The FALL *of* *the* HOUSE *of* BUSH

Also by Craig Unger

HOUSE OF BUSH, HOUSE OF SAUD:
THE SECRET RELATIONSHIP BETWEEN THE WORLD'S
TWO MOST POWERFUL DYNASTIES

BLUE BLOOD

The FALL *of* *the* HOUSE *of* BUSH

THE UNTOLD STORY OF HOW A BAND OF TRUE BELIEVERS SEIZED THE EXECUTIVE BRANCH, STARTED THE IRAQ WAR, AND STILL IMPERILS AMERICA'S FUTURE

Craig Unger

SIMON &
SCHUSTER

London · New York · Sydney · Toronto

A CBS COMPANY

First published in Great Britain in 2007 by Simon & Schuster UK Ltd
A CBS COMPANY

1 3 5 7 9 10 8 6 4 2

Simon & Schuster UK Ltd
Africa House
64–78 Kingsway
London WC2B 6AH

www.simonsays.co.uk

Simon & Schuster Australia
Sydney

A CIP catalogue for this book is
available from the British Library.

ISBN: 978-0-74328-562-9

Designed by Erich Hobbing
Text set in Stempel Garamond

Printed and bound in Great Britain by
Mackays of Chatham Ltd

And every one that heareth these sayings of mine, and doeth them not, shall be likened unto a foolish man, which built his house upon the sand. The rain fell, and the floods came, and the winds blew and slammed against that house; and it fell—and great was its fall.

—Matthew 7:26–27

CONTENTS

The FALL *of* *the* HOUSE *of* BUSH

CHAPTER ONE

Oedipus Tex

It was a cool, crisp day in the spring of 2004—a rarity for Houston—and George H.W. Bush chatted with a friend in his office suite on Memorial Drive. Tall and trim, his hair graying but by no means white, the former president was a few weeks shy of his eightieth birthday—it would take place on June 12, to be exact—and he was racing toward that milestone with the vigor of a man thirty years younger. In addition to golf, tennis, horseshoes, and his beloved Houston Astros, Bush's near-term calendar was filled with dates for fishing for Coho salmon in Newfoundland, crossing the Rockies by train, and trout fishing in the River Test in Hampshire, England.[1] He still prowled the corridors of power from London to Beijing. He still lectured all over the world. And, as if that weren't enough, he was planning to commemorate his eightieth with a star-studded two-day extravaganza, culminating with him skydiving from thirteen thousand feet over his presidential library in College Station, Texas.[2] All the celebratory fervor, however, could not mask one dark cloud on the horizon. The presidency of his son, George W. Bush, was imperiled.

One way of examining the growing crisis could be found in the prism of the elder Bush's relationship with his son, a relationship fraught with ancient conflicts, ideological differences, and their profound failure to communicate with each other. On many levels, the two men were polar opposites with completely different belief systems. An old-line Episcopalian, Bush 41 had forged an alliance with Christian evangelicals during the 1988 presidential campaign because it was vital to winning the White House. But the truth was that real evangelicals had always regarded him with suspicion—and he had returned the sentiment.

But Bush 43 was different. A genuine born-again Christian himself, he had given hundreds of evangelicals key positions in the White

House, the Justice Department, the Pentagon, and various federal agencies. How had it come to pass that after four generations of Bushes at Yale, the family name now meant that progress, science, and evolution were out and stopping embryonic stem cell research was in? Why was his son turning back the hands of time to the days when Creationism held sway?

But this was nothing compared to the Iraq War and the men behind it. George H.W. Bush was a genial man with few bitter enemies, but his son had managed to appoint, as secretary of defense no less, one of the very few who fit the bill—Donald Rumsfeld. Once Rumsfeld and Vice President Dick Cheney took office, the latter supposedly a loyal friend, they had brought in one neoconservative policy maker after another to the Pentagon, the vice president's office, and the National Security Council. In some cases, these were the same men who had battled the elder Bush when he was head of the CIA in 1976. These were the same men who fought him when he decided not to take down Saddam Hussein during the 1991 Gulf War. Their goal in life seemed to be to dismantle his legacy.

Which was exactly what was happening—with his son playing the starring role. A year earlier, President George W. Bush, clad in fighter-pilot regalia, strode triumphantly across the deck of the U.S.S. *Abraham Lincoln,* a "Mission Accomplished" banner at his back—the Iraq War presumably won. But the giddy triumphalism of Operation Shock and Awe had quickly faded. America had failed to form a stable Iraqi government. With Baghdad out of control, sectarian violence was on the rise. U.S. soldiers were becoming occupiers rather than liberators. Coalition forces were torturing prisoners.[3] As for Saddam's vast stash of weapons of mass destruction—the stated reason for the invasion—none had been found.

Bush 41* had always told his son that it was fine to take different political positions than he had held. If you have to run away from me, he said, I'll understand.[4] Few things upset him. But there were limits.

*One problem with writing about the relationship between George H.W. Bush and George W. Bush is the laborious repetition of their full names to differentiate the two men. The younger Bush is not, strictly speaking, a junior—he is missing the "H."—so referring to him as such would be inaccurate, and, perhaps as a result, reporters have sometimes referred to him as W. or, in the Texas fashion, Dubya. Because George H.W. Bush was the forty-first president of the United States and George W. Bush the forty-third, at times they will be referred to as Bush 41 and Bush 43 in this book, at times by their full names, and at times as the elder or younger Bush.

He was especially proud of his accomplishments during the 1991 Gulf War, none more so than his decision, after defeating Saddam in Kuwait, to refrain from marching on Baghdad to overthrow the brutal Iraqi dictator. Afterward, he wrote about it with coauthor Brent Scowcroft, his national security adviser, in *A World Transformed,* asserting that taking Baghdad would have incurred "incalculable human and political costs," alienated allies, and transformed Americans from liberators into a hostile occupying power, forced to rule Iraq with no exit strategy.[5] His own son's folly had confirmed his wisdom, he felt.

But now his son had not only reversed his policies, he had taken things a step further. "The stakes are high. . . ." the younger Bush told reporters on April 21.[6] "And the Iraqi people are looking—they're looking at America and saying, are we going to cut and run again?"

The unspoken etiquette of the Oval Office was that sitting and former presidents did not attack one another. "Cut and run" was precisely the phrase Bush 43 used to taunt his Democratic foes, but this time he had used it to take a swipe at his old man.

Having returned recently from the Masters Golf Tournament in Augusta, Georgia,[7] the elder Bush was eagerly looking forward to his celebrity-studded birthday bash in June.[8] But, to his dismay, the media didn't miss his son's slight of him. On CNN, White House correspondent John King characterized the president's speech as an apparent "criticism of his father's choice at the end of the first Gulf War."[9] Thanks to a raft of election season books, the press was asking questions about whether there was a rift between father and son.[10]

So on that brisk spring day, a friend of Bush 41's dropped by the Memorial Drive offices and asked the former president how he felt about his son's controversial remarks. The elder Bush was stoic and taciturn as usual. But it was clear that he was not merely insulted or offended—his son's remark had struck at the very heart of his pride. "I don't know what the hell that's about," George H.W. Bush said, "but I'm going to find out. Scowcroft is calling him right now."

The battle lines between father and son had been drawn even before the Iraq War started—a discreet, sub-rosa conflict that was both deeply personal and profoundly political. In the balance hung policies that would kill and maim hundreds of thousands of people, create millions of refugees, destabilize a volatile region that contained the largest energy deposits on the planet, and change the geostrategic balance of power for years to come.

Ultimately, it was the greatest foreign policy disaster in American history—one that could result in the end of American global supremacy.

The two men shared overlapping résumés—schooling at Andover and Yale, membership in Skull and Bones, and an affinity for Texas and the oil business. But that's about where the similarities end. From the privileged confines of Greenwich, Connecticut, where he was raised, to Walker's Point, the Bush family summer compound in Kennebunkport where his family golfed and ate lobster on the rugged Maine coast, to the posh River Oaks section of Houston after they settled in Texas, George H.W. Bush epitomized a blue-blooded, old money, Eastern establishment ethos that was abhorrent to the Bible Belt. By contrast, his son had been a fish out of water among the Andover and Yale elite, and scurried back to the West Texas town of Midland after graduating from the Harvard Business School. Nothing made him happier than clearing brush off the Texas plains.

People who knew both men tended to favor the father. "Bush senior finds it impossible to strut, and Bush junior finds it impossible not to," said Bob Strauss, the former chairman of the Democratic National Committee who served as ambassador to Moscow under Bush 41 and remained a loyal friend.[11] "That's the big difference between the two of them."

More profoundly, they epitomized two diametrically opposed forces. On one side was the father, George H.W. Bush, a realist and a pragmatist whose domestic and foreign policies fit comfortably within the age-old American traditions of Jeffersonian democracy. On the other was his son George W. Bush, a radical evangelical poised to enact a vision of American exceptionalism shared by the Christian Right, who saw American destiny as ordained by God, and by neoconservative ideologues, who believed that America's "greatness" was founded on "universal principles"[12] that applied to all men and all nations—and gave America the right to change the world.

And so an extraordinary constrained nonconversation of sorts between father and son had ensued. Real content was expressed only via surrogates. In August 2002, more than seven months before the start of the Iraq War, Brent Scowcroft, a man of modest demeanor but of great intellectual resolve, was the first to speak out. At seventy-seven, Scowcroft conducted himself with a self-effacing manner that belied his considerable achievements. Ever the loyal retainer, he was the public voice of Bush 41, which meant he had the tacit approval of the former president. "They are two old friends who talk every day," says Bob Strauss. "Scowcroft knew it wouldn't terribly displease his friend."[13]

Well aware that war was afoot, Scowcroft had tried to head it off with an August 15, 2002, *Wall Street Journal* op-ed piece titled "Don't Attack Saddam" and TV interviews. As a purveyor of the realist school of foreign policy, and as a protégé of Henry Kissinger, Scowcroft believed that idealism should take a backseat to America's strategic self-interest, and his case was simple. "There is scant evidence to tie Saddam to terrorist organizations," he wrote, "and even less to the Sept. 11 attacks."[14] To attack Iraq, while ignoring the Israeli-Palestinian conflict, he said, "could turn the whole region into a cauldron and, thus, destroy the war on terrorism."[15] A few days later, former secretary of state James Baker, who had carefully assembled the massive coalition for the Gulf War in 1991, joined in, warning the Bush administration that if it were to attack Saddam, it should not go it alone.[16]

On one side, aligned with Bush 41, were pragmatic moderates who had served at the highest levels of the national security apparatus—Scowcroft, Baker, former secretary of state Lawrence Eagleburger, and Colin Powell, with only Powell, as the sitting secretary of state, having a seat at the table in the new administration. On the other side, under the younger George Bush, were Dick Cheney, Donald Rumsfeld, Deputy Secretary of Defense Paul Wolfowitz, and Richard Perle, chairman of the Defense Policy Advisory Board Committee—all far more hawkish and ideological than their rivals.

Of course, both Scowcroft and Baker would have preferred to give their advice to the young president directly rather than through the media,[17] and as close friends to Bush senior for more than thirty years, that should not have been difficult. After all, Scowcroft's best friend was the president's father, his close friend Dick Cheney was vice president, and Scowcroft counted National Security Adviser Condoleezza Rice and her deputy Stephen Hadley* among his protégés. And James Baker had an even more storied history with the Bushes.

"Am I happy at not being closer to the White House?" Scowcroft asked. "No. I would prefer to be closer. I like George Bush personally, and he is the son of a man I'm just crazy about."[18]

But in the wake of Scowcroft's piece in the *Journal*, both men were denied access to the White House. When the elder Bush tried to intercede on Scowcroft's behalf, he met with no success. "There have been occasions when Forty-one has engineered meetings in which Forty-

*When Rice became secretary of state in Bush's second administration, Hadley took her position as national security adviser.

three and Scowcroft are in the same place at the same time, but they were social settings that weren't conducive to talking about substantive issues," a Scowcroft confidant told *The New Yorker*.[19]

Meanwhile, Bush senior did not dare tell his son that he shared Scowcroft's views. According to the Bushes' conservative biographers, Peter and Rochelle Schweizer, family members could see his torment.[20] When his sister, Nancy Ellis, asked him what he thought about his son's plan for the war, Bush 41 replied, "But do they have an exit strategy?"

In direct talks between father and son, however, such vital policy issues were verboten. "[Bush senior is] so careful about his son's prerogatives that I don't think he would tell him his own views," a former aide to the elder Bush told *New York Times* columnist Maureen Dowd.[21] When the *Washington Post*'s Bob Woodward told Bush 43 that it was hard to believe he had not asked his father for advice about Iraq, the president insisted the war was never discussed. "If it wouldn't be credible," Bush added, "I guess I better make something up."[22]

Likewise, friends who saw them together found that they had absolutely nothing to say to each other on matters of vital national importance. "I was curious to see how they related to one another, and I'll be damned," said Bob Strauss, who shared an intimate dinner with them in the White House. "They never discussed the war, never discussed politics. We talked about social things, friendships, what was going on back in Texas. It was like a couple of old friends just gossiping about the past."[23]

By 2006, however, tens of thousands of people had been killed in the Iraq War. Launched with the stated intention of eliminating Iraq's weapons of mass destructions, the war had turned up no weapons whatsoever, and had instead raised profound questions about U.S. intelligence. Likewise, it had been disastrous in terms of America's strategic ambitions. Instead of shoring up Israeli security and replacing rogue regimes in the Middle East with friendly, pro-Western allies, the war had turned Iraq into a terrorist training ground. By eliminating Saddam Hussein, the United States had sparked a Sunni-Shi'ite civil war that threatened to spread throughout and destabilize the entire Middle East. Far from creating a secular democracy, the war had empowered Shi'ite fundamentalists aligned with Iran. The Islamic Republic of Iran, America's greatest foe in the region, had, unwittingly, been empowered. Dramatic action was necessary if Bush senior's legacy was to be saved.

Enter the Iraq Study Group (ISG), a panel chaired by James Baker and former Democratic congressman Lee Hamilton and charged, in March 2006, with reassessing the deteriorating situation in Iraq and making policy recommendations. With Baker, the legendary Republican political operator and close friend of the former president, and Lawrence Eagleburger, who had also served as Bush 41's secretary of state,* on the commission, and with Brent Scowcroft a consultant to it, key figures of Bush senior's national security team finally had a politically opportune moment to present a bipartisan fig leaf that would enable the president to change course.

By now, however, Baker and Scowcroft knew that even their substantial persuasive powers would not change the president's mind, so they devised an alternative strategy.[24] The key would be to get help from one of the very few people close enough to the president who could possibly persuade him to change direction—Secretary of State Condoleezza Rice.

The forty-eight-year-old Rice was slender, perfectly coiffed, and ferociously poised—the most powerful African-American woman in the country. Having come of age working with Scowcroft and James Baker in Bush senior's administration, she had also developed a special, if somewhat strange, relationship with the younger Bush. When Rice once publicly referred to Bush as "my husband," it was widely seen as a Freudian slip that reflected how close they had become.[25] Her predecessor, former secretary of state Colin Powell, once a trusted member of Bush 41's circle, had proven remarkably ineffective in fighting the neocons, and was long gone. That left Rice, the only member of the old guard who had unalloyed access to the president, as the crucial bridge between 41 and 43.

So, in late August 2006, according to a report by Sidney Blumenthal, a former senior adviser to Bill Clinton, in *Salon,* Scowcroft met with Rice to explain that a comprehensive new approach to the Middle East was in order, including a focus on the Israeli-Palestinian conflict. Rice seemed to agree. "How are we going to present this to the president?" she asked Scowcroft.[26]

"Not we," replied Scowcroft. "You." He emphasized that only she was in a position to get Bush to change his policies.

* * *

*Another high-level Bush 41 appointee, former CIA director Robert Gates, served on the panel until he was nominated to be secretary of defense in 2006. Eagleburger replaced him.

About two weeks later, there were signs from the State Department that Scowcroft's meeting with Rice had paid off. On September 15, Philip Zelikow, Rice's closest aide, gave a speech asserting that the Israeli-Palestinian conflict must be addressed if Arab moderates and Europeans were to cooperate with the United States in the Middle East.[27]

Zelikow's talk was widely seen as the first sign of a dramatic shift in administration policy. As the November 2006 midterm elections approached, thanks to growing antiwar sentiment, the hawks were finally in retreat. Deputy Secretary of Defense Paul Wolfowitz and Under Secretary of Defense Douglas Feith, two highly controversial neoconservative architects of the war, had left the administration the previous year under fire. And eight retired generals had demanded Donald Rumsfeld's resignation.[28]

But the neocons were not dead yet. Immediately after Zelikow's speech, Cheney's office responded with fierce attacks on Zelikow from inside the bureaucracy.[29] Reports surfaced in the *Jerusalem Post*[30] and the *New York Sun*,[31] two neocon papers, undermining Zelikow's message. Faced with the prospect of battling Cheney, Condoleezza Rice caved instantly. The State Department assailed Zelikow. "The issues of Iran and Israeli-Palestinian interaction each have their own dynamic, and we are not making a new linkage between the two issues," State Department spokesman Sean McCormack announced.[32] "Nothing in Philip's remarks should be interpreted as laying out or even hinting at a change in policy." On November 27, Zelikow abruptly resigned.[33]

But, most important, Rice never followed through after her meeting with Scowcroft. She never stepped up to the plate to try to persuade Bush to change course. Once again, the neocons held sway. As a result, Bush 41's moderates were in a much weaker position than they had anticipated as the Iraq Study Group prepared to make its presentation.

The ritualized pomp and circumstance began at dawn on December 6, 2006, with the motorcade of long black sedans carrying the esteemed Wise Men (and one woman) to the White House. At 7:00 a.m., all ten members of the Iraq Study Group—James Baker, Lee Hamilton, former Supreme Court justice Sandra Day O'Connor, former secretary of state Lawrence Eagleburger, Clinton friend and adviser Vernon Jordan, former attorney general Edwin Meese, former White House chief of staff Leon Panetta, former secretary of defense William Perry, and former senators Charles Robb and Alan Simpson—arrived to hand-deliver signed copies of their report, "The Way Forward—A New Approach," to the president.[34]

Bush formally received the group, thanked Baker and Hamilton, congratulated them on their work, and made a brief, pro forma statement to the press: "We applaud your work. I will take it very seriously. And we will act on it in a timely fashion."[35]

Then the dignitaries climbed back in their motorcade, which made its way up to the Capitol. With police sirens signaling their arrival, they met first with the House leadership behind closed doors, and handed out copies of the report.[36] Book bearers in tow, Baker and Hamilton led the way down a basement corridor, through a banquet kitchen and a locker room where waiters had donned bow ties. Dozens of photographers and about two hundred journalists were present to document the delivery of the report to Senate leaders Bill Frist and Harry Reid.

Bipartisan commissions, by their nature, tend to be bland affairs, but in the urgency of the political moment, the Iraq Study Group was different. Congressional Republicans had just been swept out of power in the 2006 midterm elections; Rumsfeld had been tossed overboard. Now came James Baker, the Bush family's longtime friend and consigliere, to talk some sense into the president. With his steely-eyed toughness, Baker was the neocons' worst nightmare. But to war-weary Americans, his presence signaled a moment of hope when it seemed that the president might finally accept the failure of his policies and try something new.

A reporter asked Baker if, given his close relationship to the Bush family, he thought the president could "pull a 180." "I never put presidents I worked for on the couch," he replied. "So I'm not going to answer that, because that would mean I'd have to psychologically analyze the inner workings of his mind. And I don't do that."[37]

In fact, Bush had no intention of acting on the report's recommendations at all. Much of its content had leaked out beforehand, and Bush did not like what it said. With phrases like "grave and deteriorating" and "pessimism is pervasive," its verdict was clear: America's policies had failed. It was time to cut losses. The report highlighted the basic fallacy behind the administration's strategy: the new Iraqi army, the police force, and even Prime Minister Nuri Kamal al-Maliki often showed greater loyalty to their ethnic identities than to the ideal of a nonsectarian, democratic Iraq. Ultimately, military solutions—that is, sending more soldiers to Iraq—could not resolve what were fundamentally political problems.

Journalists from the *Washington Post* and the *New York Times* said that, in rejecting all of its substantive recommendations, President Bush had, in private, used especially colorful language, calling the ISG study "a flaming turd."[38] Even if that account was exaggerated, it seemed to convey the feelings that led him to dismiss the report so brusquely.

As for the elder George Bush, he was in the news, too. On December 5, less than twenty-four hours before the Iraq Study Group report was released, the former president addressed the Florida state legislature in Tallahassee, where his son Jeb was governor. In a speech about leadership—run-of-the-mill stuff for a former chief of state—Bush told the legislators how proud he was about Jeb's brave reaction to his defeat in the 1994 Florida gubernatorial election. "When it came down the homestretch," Bush senior said, "[Jeb] saw some unpleasant things happen, unfair stuff, but he didn't whine about it, he didn't complain."[39]

Then the former president grabbed the podium as if to steady himself. He paused, obviously shaken. "Barbara will bawl me out. . . ." he said wanly.

Now near tears, he continued. "A true measure of a man is how you handle victory," he said, his voice wavering. Again, Bush grasped the podium and hesitated before going on. "And also defeat."[40]

Then his voice cracked. "So in '94 Floridians chose to rehire the governor," he said, "but they took note of his worthy opponent, who showed with not only words but with actions what decency he had." Fighting to keep his composure, he collapsed weeping as Jeb rushed over to comfort him.

In an earlier epoch, an inconsequential speech made in Tallahassee by a former president might not have even made the evening news. But in the era of YouTube and Internet video, it circled the globe instantly, not just for one news cycle but forever.

"It is not fully right, or fully fair, to guess about another's emotions," Peggy Noonan, Bush senior's former speechwriter, wrote afterward in the *Wall Street Journal*.[41] "But no one who knows George H.W. Bush thinks that moment was only about Jeb. It wasn't only about some small defeat a dozen years ago. It would more likely have been about a number of things, and another son, and more than him."

Noonan pointed out that Bush senior must have known the contents of the Iraq Study Group's report that was being released the next day, and its damning judgment of his son's presidency. "Surely Mr. Bush knew—surely he was first on James Baker's call list—that the report would not, could not, offer a way out of a national calamity, but only suggestions, hopes, on ways through it. To know his son George had (with the best of intentions!) been wrong in the great decision of his presidency—stop at Afghanistan or move on to Iraq?—and was now suffering a defeat made clear by the report; to love that son, and love your country, to hold these thoughts, to have them collide and come together—this would bring not only tears, but more than tears."[42]

Less noticed, but just as striking as the former president's tears, was the fact that the son who had created this catastrophe was at the other end of the emotional spectrum. Far from showing signs of anguish at the horrors he had unleashed, George W. Bush displayed what Noonan called "a jarring peppiness."[43]

And consider the context. A bipartisan panel had just eviscerated the centerpiece of his entire presidency. Moreover, by this time, there had been so many astounding revelations about Iraq that it was difficult to process them all. From Saddam's phantom WMDs to the "Mission Accomplished" photo op, from the fairy-tale pluck of Jessica Lynch to the heroic martyrdom of NFL star Pat Tillman, who had been killed by friendly fire, the Pentagon had trumpeted one Hollywoodized saga of the Iraq War after another. By and large, most of them had been revealed as lies. No longer was the Bush White House able to maintain control of the narrative. The carefully managed perceptions of the Iraq War were vanishing.

How could one believe in the noble ideal of democratizing the Middle East when American soldiers—and even the Iraqi government itself—hid out in the Little America bubble of the Green Zone, the so-called Emerald City, with its discos, fast food, porno shops—and thousands of contractors from Halliburton? How could one see American soldiers as liberators after the reports of torture and horrifying abuses at Abu Ghraib that drove thousands of Iraqis not just to join the insurgency, but to cheer as the charred, mutilated bodies of dead Americans were dragged through the streets of Fallujah?[44] How could one celebrate the rebuilding of Iraq's infrastructure when at least $12 billion in cash was flown to Baghdad, shrink-wrapped in plastic, and $9 billion of it vanished under highly suspicious circumstances.[45] Or when untold billions went to virtually unregulated private security firms, which brought in tens of thousands of mercenaries who were paid enormous sums.

Meanwhile, in terms of blood and treasure, the costs of war had soared beyond anyone's worst nightmare. Billed as likely to last only a few months,[46] this was a war that was to have practically paid for itself, officials had said.*[47] But with no end in sight, according to the ISG report, the war's price tag exceeded $400 billion and Nobel Prize–winning economist Joseph Stiglitz put its "true cost" much, much higher—at more than $2 trillion.[48] Far from funding the war as promised, Iraq's oil industry was being systematically sabotaged, its oil hijacked, with billions of dollars going to subsidize terrorists.[49] No wonder oil prices had more than tripled since 2002 to well over $70 a barrel.[50]

And then there were the gruesome and horrific human losses. Walter Reed Army Medical Center, itself a scandalous victim of neglect, was teeming with soldiers who had lost hands and arms, feet and legs, whose faces had been burned off, who had been paralyzed, who confronted lives very different from what they had imagined scant months earlier.

Meanwhile, at home, the world's greatest constitutional democracy had implemented unprecedented secrecy and spying on its own citizens. There had been a dramatic erosion of civil liberties. The creation of a Soviet-style gulag at Guantánamo made a mockery of America's Constitution by suspending habeas corpus and embracing the detention of prisoners—allowing them no rights whatsoever. The presidency itself had become an "imperial presidency," consolidating enormous powers far beyond those intended by the founding fathers, effectively gutting the concepts of checks and balances.

None of which took into account the unforeseeable consequences that lay ahead for America thanks to the strategic disaster that was unfolding. Indeed, the Iraq War had accomplished precisely the opposite of its intentions. Rather than end terrorism, it created blood-drenched killing fields and vast new training grounds for tens of thousands of jihadists and Islamist militias. It created a new Iraqi state dominated by Shi'ites who saw Israel and America as their enemy. If there had been any winner at all, it had been the Islamic Republic of Iran. America's military was being stretched thin, its troops overburdened. Ultimately, the war had resulted in a historic decline in American power and prestige.

*In testimony before Congress, Deputy Secretary of Defense Paul Wolfowitz said, "It's hard to conceive that it would take more forces to provide stability in post-Saddam Iraq than it would take to conduct the war itself." He added, "The oil revenue of that country could bring between 50 and 100 billion dollars over the course of the next two or three years. We're dealing with a country that could really finance its own reconstruction, and relatively soon."

And yet, in the midst of all this, George W. Bush was, as Noonan put it, "resolutely un-anguished." How could he be so free of doubt in the face of such a cataclysm? As his father wept, how could he remain so serene?

As the situation in Iraq continued to deteriorate in the summer of 2007, the larger question of how America arrived at this moment, and precisely what that moment meant, was even less clear or understood. In the prosecution of the war, and the implementation of Bush's broader vision, many of America's most sacred institutions, from its judiciary to its national security apparatus, had been sabotaged and subverted. When it came to the constitutional checks and balances, to the powers of the executive branch, lines had been crossed, fundamental principles violated, putting at risk precisely what made America so special. Dick Cheney had led Donald Rumsfeld and the neocons in creating a separate, shadow national security apparatus to create a disinformation pipeline putting forth its own wished-for reality as a mechanism to start the war. As the summer of 2007 drew to an end, there was even reason to believe that the Bush administration would "double down" by bombing Iran, a potentially disastrous move that could ignite a global oil war and might spell the end of American supremacy forever. How had Americans been tricked into allowing these radical policies to be implemented? What were the deep cultural forces that had led the country to this historic catastrophe?

A hint could be found in Bush's ready explanation as to why he had not gone to his father for advice on Iraq. "You know," he said, "he is the wrong father to appeal to in terms of strength. There is a higher father that I appeal to."[51]

Indeed, it was precisely this faith—as opposed to reason—that had made Bush such an ideal vehicle through whom to implement a revived vision of American exceptionalism, a vision shared by neoconservatives and the Christian Right, asserting America's right to fight tyranny all over the world. But how had this shared vision, this strange alliance of faith and ideology, come to be?

Astonishingly enough, the story of how neoconservative ideologues banded together with the Christian Right to forge these radical policies under Bush has never been fully told. In part, that may be because the religious sensitivities of both evangelical Christians and Jews make deep criticism of America's Middle East policies the third rail of American politics. Indeed, the entire topic is at odds with the way the discourse about the Middle East conflict has been framed in the United States, and so taboo that it rarely appears in the American press in any context whatsoever.

To truly understand the scope and meaning of the relationship between the neocons and the Christian Right, however, one would have to delve into subjects as varied and seemingly unrelated as theology and espionage, ancient history and the geopolitics of oil, biblical prophecy, political assassinations, and the secret ties between the Pentagon and Israel. One would have to travel in time from the exile of the Jews from the Temple in the pre-Christian era to the days of the early Puritan settlers in New England to the unending intrigue in Washington and the Middle East today. One would have to interview messianic Jews in Jerusalem and settlers in the West Bank; Likudnik politicians of the Israeli right and Rapturite fundamentalists from the Bible Belt; leaders of the Christian Right in Lynchburg, Virginia; neoconservative ideologues in Washington think tanks; CIA intelligence operatives in Langley, Virginia, and their Israeli counterparts in Mossad; and the many military officers and intelligence officials who rebelled against the Bush administration.

Finally, one would conclude that the most significant "clash of civilizations" today is not between Islam and the West at all, as the conflict is usually framed, but between fundamentalists—not just Islamists, but Christian and Orthodox Jewish fundamentalists as well—and the modern world. In other words, the most powerful enemies of our modern, humanist post-Enlightenment world may not be militant Islamists more than an ocean away, but Christian fundamentalists and their neoconservative allies who have been waging a ferocious war against "militant secularists," and who finally became influential enough to install, for the first time, a powerful leader of the Christian Right, George W. Bush, in the White House.

CHAPTER TWO

Redeemer Nation

On a scorching afternoon in late May 2005, Tim LaHaye, the seventy-nine-year-old coauthor of the Left Behind series of apocalyptic thrillers, leads several dozen of his acolytes up a long, winding path to a hilltop in the ancient fortress city of Megiddo, Israel. LaHaye is not a household name in the secular world, but in the parallel universe of evangelical Christians he is a potent cultural icon. The author or coauthor of more than seventy-five books, LaHaye in 2001 was named the most influential American evangelical leader of the past twenty-five years by the Institute for the Study of American Evangelicals—over Billy Graham, Jerry Falwell, and scores of other famous evangelists.[1] With more than 63 million copies of his Left Behind novels sold, he is one of the bestselling authors in all of American history. Here, a group of about ninety American evangelical Christians who embrace the astonishing theology he espouses have joined him in the Holy Land for the "Walking Where Jesus Walked" tour.

Megiddo, the site of roughly twenty different civilizations over the last ten thousand years, is among the first stops on the pilgrimage. Given that LaHaye's field of expertise is the apocalypse,* it is also one of the most important. Alexander the Great, Saladin, Napoleon, and other renowned warriors all fought great battles here. But if Megiddo is to go down in history as the greatest battlefield on earth, its real test is yet to come. According to the Book of Revelation, the hill of Megiddo,

*Perhaps because of its association with End Times, the word *apocalypse* is widely thought to refer to the end of the world. However, translated literally from the Greek, it means "the lifting of the veil"—that is, "revelation"—and refers to the disclosure to a few privileged persons of something hidden. The last book of the *New Testament* is sometimes referred to as the Apocalypse of John, and, in English, is known as the Revelation of St. John the Divine or the Book of Revelation.

better known as Armageddon—will be the site of the cataclysmic battle between the forces of Christ and the Antichrist.

To get a good look at the battlefields of the apocalypse, the group takes shelter under a makeshift lean-to at the top of the hill. Wearing a floppy hat to protect him from the blazing Israeli sun, LaHaye yields to his colleague Gary Frazier to explain what will happen during the Final Days. The tour organizer and founder of Discovery Ministries in Arlington, Texas, near Dallas, Frazier has the demeanor of a high school football coach herding his charges on and off the bus during a road trip.

"How many of you have read the Left Behind prophecy novels?" asks Frazier.[2] Almost everyone raises a hand.

"The thing that you must know," Frazier tells them, "is that the next event on God's prophetic plan, we believe, is the catching away of the saints in the presence of the Lord. We call it 'the Rapture.'"

Frazier is referring to a key biblical passage, in the first book of Thessalonians, that says the Lord will "descend from heaven with a shout. . . . The dead in Christ shall rise first. Then we which are alive and remain shall be caught up together with them in the clouds, to meet the Lord in the air."[3] Because the Greek word *harpazo* in the original text is sometimes translated as "caught up" or "raptured," adherents cite this as the essential biblical reference to the Rapture.

"Christ is going to appear," Frazier continues. "He is going to call all of his saved, all of his children, home to be with him."

In other words, "in the twinkling of an eye,"* as the Rapturists often say, millions of born-again evangelicals will suddenly vanish from the earth—just as they do in LaHaye's Left Behind books. They will leave behind their clothes, their material possessions, and all their friends and family members who have not accepted Christ—and they will join Christ in the Kingdom of God.

Frazier continues. "Jesus taught his disciples that he was going to go away to his father's house, but that he was not going to abandon them. . . . Jesus is going to come and get his bride, which comprises all of us who are born again."

Frazier is a fiery preacher, and as his voice rises and falls, his listeners respond with cries of "Amen" and "That's right."

"I'm going to tell you with zeal and enthusiasm and passion Jesus is

*The phrase comes from the first book of Corinthians, 15:52: "In a moment, in the twinkling of an eye, at the last trump: for the trumpet shall sound, and the dead shall be raised incorruptible, and we shall be changed."

coming on the clouds of glory to call us home. . . . Now, ladies and gentlemen, I want you to know . . . that Christ is coming. And we believe that that day is very, very near."

For miles around in all directions the fertile Jezreel Valley, known as the breadbasket of Israel, is spread out before them, an endless vista of lush vineyards and orchards growing grapes, oranges, kumquats, peaches, and pears. The sight LaHaye's followers hope to see here in the near future, however, is anything but bucolic. Their vision is fueled by the Book of Revelation, the dark and foreboding messianic prophecy peopled with grotesque monsters—the whore of Babylon, the Beast with ten horns and seven heads—that foresees a gruesome and bloody confrontation between Christ and the armies of the Antichrist at Armageddon.

Addressing the group from the precise spot where the conflict is to take place, Frazier turns to Revelation 19, which describes Christ going into battle. "It thrills my heart every time that I read these words," he says. Then he begins to read: "'And I saw heaven standing open. . . . And there before me was a white horse, whose rider is called Faithful and True. With justice he judges and makes war. His eyes are like blazing fire.'"

Frazier pauses to explain the text. "This doesn't sound like compassionate Jesus," he says. "This doesn't sound like the suffering servant of Isaiah 53. This is the Warrior King. He judges and makes war."

Frazier returns to the Scripture: "He has a name written on him that no one but he himself knows. He is dressed in a robe that is dipped in blood and his name is the word of God."

What is to happen next is the moment the Rapturists fervently await. Light-years away from the lamblike, turn-the-other-cheek Christ of the Sermon on the Mount, this Jesus is an angry and merciless God, and the magnitude of death and destruction he wreaks will make the Holocaust seem trifling. Finally, when the battle begins, those who remain on earth are the unsaved, the left behind—many of them dissolute followers of the Antichrist, who is massing his army against Christ. Accompanying Christ into battle are the armies of heaven, riding white horses and dressed in fine linen.

"This is all of us," Frazier says.

Frazier points out that Christ does not need high-tech weaponry for this conflict. "'Out of his mouth comes a sharp sword,' not a bunch of missiles and rockets," he says.[4]

Once Christ joins the battle, both the Antichrist and the False

Prophet are quickly captured and cast alive into a lake of fire burning with brimstone.[5] Huge numbers of the Antichrist's supporters are slain.

Meanwhile, an angel exhorts Christ, "Thrust in thy sickle, and reap."[6] And so, Christ, sickle in hand, gathers "the vine of the earth."

Then, according to Revelation, "the earth was reaped." These four simple words signify the end of the world as we know it.

Grapes that are "fully ripe"—billions of people who have reached maturity but still reject the grace of God—are now cast "into the great winepress of the wrath of God."[7]

Here we have the origin of the phrase "the grapes of wrath." In an extraordinarily merciless and brutal act, Christ crushes the so-called grapes of wrath. In the process, he is killing billions of people because they did not accept him as their savior. Then, Revelation says, blood flows out "of the winepress, even unto the horse bridles, by the space of a thousand and six hundred furlongs."*

With its highly figurative language, Revelation is subject to profoundly differing interpretations. Nevertheless, LaHaye's followers insist on its literal truth and accuracy, and they have gone to their calculators to figure out exactly what this passage of Revelation means. As we walk down from the top of the hill of Megiddo, one of them looks out over the Jezreel Valley. "Can you imagine this entire valley filled with blood?" he asks.[8] "That would be a 200-mile-long river of blood, four and a half feet deep. We've done the math. That's the blood of as many as two and a half billion people."

When this will happen is another question. The Bible says that "of that day and hour knoweth no man." Nevertheless, LaHaye's disciples are certain these events, variously known as the End of Days, the Final Conflict, and Armageddon, are imminent.

In fact, one of them has especially strong ideas about when they will take place. "Not soon enough," she says. "Not soon enough."

If such views sound astonishing, the people who hold them are decidedly not. For the most part, the group could pass for a random selection culled from almost any shopping mall in America. There are warm and loving middle-aged couples who hold hands as they stroll through the Holy Land. There is a well-coiffed Texas matron with an Hermès scarf. There's a duck-tailed septuagenarian and a host of post-teen mall

*A furlong is one-eighth of a mile. The phrase "unto the horse bridles" has been interpreted to mean a depth of about four and a half feet.

rats. There are young singles. One couple even chose this trip for their honeymoon. A big-haired platinum blonde with a white sequined cowboy hat adds a touch of Dallas glamour. There is a computer-security expert, a legal assistant, and a real estate broker; a construction executive, a retired pastor, a caregiver for the elderly, and a graduate student from Jerry Falwell's Liberty University. They hail from Peoria, Illinois, and Longview, Texas, as well as San Diego and San Antonio. These are the kind of people who made George W. Bush president.

Many have attended the Left Behind Prophecy Conference on one of its tours of the United States, and almost all of them are fans of the Left Behind books. In their unquestioning religiosity, however, they seem stunningly oblivious to the genocidal nature of the series' theology.*[9] Like the angry warrior Christ in the Book of Revelation, they seem to have no problem with the slaughter of billions of people who have not accepted Christ as their savior.

Such beliefs may seem astounding to secular Americans, but they are not unusual. According to a *Time*/CNN poll from 2002, 59 percent of Americans believe the events in the Book of Revelation will take place.[10] In addition, a January 2007 study by the Barna Group, a Christian research firm, found that there are as many as 84 million[11] adult evangelicals in the United States—about 38 percent of the population.

Exactly what such surveys mean, however, is a different story. The same Barna Group poll, for example, found that when a "theological filter" of nine rigorous questions was used to find out if respondents were *really* "evangelical," only 8 percent of the adult population—18 million people rather than 84 million—passed the test.[12] Consequently, the actual number of evangelicals varies widely from one survey to another, and terms such as "evangelical," "fundamentalist," and "born-again Christian" are open to a variety of interpretations.† Perhaps most

*The Left Behind video game actually invites teenagers to become virtual soldiers in a Christian Taliban. As Talk to Action, a website about the Christian Right, puts it, you are asked to join "a paramilitary group whose purpose is to remake America as a Christian theocracy, and establish its worldly vision of the dominion of Christ over all aspects of life. You are issued high-tech military weaponry, and instructed to engage the infidel on the streets of New York City. You are on a mission—both a religious mission and a military mission—to convert or kill Catholics, Jews, Muslims, Buddhists, gays, and anyone who advocates the separation of church and state."

†Nevertheless, for those who are unfamiliar with such terms, shorthand definitions can be worthwhile. The Barna Group defines "born-again Christians" as those who have made a personal commitment to Jesus Christ and believe when they die they

important, many millions of evangelicals belong to more than 200,000 churches, most of which are run by pastors who belong to conservative political organizations that make sure their flocks vote as a hard-right Republican bloc.

A fascination with Revelation and America's specially ordained place in the divine plan has always been a powerful and elemental, if often unseen, component of the American consciousness, predating America itself and playing a vital role in its creation. Nor has it been confined to the pulpit. Imagery from the Book of Revelation has inspired poets and writers from William Blake and William Butler Yeats to Joan Didion and Bob Dylan. Elements of Revelation, secularized or otherwise, turn up in movies such as the *Star Wars* trilogies, *High Noon* starring Gary Cooper, *The Omen* starring Gregory Peck and its various sequels, Clint Eastwood's *Pale Rider,* and countless others.

Likewise, the myth that Americans are a Chosen People specially ordained by God has remained alive and well in various forms over the centuries, evolving through time to fit changing economic, military, political, and historic circumstances. It has taken the shape of the Puritanical assertion that America is a Redeemer Nation with a special millennial mission. It has been called Manifest Destiny in reference to westward expansion through which the early settlers tamed the frontier and Christianized heathen savages. It has taken on the guise of American exceptionalism when the U.S. military ventured forth to make the world safe for

---

will go to Heaven because they have confessed their sins and have accepted Jesus Christ as their savior.

As for "evangelicals," in the American Religious Report of 2004, John C. Green of the University of Akron, in Ohio, one of the leading demographers in the United States, attributed four key beliefs to evangelicals—biblical inerrancy, salvation through faith in Jesus rather than through good works, a *personal* trust in Jesus Christ as savior, and a commission to evangelize and to be publicly baptized as an act of faith.

As for fundamentalism, theologian and religious historian George C. Marsden has famously described fundamentalists as "angry evangelicals." Core fundamentalist beliefs include the inerrancy of Scripture, the virgin birth and deity of Jesus, the notion of substitutionary atonement—that is, salvation through faith in Christ because he died for our sins; the resurrection of Jesus; and the authenticity of Christ's miracles, including his Second Coming.

Fundamentalists sometimes criticize evangelicals for a lack of doctrinal purity. But in general, for the purposes of this book, insofar as fundamentalism is a subset of evangelicalism, the difference between the two has less to do with diverging theological doctrines than with Christian fundamentalists' more aggressive political posture in fighting modernism, especially with regard to the antiabortion movement, stem cell research, gay marriage, and other political issues.

democracy. It has been secularized and transformed into various expressions of romantic nationalism in one era after another, from the days of the Puritans to the Wild West, from the Spanish-American War when American soldiers Christianized the Philippines to Woodrow Wilson's notion of America as the servant of mankind, from fighting Hitler in World War II to the war against godless communism in Vietnam. It has been used to rationalize a lust for power, energy security, oil, and wealth. Today it means "democratizing" the Middle East and fighting terrorism.

The notion of American exceptionalism began in the late fifteenth century when America itself was no more than a gleam in Christopher Columbus's eye. A student of biblical prophecy as well as an explorer, Columbus asserted that his role in crossing the ocean was to expedite the messianic fulfillment of God's plan for the millennial kingdom on earth. "God made me the messenger of the new heaven and the new earth of which he spoke in the Apocalypse of St. John . . . and he showed me the spot where to find it," Columbus wrote in his diary.[13]

More than a century later, English theologians espoused similar messianic ideals and set about to fulfill them. With the advent of the Protestant Reformation and the first English-language Bibles came the early stirrings of Christian Zionism, the belief that the return of Jews to the Holy Land is in accordance with biblical prophecy. By the sixteenth century, theologians such as Thomas Brightman, Thomas Draxe, Edmund Bunny, and Francis Kett began calling for the restoration of the Jews in Palestine.[14]

Initially, such ideas were deemed so heretical that some theologians, including Kett, were burned at the stake. Nevertheless, the identification of the Puritans with the Jews was so deep that some Puritans even wanted to use Hebrew in their prayers and to have Mosaic law enforced.[15] Ultimately, the most radical among them split off and, like the Israelites themselves, fled England to seek a new Zion, a new Promised Land, in America.

Then, in 1630, a wealthy attorney named John Winthrop took a group of fellow Puritans across the Atlantic to build a New Jerusalem in the New World, one that would provide the foundation for a millennial Kingdom of Christ. Just before his ship, the *Arbella*, made its way to Salem,[16] Massachusetts, Winthrop, who later became the first governor of Massachusetts Bay Colony, addressed his fellow pilgrims with his historic sermon, "A Model of Christian Charity." Borrowing a phrase from the Book of Matthew's Sermon on the Mount, Winthrop characterized their new homeland as a New Jerusalem by describing it as a "city on a hill," thereby searing in the national consciousness forever an iconic phrase to define America's special role in the world. Famously quoted by President-elect John F. Kennedy in 1961 and President Ronald W. Reagan in his 1989 farewell address, cited by countless politicians and pastors, Winthrop's sermon became one of the essential texts that shaped America, and the first to put forth the notion that America was a Promised Land. The term *Christian Zionism* is rarely used to describe the seminal myths on which America was founded. Yet the enduring power of Winthrop's celebrated sermon lay in its ability to remake Christian Zionism into a uniquely American myth. A city on a hill, after all, meant Jerusalem—and Jerusalem was Zion.[17] Christian Zionism had been transformed. America was God's new Israel. It had inherited the spiritual legacy of the Jews.

Specifically, Winthrop told his followers that, like the Israelites being driven from the Temple, they had been persecuted and sent into exile. But God's providence was such, he said, that they had been entrusted with enacting the final chapter in human history. They were the Chosen People and they were now on their way to the Promised Land.*[18] Winthrop's metaphor became so deeply rooted among the early settlers that Puritan minister Cotton Mather[19] preached that Christ would return once America was built into a truly righteous millennial kingdom.[20] This doctrine became known as postmillennialism because it posited that the Second Coming of Christ would take place *after* the millennium.

As a result, the Puritans had created a powerful and lasting sense of America as part of a utopian mission in the war of good against evil, freedom against tyranny. The Puritan imprint was indelibly stamped on the map of America. Salem, Massachusetts, where Winthrop had landed, and later Salem, Oregon, Illinois, and more than twenty other states, was the New World name for Jerusalem.† There were Goshens, Canaans, and

*The passage to which Winthrop was referring comes from Matthew 5:14: "You are the light of the world. A city set on a hill cannot be hid." In part, he told his fellow pilgrims: "Wee shall finde that the God of Israell is among us, when ten of us shall be able to resist a thousand of our enemies; when hee shall make us a prayse and glory that men shall say of succeeding plantations, 'the Lord make it likely that of New England.' For wee must consider that wee shall be as a citty upon a hill. The eies of all people are uppon us."

†At least twenty-three states have towns named Salem. Three ships in the U.S. Navy have been called the USS *Salem,* and it is the name of a cigarette brand and a nuclear power plant.

New Canaans, not to mention various Bethlehems, Zions, and Hebrons throughout the nation.[21] From Increase Mather (Cotton Mather's father)[22] in the late seventeenth century, to Ezra Stiles and Timothy Dwight,[23] both of whom became presidents of Yale University, countless other theologians referred to the United States as the American Israel.[24]

This notion of America as the new Israel was so powerful that it transcended Puritan theology and even crossed over into the world of the far more secular, post-Enlightenment founding fathers, who, without embracing Puritan theology, found enormous value in using these biblical myths and symbols to give meaning to the colonial experience. In fact, on September 7, 1774, with the outbreak of the Revolutionary War, George Washington attended the First Continental Congress in Philadelphia and heard an Episcopal priest read Psalm 35, clearly suggesting that Americans, like the Jews before them, had earned the right to be called God's Chosen People, and that God would fight for America just as He had fought for Israel.*[25]

The analogy was clear. The British were tyrants like the Egyptian pharaohs; the American rebels seeking their freedom were akin to the Israelites in Exodus; and America was Zion, the New Jerusalem, the Promised Land. Even to committed secularists, such as Adams, Franklin,

*Likewise, before it adjourned on July 4, 1776, having passed the Declaration of Independence, the Continental Congress put together a special committee consisting of John Adams, Benjamin Franklin, and Thomas Jefferson to design a seal for the new nation. Inspired by the Bible, the committee came up with the idea of depicting "The Children of Israel in the Wilderness" by having a pharaoh riding in an open chariot with a sword in his hand as he and his men pursued the Israelites into the divided Red Sea. Moses was to be standing on the shore, his people behind him, extending his hand over the sea so it would destroy the Egyptians. "Rebellion to Tyrants is Obedience to God," read the text surrounding the image.

The Great Seal of the United States was subsequently changed, but even in its final form, which can be seen on the one-dollar bill, the thirteen stars, representing each of the original colonies, are arranged in the shape of the six-pointed Star of David.

Exactly why the thirteen-star constellation is arranged in the pattern of a Star of David–like hexagram is a matter of considerable dispute. Preliminary versions of the design show the constellation taking a random pattern, which is consistent with heraldic tradition. Historians have speculated on various possible reasons for the change, including George Washington's friendship with Haym Solomon, a Jewish banker who helped finance the American Revolution. In his description of the shield, Charles Thomson, the secretary of Congress who designed the final 1782 version of the seal, failed to answer why it was changed to a hexagram and simply wrote that the constellation "denotes a new State taking its place and rank among other sovereign powers."

and Jefferson, these images and biblical myths had an undeniable attraction in that they gave narrative coherence to the American Revolution, the meaning of which was indisputable: America was about freedom. It was the enemy of tyranny.

Of course, the Puritans were not the dominant force colonizing America. Church membership during the colonial period never exceeded 20 percent of the population,[26] which also included Catholics, Quakers, Lutherans, Congregationalists, Baptists, Presbyterians, and Jews, many of whom, like the Puritans, sought their own brand of religious expression. Other early settlers included those seeking economic opportunity in the New World and escape from criminal punishment in the Old. And the more worldly Founding Fathers were the rationalist men of the Enlightenment—George Washington, John Adams, Benjamin Franklin, Thomas Jefferson, John Jay, James Madison, Thomas Paine, and Alexander Hamilton—who envisioned a Republic that was very different from that imagined by the Puritans.[27] They rejected the idea of society's redemption through Jesus Christ, but were able to find common cause with the Puritans in fighting British tyranny.

Meanwhile, the Puritans saw the American Revolution itself as the birth of the new millennial age.[28] By asserting their identity as the Chosen People against the colonial British, they, too, played a key role in the Revolutionary War. As a result, the American struggle for independence was powered by a variety of forces.

On the one hand, the proponents of a rational, post-Enlightenment America advocated the separation of church and state, freedom of and *from* religion, and they explicitly asserted that "the United States of America is not in any sense founded on the Christian Religion."[29] On the other hand, the Puritans believed in a fundamentalist, theocratic state. In parts of colonial New England, they had gone so far as to make biblical law the law[30] of the land—just as Islamic fundamentalists later did in making Sharia, Islamic law, the rule in some Muslim countries in the Middle East and Africa.* Reason was their enemy. In *Memorable Providences and Wonders of the Invisible World*, Cotton Mather even

*Records from the Colony of New Haven in 1644, for example, say "Itt [*sic*] was ordered that the judicial lawes of God, as they were delivered by Moses, . . . be a rule to all the courts in this jurisdiction in their proceeding against offenders. . . ." Likewise, the Massachusetts Bay Colony under Winthrop was effectively a Puritan theocracy based on his religious beliefs.

argued passionately for mass executions of women for witchcraft.[31] Even at America's birth, we see the split between the religious conservatives and the secularists that would forever divide America.*

When it came to the founding of the new Republic—and writing the Constitution, the Bill of Rights, the Declaration of Independence—the post-Enlightenment Founding Fathers held sway. Because there was the potential for enormous discord among the thirteen colonies, they chose to accommodate religious values by stressing religious freedom over purity. "The founders' professed goal was to establish a nation true to the spirit of divine law, a spirit some understood in Christian terms and others according to the canons of Enlightenment philosophy," writes Forrest Church in *The Separation of Church and State*.[32] Church concludes that "these seemingly opposite world-views collaborated brilliantly and effectively to establish the separation of church and state in America."[33]

In the last half of the nineteenth century, American evangelicalism underwent a dramatic transformation and advanced what has become the most popular doctrine embraced by the Christian Right today: premillennial dispensationalism, a canon championed by evangelicals from Billy Graham to the late Jerry Falwell, popularized in apocalyptic bestsellers such as Hal Lindsey's *The Late Great Planet Earth* and Tim LaHaye's and Jerry Jenkins's Left Behind series, the same doctrine espoused by LaHaye and Gary Frazier on the hill in Megiddo, Israel.

The key agent of change was a renegade Irish Anglican theologian, author, and mystic named John Nelson Darby, sometimes known as the father of premillennialism.[34] As the leader of a strict fundamentalist sect known as the Plymouth Brethren, or Darbyites,[35] Darby won a modest following in England.[36] But when he began a series of American tours in 1859 his fame spread far and wide. Crippled and burdened by a homely, deformed face, Darby was nevertheless an intensely charismatic figure whose personality "could enslave by its sheer attractiveness."[37] He made his mark by systematically assembling the prophetic passages in the Bible into one overarching, grandiose, and intoxicating apocalyptic vision of End Times that combined divine redemption and millennial

*There were important exceptions, of course. In the nineteenth century, evangelicals who supported the Social Gospels by fighting poverty, crime, racism, and the like strayed to the left side of the political spectrum, as did Abolitionists who fought slavery before the Civil War, and the black church in the civil rights movement of the 1960s.

bliss with brutal justice and gruesome catastrophes into a horrifying, nihilistic fantasy.

At the heart of Darby's doctrine was the precept that biblical prophesies from even the most convoluted and abstruse part of the New Testament, the Book of Revelation, would come to pass exactly as predicted. Specifically, Darby preached that there are seven epochs, or dispensations, in human history, starting with the Garden of Eden and ending with the thousand-year reign of the saints after the Second Coming of Christ.[38] He asserted that we are now living near the end of the sixth dispensation, the period immediately preceding the Second Coming of Christ. Because Christ would return *before* the new millennium, this doctrine was known as premillennial dispensationalism and was a distinct break from the Puritan theology of postmillennialism, which posits that Christ will return only *after* a millennium of peace on earth.[39] But Darby's greatest fame came from popularizing the Rapture. The current era would end, he preached, when those who have accepted Christ were suddenly "caught up" from the earth to join the Lord in the air, thereby escaping the horrifying Tribulation that was to follow.

Much of Darby's theology was old hat. As author Paul Boyer points out in *When Time Shall Be No More,* various forms of it had been around for hundreds of years, and the Rapture doctrine had been expounded in America by Increase Mather two centuries earlier.[40] But Darby systematically wove diverse texts into a rich theological fabric that he diligently promoted through his writing and nationwide tours. He also revived the Christian Zionist notion of the restoration of the Jews into Palestine, an event to which he gave great prominence in his dispensational system. Once again, Christian Zionism was alive as part of American evangelicalism.

The times could not have been more propitious for America to embrace a new apocalyptic vision. Between 1780 and 1860, the number of congregations in the United States soared from 2,500 to 52,000.[41] Meanwhile, America was reeling from the Civil War, a conflict so bloodily apocalyptic that the lyrics for its anthem had been cribbed from the Book of Revelation. "Mine eyes have seen the glory of the coming of the Lord" went the "Battle Hymn of the Republic." "He is trampling out the vintage where the grapes of wrath are stored."

In the war's aftermath, a shattered America rapidly transformed itself into the greatest industrial power in the world. "Instead of utopia, the northern states experienced the rapid and painful transition from an agrarian to an industrialized society," writes Karen Armstrong in *The*

Battle for God.[42] "New cities were built, old cities exploded in size. . . . New immigrants poured into the country. Capitalists made vast fortunes from the iron, oil and steel industries, while workers lived below subsistence level."

Meanwhile, even the most basic tenets about man and society were being questioned. In 1859, the publication of Darwin's *The Origin of Species,* and, later, *The Descent of Man,* seemed to say that humans were just like animals and were not created by God. Liberal theologians in America and historians in Germany challenged the dogma that the Bible was divinely inspired.[43] In 1882, Friedrich Nietzsche proclaimed that God was dead.

Many of the institutions founded to promote Puritan orthodoxy gradually became something else entirely and promoted this new brand of post-Enlightenment humanism. Among them was Yale University, which was founded with the purpose of sending its graduates on as missionaries to spread Christianity throughout the world,[44] but which evolved into a training ground for a secular elite, including four generations of Bushes.*

But each step forward taken by modernism provoked an equal and opposite reaction from fundamentalists. Far from making man more virtuous, Darbyites believed, the Enlightenment, science, and modernism had left mankind so wanton and dissolute that God had to intervene and inflict untold misery upon the human race. And part of the seductive beauty of Darbyism was that it could be interpreted to explain away contemporary social and political forces in terms of Christ and the Antichrist, that it could frame the temptations of the modern world as going hand in glove with Satan.

In response to the shattering uncertainties wrought by the upheavals of Civil War and modernism, millions of Americans accepted Darby's emphasis on the inerrancy of the Bible and on the final conflict between God and Satan. In that context, Darby's version of the Rapture was an alluring revenge fantasy that allowed this aggrieved class of fundamen-

*Founded in 1701 as the Collegiate School, Yale won funding, and its current name, after Increase Mather, the president of Harvard, and his son, Cotton Mather, became disillusioned with Harvard's liberalism and contacted a businessman named Elihu Yale on behalf of the new college. More than two centuries later, in his 1951 book *God and Man at Yale,* the founding father of the modern conservative movement, William F. Buckley, took on his alma mater for promoting secular liberalism and undermining Christianity.

talists to believe that ultimately they would be able to look down from the Kingdom of Christ at elite secularists who had once mocked their beliefs, but would suffer through an eternity of hellfire and damnation.

Thanks to a series of highly publicized tours across the United States by Darby between 1859 and 1877, millions of Americans, particularly Baptists and Presbyterians, eagerly embraced his apocalyptic vision.[45] Dwight Moody, one of the fathers of modern American fundamentalism, became a Darbyite, and spread the word through huge tours all over America.[46] Darby's theology became an essential part of the curriculum at the Moody Bible Institute and other Bible schools that trained thousands of pastors who in turn spread the word across the land. In 1909, Congregationalist minister Cyrus Scofield published the *Scofield Reference Bible,* sometimes called "the most important single document in all of fundamentalist literature,"[47] annotating both the Old and New Testaments with thousands of extended footnotes explicating the Scriptures with Darby's system. It sold more than 10 million copies, and made Darby's premillennial dispensationalism the dominant theology in American fundamentalism.

At the turn of the century, the growing sense that America must fulfill its destiny as a Redeemer Nation became a driving force in America's first military adventure overseas. On February 15, 1898, the battleship *Maine* blew up in Havana harbor, killing 266 Americans. The explosion was probably accidental, but, spurred on by the cry "Remember the *Maine*!" the American military seized the Spanish colonies of Cuba and Puerto Rico before moving on to Guam and the Philippines.

"Evangelicals who [had] cheered the liberation of Cuba and the Philippines, now suddenly embraced their colonization," writes John Judis in *The Folly of Empire.*[48] The pro-war press, he adds, assured readers that "Americans were not acting like exploiters but in their special role as redeemers of the world." Or, as Republican legislator Albert J. Beveridge proclaimed, the end result of the American occupation "will be the empire of the Son of Man."[49] At last, America had projected its military might abroad. And as Teddy Roosevelt demonstrated in 1908 by sending America's Great White Fleet of sixteen ships on an unprecedented 43,000-mile voyage around the world, it was ready, willing, and able to act globally.

But the evangelicals' victories were often pyrrhic. In 1925, fundamentalist William Jennings Bryan won the historic Scopes Monkey Trial when a Tennessee court ruled that biology teacher John Scopes was guilty of teaching evolution.*[50] In the process, however, Clarence Darrow, the celebrated civil liberties lawyer, humiliated Bryan. H. L. Mencken derided Bryan's fundamentalist supporters as "gaping primates of the upland valleys."[51] The South and the entire fundamentalist movement emerged bruised and battered, characterized as credulous rubes and hicks, hate-filled yokels who were enemies of science and had no business in the modern world. Likewise, in 1933, Prohibition was repealed, overturning Bryan's successful campaign to pass the Eighteenth Amendment in 1918.[52] The fundamentalist foray into the national debate had turned into a devastating defeat.

In response, the religious right began a complex process of disengagement from formal national politics that lasted late into the Cold War. Increasingly, parishioners left mainstream Christian denominations for smaller sects. They regrouped and realigned themselves.[53] The religious right began to see itself as a beleaguered minority,[54] and embraced various forms of anti-Semitism, anti-Catholicism, and anticommunism.

In the fifties, Carl McIntire's American Council of Churches linked godless communism to the satanic beast prophesied in the Book of Revelation.[55] In 1962, Billy James Hargis, leader of the anticommunist Christian Crusade, declared that "the primary threat to the United States is internationalism." Along with McIntire, he had been an ardent supporter of Senator Joseph McCarthy's anticommunist witch hunts.[56] There was a new sense of anger and militancy. Fundamentalists felt marginalized—and with good reason. Their faith was under attack by mainstream secular culture.[57]

By midcentury, most secularists thought such right-wing fundamentalism had been consigned to the dustbin of history. The Atomic Age had begun. For better or for worse, man had seized control over his own fate. On the one hand, with the horrors of the Holocaust and Hiroshima still fresh, both the United States and the Soviet Union were developing the H-bomb, and with it the power to destroy humanity. On the other hand, the cosmos was being demystified at every level. Thanks to *Sput-*

*In 2006, right-wing polemicist Ann Coulter tried to link Hitler's atrocities to Darwinism, just as William Jennings Bryan asserted that there were close ties between the theory of evolution and German militarism and that Darwinism played a key role in starting the First World War. "Eugenics is applied Darwinism," Coulter said. "And it sticks out like a sore thumb that all of these German eugenicists preceding the Nazi regime were enthusiastic Darwinists."

nik and the space race, outer space was being conquered. Jonas Salk had developed a vaccine for polio. And in 1959, James D. Watson and Frances Crick decoded the basic building block of life, DNA. By the dawn of the sixties, man could go to the moon or wage war on poverty. Reason was triumphant.

Coming as they did in the postwar boom, these scientific marvels and technological advances gave birth to and fueled a new, insatiable, and extraordinarily powerful consumer culture that dramatically transformed everyday lives. For the first time, America's wide-open spaces were connected by newly paved interstate highways. The automobile of the fifties, festooned with extravagant fins inspired by jet-age aviation technology, became an essentially American metaphor for mobility, independence, freedom, and sexuality.

Thanks to radio and television, a vast continent suddenly became a global village. On the radio, Ray Charles transformed gospel into the devil's music, the secular—and sexual—beat of rhythm and blues. Rock and roll was born. Elvis made its sexuality suitable for mass-market consumption. For the martini set, Frank Sinatra, Dean Martin, Joey Bishop, and Sammy Davis Jr. led the Rat Pack in Vegas. Hugh Hefner's *Playboy* magazine was newly ascendant. In 1960, the FDA approved the birth control pill, enabling tens of millions of Americans to separate sexual behavior from procreation. The sexual revolution was under way.

And with the sixties came Bob Dylan, the Beatles and the Rolling Stones, marijuana and LSD. Powerful social movements—the antiwar movement, the civil rights movement, environmentalism, feminism, and gay rights—divided the country. With the notable exception of civil rights, which was rooted in the black church, these movements were largely secular in nature.

By 1965, secularism's triumph seemed so complete that one of America's best-known theologians—albeit a liberal one—was conspicuously waving the white flag of surrender. "God is teaching man to get along without Him, to become mature. . . ." wrote Harvard professor Harvey Cox, a Baptist minister, in *The Secular City.* "It may well be that the English word God will have to die. . . ."[58]

Given the cataclysmic tenor of the times in the sixties, it was not surprising how easily many Americans dismissed evangelicalism as nothing more than a gaudy, irrelevant, Bible Belt sideshow. After all, with the likes of JFK and RFK, Martin Luther King Jr., and Bob Dylan as the powerful voices of a new generation, who could take seriously fearmongering evangelicals railing about godless communism? The fact that evangelicals staged mass burnings of Beatles albums—after John Lennon said the band was more popular than Jesus—showed how astoundingly out of step they were with the times. While their hip cohorts lined up to see *Easy Rider,* students at evangelical Wheaton College had to lobby college administrators—just to see the film *Bambi.*[59] No wonder secularists saw evangelicals as quaint throwbacks to an America of country bumpkins and one-horse towns.

A case in point was *Time* magazine's historic April 1966 cover[60] that consisted of a black background and just three words: *Is God Dead?* Citing Harvey Cox and a host of European intellectuals, *Time* concluded that modern, secular man had realized he "did not need God to explain, govern or justify certain areas of life."[61] In other words, the game was over. Secularists had won. God was dead—except for a few Bible-thumping rednecks in the hinterlands.

In proclaiming the demise of God, however, *Time* had glossed over one extraordinary detail: no fewer than 97 percent of Americans believed in God.[62] True, the mainline churches with which *Time*'s editors were familiar—Presbyterian, Methodist, Episcopal, Congregational, American Baptist, and so on—were fading. But as the country's center shifted demographically toward the South and West, old-line religion gave way to "that old-time religion"—Southern Baptists, Pentecostalists, Assemblies of God, and a wide variety of independent evangelical and fundamentalist congregations.[63]

To these evangelicals, the bold scientific advances and highly prized consumer goods hailed by secularists, far from being genuine achievements, were daggers poised to strike at all that was sacred. The permissiveness of Dr. Spock's liberal child-rearing practices combined with the newly developed birth control pill and the backseat of a readily available Chevy to make a perfect Satanic recipe for godless promiscuity. Even the most profound scientific discoveries—DNA—posed daunting threats to evangelicals, by undermining the theology of Creation.

As the sixties wore on, the burgeoning counterculture continued to assault the values of evangelicals at every level. There was drug use, the sexual revolution, and feminism. Draft cards and American flags were burned. There were militant Black Panthers, "Burn, baby, burn" riots in the ghetto, the radical Weather Underground, and one militant antiwar demonstration after another. Revered institutions such as schools, the government, the military, and the church were in turmoil. Even the family was under attack thanks to liberation movements for women and gays.

At times it was as if the United States consisted of parallel universes that overlapped, but often didn't talk to each other, inhabited by two distinctly different peoples with different values, cultures, myths, heroes and villains, and history, one of which sent men to the moon and unraveled the human genome, the other which believes that the universe started six thousand years ago and will come to an end at any moment. Ultimately, both sides sought political power as a means of accomplishing their goals.

What had not been in the headlines, however, was that in the five decades since the Scopes trial, the religious right had quietly begun building a vast infrastructure of Christian colleges and Bible institutes, magazines, broadcast outlets, crusades to convert the unsaved, and thousands of new churches. Bob Jones University, founded in 1927, had become the biggest producer of fundamentalist preachers in the United States.[64] By 1930, there were at least 50 fundamentalist Bible colleges.[65] Two dozen more were founded during the Depression, and eventually there were more than 130 in the United States[66] that were effectively "separatist" educational institutions, teaching an entirely different system of beliefs and values from that taught in secular universities.

Fundamentalist publishing and broadcasting empires took root. As early as 1934, Gerald Winrod's *Defender Magazine* boasted 600,000 subscribers. McIntire's *Christian Beacon* reached 120,000 homes, and countless more in his *Twentieth Century Reformation Hour* radio show.[67] In 1950, Billy Graham, the relatively moderate face of evangelicalism during the postwar era, drew 50,000 people to Boston Common,[68] and proceeded to lead successful crusades all over the world.

In 1960, Pat Robertson founded the Christian Broadcasting Network, which televised *The 700 Club,* eventually reaching 96 percent of the television markets in the country.[69] Carl McIntire's *Twentieth Century Reformation Hour* was said to have been broadcast on as many as six hundred radio stations. In 1973, Paul and Jan Crouch founded the Trinity Broadcasting Network, which ultimately became the world's biggest Christian network, reaching six thousand stations in the United States and seventy-five countries around the world.[70]

For leading evangelists, the temptation to enter the political fray was overpowering. As early as 1951, Billy Graham had announced that the "Christian people of America will not sit idly by during the 1952 presidential campaign. [They] are going to vote as a bloc for the man with the strongest moral and spiritual platform. . . . I believe we can hold the balance of power."[71] In the end, however, he refrained from endorsing any candidate. Likewise, in 1972, under siege by the antiwar movement, Richard Nixon called out for the great "silent majority" to come to his aid. Graham tacitly supported him—but ultimately refrained from giving Nixon an explicit endorsement.[72]

At the same time, Jerry Falwell and other pastors of his generation were very much heirs to the unsavory racist legacy of the fundamentalists who had preceded them. Baptist preacher Gerald Winrod had traveled to Nazi Germany in the thirties and returned to denounce the satanic "Jewish New Deal."[73] In 1936, missionary Francis X. Buchman told the *New York World Telegram*, "I thank heaven for a man like Adolf Hitler, who built a front line of defense against the anti-Christ of Communism."[74] And members of the Ku Klux Klan and the racist White Citizen Councils were key parts of the fundamentalist constituency well into the fifties.[75]

Likewise, in 1958, Falwell, then a twenty-five-year-old pastor who had just founded the Thomas Road Baptist Church, told his congregation that if "Chief Justice Warren and his associates had known God's word," the 1954 Supreme Court decision desegregating schools, *Brown v. Board of Education*, "would never have been made. The facilities should be separate. When God has drawn a line of distinction, we should not attempt to cross that line. The true Negro does not want integration. . . . [H]is potential is far better among his own race."*[76]

But still wounded by the humiliation of Prohibition and the Scopes trial, most evangelicals stayed on the sidelines when it came to electoral politics. "The failure of the Prohibition movement discouraged conservative religions so much that they withdrew from the battle," Falwell said in an interview before his death in 2007.[77] "Most conservative evangelicals pulled out of the political scene totally."

In that context, nothing was more galling to Falwell than Reverend Martin Luther King Jr.'s electrifying leadership of the civil rights movement, and in 1965, Falwell took King on for crossing the inviolable

*In 1967, Falwell founded the Lynchburg Christian Academy, which was described by the *Lynchburg News* as "a private school for white students"—one of the scores of segregation academies that arose in the South after the Supreme Court decision. Blacks were admitted to the Lynchburg school less than two years after its opening, and for years Falwell took issue with the *Lynchburg News* report. According to Jonathan Wright, author of *Shapers of the Great Debate on Freedom of Religion*, in founding the school, Falwell may have been motivated more by the Supreme Court rulings against school prayer than by its desegregation ruling.

barrier between religion and politics. "The only purpose on this earth [for ministers] is to know Christ and to make him known," Falwell said. "Believing the Bible as I do, I would find it impossible to stop preaching the pure saving Gospel of Jesus Christ and begin doing anything else—including the fighting of Communism or civil rights reform. . . . Preachers are not called to be politicians, but to be soul winners."[78]

On the other hand, if Falwell and his colleagues decided to tear down the wall that kept fundamentalists out of politics, all bets were off.

CHAPTER THREE

Birth of the Neocons

Christian evangelicals weren't the only ones to react violently to the antiwar movement and the sixties counterculture. During the same period, when the antiwar movement took hold of the Democratic Party, a group of formerly leftish Democrats began to seek other political outlets. As characterized in *The Rise of the Counter-Establishment,* Sidney Blumenthal's 1986 chronicle of their ascent, the first neoconservatives were "mostly second-generation Jews torn between cultures,"[1] intellectuals grounded "in the disputatious heritage of the Talmud."[2] They came of age as part of a rarefied circle peopled with liberal New York thinkers, internationally known novelists, Beat poets, playwrights, political theorists, academics, and literary critics—Norman Mailer, Allen Ginsberg, Lillian Hellman, Lionel Trilling, Hannah Arendt, Paul Goodman, the *Partisan Review* crowd, and the like.[3] Their world was a cerebral hothouse of ferocious intellectual brawls in which ideas mattered. So did politics.

When it came to policy making, this was an era in which Ivy League scholars such as John Kenneth Galbraith and Arthur Schlesinger Jr., policy wonks at the Brookings Institute, Ford Foundation, and the Council on Foreign Relations, held sway. Within that elite, secular world, two of the founding fathers of neoconservativism, Irving Kristol, the managing editor of *Commentary* magazine from 1947 to 1952 and the father of *Weekly Standard* editor William Kristol, and Norman Podhoretz, the editor in chief of *Commentary,* carved out comfortable places as arbiters of taste and power. But in response to the turmoil of the sixties, in Kristol's memorable phrase, they became "liberals mugged by reality," made a sharp right turn—and neoconservatism was born.

It began modestly enough. In 1965, Kristol founded *The Public Interest* with Harvard sociologists Daniel Bell and, later, Nathan Glazer as co-editors. Podhoretz and Kristol published scores of articles questioning the conventional wisdom of Lyndon Johnson's Great Society and assaulting the shibboleths of the New Left by a host of like-minded intellectuals including Daniel Patrick Moynihan, Jeane Kirkpatrick, Diana Trilling, Seymour Martin Lipset, and Podhoretz's wife, Midge Decter. Not every writer who published in these journals identified himself as a neoconservative, but by and large they constituted a growing but tightly knit clique with a shared political sensibility that, over time, acquired real clout.[4]

In part, their apostasy could be attributed to angst about their careers and social standing. Podhoretz, in particular, created a veritable cottage industry by putting his status anxiety nakedly on display in one confessional memoir after another. In *Making It,* he unveiled his "dirty little secrets" about social climbing and corruption in the literary world; in *Breaking Ranks,* he codified his split with his left-wing past in ideological terms; and finally, in *Ex-Friends,* he recounted an assortment of hurts, slights, and betrayals by literary hipsters and friends next to whom Podhoretz had become hopelessly retrograde. "I have often said, if I wish to name drop, I have only to list my ex-friends,"[5] Podhoretz begins *Ex-Friends,* a memoir in which he chronicles his disaffection with Allen Ginsberg, Lillian Hellman, Hannah Arendt, and Norman Mailer.[6]

Once an acolyte of Ginsberg, who was the raw, emotive voice of the Beat generation and the author of *Howl,*[7] Podhoretz began attacking the "know-nothing Bohemians" for leading a "revolt of the spiritually underprivileged and crippled of soul" against "normal feeling and the attempt to cope with the world through intelligence." He published articles in *Commentary* such as "Boys on the Beach" by his wife, Midge Decter, who ruminated about the "hairless bodies . . . and smooth feminine skin" of gay men on the beaches of Fire Island, and asked why lesbians had "a marked tendency to hang out in the company of large and ferocious dogs."*[8]

At one point, Podhoretz even tried to get Allen Ginsberg to abandon his fellow Beats, an attempt Ginsberg recalled as "an epiphanous moment in my relation with Podhoretz and what he was part of—a large, right wing, protopolice surveillance movement."[9] Once part of the

*In response to which author Gore Vidal wrote: "Well, if I were a dyke and a pair of Podhoretzes came waddling toward me on the beach, copies of Leviticus and Freud in hand, I'd get in touch with the nearest Alsatian dealer pronto."

cultural vanguard, Podhoretz and his colleagues had become cultural cops, antihipsters.*[10]

Such cultural skirmishes were inconsequential, however, compared to the political upheavals in the works. Faced with the ascendancy of the liberal McGovern wing of the Democratic Party in the early seventies, the neoconservatives turned to Senator Henry "Scoop" Jackson (D-Wash.). A self-styled "muscular Democrat" who was sometimes known

*Two episodes reported about Podhoretz suggest that his sharp neocon shift to the right may have been partially facilitated by the treacherous, high-altitude social climbing he engaged in during the post-Camelot era. As reported in Sidney Blumenthal's *The Rise of the Counter-Establishment* (and denied by Podhoretz), on one occasion in the middle of that turbulent decade, Podhoretz invited Senator George McGovern to dinner at a French restaurant. McGovern, who apparently had never met Podhoretz's wife, Midge Decter, arrived early and sat alone.

When Podhoretz finally arrived and took his seat opposite McGovern, the senator sought to break the ice by pointing out some attractive women at nearby tables, as well as a couple who were less appealing. "Norman, you get to look at those good-looking [women], while I have to look at those turkeys," McGovern said, pointing to a woman in the latter category.

After a moment of uneasy silence, the flustered Podhoretz finally responded. "That's my wife," he said.

The other episode involved Jacqueline Kennedy, who had moved to New York after her husband's assassination and, thanks to her friendship with John Kenneth Galbraith, was emerging from the trauma that shook the entire country and devastated her family. As Podhoretz relates in *Ex-Friends,* he first met the former first lady when Richard Goodwin, the former Kennedy aide, called and asked if he could drop by with an unnamed friend who wanted to meet Podhoretz. "Within minutes, [Goodwin] showed up at my door with a jeans-clad Jackie Kennedy in tow," Podhoretz wrote.

At the time, Jackie, still in her thirties, was the stunningly beautiful and glamorous but fragile widow of John F. Kennedy and, arguably, the most-sought-after woman in the world. Podhoretz, by contrast, was the sometimes witty and charming but less-than-stunningly-handsome husband of Midge Decter, lived on the less-than-fashionable Upper West Side, and, when it came to stylish attire, was given to dowdy brown suits and brown shoes. Nevertheless, according to Podhoretz, they struck an "instant rapport" and "at her initiative" had tea regularly alone in her Fifth Avenue apartment—which was enough, apparently, to put ideas in his head.

"He thought she was coming on to him and he had a terrible crush on her," says someone who knew them both. "She was just trying to make new friends," says another source with knowledge of the episode. "It was a very fragile time in her life. She wasn't looking to have an affair. She was just trying to have a normal life. But he thought she wanted one and he was telling *everyone.*"

When Podhoretz finally cornered Jackie at a cocktail party and made his feelings known, the source says, she looked at him with an icy gaze the meaning of which was unmistakable. "Why, Mr. Podhoretz," she said. "Just who do you think you are?"

as "the senator from Boeing," Jackson fused strong support of labor and civil rights with his staunch Cold War opposition to the Soviet Union.[11] To the delight of the neoconservatives, his hawkish internationalist stance put American military power behind "moral realism" and support for democracy and human rights abroad.[12]

Jackson's other great contribution to neoconservatism was the bright young staff he assembled, five members of which would later become key figures in George W. Bush's war in Iraq and the grand neoconservative strategy to "democratize" the Middle East. They included Richard Perle, the so-called Prince of Darkness who later served as chairman of the Defense Policy Board Advisory Committee under George W. Bush; Douglas Feith, who became under secretary of defense for policy under Bush; Elliott Abrams, Bush's deputy national security adviser;[13] Abram Shulsky,[14] who headed the Pentagon's Office of Special Plans in the Bush administration and later its Iranian Directorate; and Paul Wolfowitz, the future deputy secretary of defense, who did not serve directly under Jackson but who worked with him to help persuade the Senate to fund an antiballistic missile system, and who became one of the principal architects of the Iraq War.[15]

Known as "the bunker" to staffers, Jackson's office became home to a cadre of ambitious young ideologues imbued with a powerful sense of mission. "We had a vision of fighting the lonely battle against the forces of darkness," one former Jackson aide told the *Washington Post*.[16] Dorothy Fosdick, Jackson's foreign policy adviser,[17] was den mother to the group—staffers referred to her as *bubbe* (Yiddish for grandmother)—and Jackson himself played the doting father. "He was paternalistic in every sense," Perle said.[18] "He was unusually so with me because my father had just died. He felt every young person ought to have a parent. He came naturally to that protective role. I get choked up talking about Scoop even now."

Jackson engendered such fierce loyalty because he and his acolytes shared a grandiose missionary belief that American values and principles were both virtuous and universal. This was the secular version of American exceptionalism, a romantic nationalism that saw America as a savior nation whose democratic values could save the rest of the world from communism and other sorts of tyranny—and had the moral duty to do so.

Few neoconservatives had doubts about the righteousness of their mission, in large part because many of them came of age fighting two men who embodied evil—Adolf Hitler and Joseph Stalin. As Douglas

Feith explained to Jeffrey Goldberg in the *New Yorker*, "[My] family got wiped out by Hitler, and . . . all this stuff about working things out—well, talking to Hitler to resolve the problem didn't make any sense."[19]

As a result, when it came to adversaries of any sort—not just Nazis, but communists and, later, Islamist fundamentalists—neoconservatives came up with policies in which military force was the first resort, not last. The mere thought of compromise brought forth the abhorrent notion of appeasement.*[20] "They go right back to Munich," said Stefan Halper, a White House and State Department official in the Nixon, Ford, and Reagan administrations and coauthor of *America Alone*, a highly regarded account of the rise of neoconservatism. "There is not a neocon you will ever meet who won't remind you of the tap, tap, tap of [British prime minister Neville] Chamberlain's umbrella. It is use force first and diplomacy down the line."[21]

On the other hand, Dorothy Fosdick,† Richard Perle, and many oth-

*As journalist Jim Lobe has noted, it need not take an existential crisis for neoconservatives to fall back on the Munich-appeasement trope. Lobe points out that Donald Kagan, a classicist who is the father of the *Weekly Standard*'s Robert Kagan, attributed his disillusionment with liberalism to an episode in the late sixties when Cornell University decided to negotiate with black students who were pressuring it into starting a black studies program. As Kagan put it, "Watching administrators demonstrate all the courage of Neville Chamberlain had a great impact on me and I became much more conservative." In other words, in Kagan's Manichaean, neoconservative worldview, black students, by analogy, were Nazis, and their goal of studying black history was the equivalent of world conquest and exterminating the Jews.

†Dorothy Fosdick was the daughter of Harry Fosdick, the famous liberal pastor of New York's Riverside Church who was a central figure in the modernist battle against fundamentalism in the twenties and thirties. The prevalence of left-wing backgrounds among neoconservatives was not confined to Jewish intellectuals. Even Jeane Kirkpatrick, who grew up in Duncan, Oklahoma, and attended Stephens College in Columbia, Missouri, managed to join the Trotskyist Young People's Socialist League as a college freshman in 1945. "It wasn't easy to find the YPSL in Columbia, Missouri," she recalled at a symposium in 2002. "But I had read about it and I wanted to be one. We had a very limited number of activities in Columbia, Missouri. We had an anti-Franco rally, which was a worthy cause. You could raise a question about how relevant it was likely to be in Columbia, Missouri, but it was in any case a worthy cause. We also planned a socialist picnic, which we spent quite a lot of time organizing. Eventually, I regret to say, the YPSL chapter, after much discussion, many debates, and some downright quarrels, broke up over the socialist picnic. I thought that was rather discouraging."

In addition to the elder Podhoretz and Kristol, other Trotskyist neocons included nuclear policy guru Albert Wohlstetter; Penn Kemble, a former chairman of the Young People's Socialist League (YPSL, pronounced Yipsel) who went on to found a host of hard-line lobbying groups during the Cold War; Joshua Muravchik, who also

ers, not unlike Podhoretz and Kristol, came to neoconservativism from the Left.[22] More specifically, many of them were Trotskyists—communist followers of Leon Trotsky, the Bolshevik revolutionary who lost a power struggle with Stalin and was eventually assassinated by a Soviet agent. An orthodox Marxist and Bolshevik-Leninist, Trotsky opposed both capitalism and Stalinism, and asserted that the Marxist promise of a proletarian revolution had failed in the Soviet Union because of Stalin's treachery. "The Trotskyists despised Stalin because he betrayed socialism," says Harvard Sovietologist Richard Pipes, a hard-liner himself who worked closely with Richard Perle, Paul Wolfowitz, and other neoconservatives, and whose son, Daniel Pipes, is a staunch neocon.[23] "I can see psychologically why it would not be difficult for them to become hard-liners. It was in reaction to the betrayal."

Critics who have pointed out the prevalence of Jewish intellectuals and ex-Trotskyists among neoconservatives, however, have done so at considerable peril. David Brooks, a neoconservative columnist at the *New York Times*, for example, derided detractors of the neoconservatives as anti-Semitic conspiracy nuts, "full-mooners" who believe there is "sort of a Yiddish Trilateral Commission" and for whom "Neo is short for 'Jewish.'"[24] "I have been amazed by the level of conspiracy-mongering around neocons," he told the *New York Observer.*[25] "I get it every day—'the evil Jewish conspiracy.' . . . We actually started calling it the Axis of Circumcision." Likewise, in *The Chronicle of Higher Education*, Robert Lieber asserted that theories about right-wing "Jewish masterminds" with "a Trotskyist legacy" echo "classic anti-Semitic tropes linking Jews to both international capitalism and international communism."[26]

Yet many neoconservatives honed their rhetorical skills and ideology as Trotskyists in Marxist cells of the postwar era—and were proud of it. As Irving Kristol himself put it, on graduating from City College in 1940, the "honor I most prized was the fact that I was a member in good standing of the [Trotskyist] Young People's Socialist League (Fourth International)."[27]

---

served as national chairman of the Young People's Socialist League from 1968 to 1973 and later became a prominent neocon Middle East scholar at the American Enterprise Institute; Stephen Schwartz, a former labor organizer and member of Social Democrats USA who became a neocon journalist; *Vanity Fair*'s Christopher Hitchens, a former Trotskyist who ended up supporting the Iraq War and becoming friends with Paul Wolfowitz; and Iraqi exile Kanan Makiya, who is best known as the author of *Republic of Fear.*

"To a great extent, I still consider myself to be [one of the] disciples of L.D," said Stephen Schwartz, a writer for the neoconservative *Weekly Standard,* referring to Trotsky affectionately by his birth name, Lev Davidovich Bronstein.[28] Writing on the website of the *National Review* (whose early writers included former leftists such as James Burnham, Frank Meyer, and Whittaker Chambers),[29] Schwartz noted that many first-generation neocons had strong ties to Trotskyism via Max Schactman,[30] Trotsky's leading American disciple, who stayed loyal to Trotsky through the thirties, but veered right afterward and became a supporter of Scoop Jackson.[31] It was not a lengthy journey from Schactman to Richard Perle, who called himself a socialist when he joined Scoop Jackson's staff,[32] and as a practitioner adopted an insistent, uncompromising, hard-line Bolshevik style.

Ultimately, the neoconservative ties to the Trotskyists had nothing to do with Jewish conspiracies or fantasies about proletarian uprisings. But they are worth noting because the neoconservatives saw themselves as the intellectual vanguard of a revolutionary movement. Just like the Trotskyists, they were visionaries, hardened ideological warriors on the cutting edge of history.

And just like the Trotskyists, they were highly skilled in navigating Byzantine sectarian political disputes and bitter internecine battles. For Perle and his associates, that meant they knew how to insinuate themselves masterfully into the bureaucracy, when and how to grease its wheels—and when to throw wrenches into the bureaucracy if necessary. It meant they had learned how to create a network of allies on whom they could always rely, how to recruit true believers who would never deviate from the ideological line. It meant they knew how to identify adversaries—and how to destroy them. "Whenever [Perle] saw someone as a political problem, he'd identify a vulnerability," an associate who had worked closely with Perle for several years told the *Washington Post.* "If he disagreed with a person, it was a matter of life and death. There was a willingness to play very rough. . . . There was a constant discussion of people to be promoted, people to be helped, people to be gotten rid of. This is a good Bolshevik principle. We have to build our own cadres, people who support our philosophy."[33]

As it happened, the most influential figure in the neoconservative orbit was a Trotskyist, but not a member of Scoop Jackson's staff. Back in the fifties when Richard Perle was in his junior year at Hollywood High

School,*[34] he met the man who became his guru when he was invited over to swim at the house of a classmate named Joan Wohlstetter.[35] Out by their pool, Perle was introduced to her father, Albert Wohlstetter, an analyst at the RAND Corporation, the mammoth global policy think tank, who gave Perle an article to read, "The Delicate Balance of Terror," about the strategic relationship between the United States and the Soviet Union.[36] "Perle's relationship with Joan Wohlstetter wilted, but her father became his close friend and intellectual mentor. [37]

A protégé of the influential German-born academic Leo Strauss, who is often said to be the intellectual godfather of neoconservatism, Wohlstetter, who died in 1997, was an even more important force in the movement. Thanks in large part to Wohlstetter, to his methodology, his demeanor, his political know-how, proto-neocons learned how to turn their ideas into political action. "Albert Wohlstetter was one of the most important unknown men of the twentieth century, and he liked it that way," said former *Wall Street Journal* economics columnist Jude Wanniski. "He was essentially, from the 1950s on, the man who played chess for our political establishment against the Soviet Union. There really was only one guy—master chess player—who was taking the hawk position. . . . Wohlstetter . . . was the mastermind. Maggie Thatcher didn't make a move on national security unless it was cleared by Albert. . . . He was truly a genius."[38]

Wohlstetter was often accompanied by his wife, Roberta, another RAND analyst who was the author of a Bancroft Prize–winning history of Pearl Harbor and was known as "the first lady of intelligence."[39] Wohlstetter was tall, animated, and self-assured, wore a white goatee, and carried himself with "a very European, very aristocratic demeanor," said Fred Kaplan, author of *Wizards of Armageddon,* a 1983 study of Wohlstetter and other thinkers at RAND.[40] A gourmet in a prefoodie era, he and Roberta famously served such exotic fare as fondue at their lavish Hollywood Hills dinner parties, accompanied by elaborate instructions on the optimal ratio of fondue dipping necessary to achieve maximum flavor.

Wohlstetter was as rhapsodic about his work as he was about food.

*Perle's fellow alumni at Hollywood High included Judy Garland, Mickey Rooney, Fay Wray, Lon Chaney Jr., Carole Lombard, Alan Ladd, Barbara Hershey, James Garner, Carol Burnett, Ricky Nelson, Laurence Fishburne, Keith and Robert Carradine, John Huston, Tuesday Weld, Sharon Tate, Jill St. John, many other stars—and last, but not least, former *New York Times* journalist Judith Miller.

"Albert waxed and never waned [when he spoke in public]," one colleague recalled. "He was an uncontrollable missile. He wouldn't accept notes, he wouldn't take glances, he wouldn't take nudges." At a talk in Los Angeles, Wohlstetter rambled on endlessly through the entire morning session and into the afternoon.[41]

Richard Holbrooke, who later became ambassador to the United Nations, recalled editing an article Wohlstetter had written for *Foreign Policy*. "It was the most difficult single editing job of my entire life," said Holbrooke. "[Wohlstetter's articles] were hell to edit, because . . . I kept trying to make them more accessible to the general reader, and he kept arguing the precision of language."[42]

When he studied at Columbia University, Wohlstetter had been a member of an obscure Trotskyist sect known as the Fieldites that was adept at conducting run-throughs of world historical clashes in their heads.[43] One of several analysts who is said to have been the model for the title character in Stanley Kubrick's *Dr. Strangelove* (the original working title for which came from Wohlstetter's most famous paper, "The Delicate Balance of Terror"),*[44] Wohlstetter evolved into an expert in game theory and systems analysis. The high priest of nuclear strategy who added phrases like "fail-safe" and "second strike" capability to the nuclear lexicon,[45] Wohlstetter told the national security establishment that our Strategic Air Command bombers and ICBMs were vulnerable to Soviet attack.[46]

Intellectually, his legacy went far beyond merely applying systems analysis to weapons deployment. "It was his analytic demeanor and his style," said Fred Kaplan. "Wohlstetter was politically very savvy, much more so than his brethren. He would not just do a study, he would give briefings on it, hundreds of times. Nobody did this back then."

Thanks to his briefings and his rigorous methodology, Wohlstetter was able to project the illusion of greater certainty, even when it wasn't justified. "The veneer of scientific analysis can be very misleading," said Kaplan. "At times, his numbers were based on extremely inaccurate intelligence, so it could be much less accurate than thought. It's a case of garbage in, garbage out. If the inputs are all screwed up, the results will be screwed up too."[47]

Wohlstetter diverged most sharply from the realists when it came to

*Others who have vied for the honor include hydrogen bomb designer Edward Teller, Herman Kahn of the Hudson Institute, rocket scientist Werner von Braun, and Henry Kissinger.

the doctrine of "mutual assured destruction" (MAD)—the notion that both sides would be wiped out in a nuclear war, that nuclear weapons had made war obsolete for all but the insane. Unlike the realists, Wohlstetter believed that we could not afford to be so Pollyannaish as to believe that the Soviet Union would behave rationally. Ultimately, his conclusion that deterrence was a much more delicate proposition provided fodder for a new breed of Cold War hawk.

As he won fame in the national security world, Wohlstetter earned a reputation as a "mad genius" of sorts, a guru who collected creative young minds. Perle was hardly his only acolyte. In 1965, when Wohlstetter taught at the University of Chicago, a young grad student named Paul Wolfowitz became his protégé,[48] as did Zalmay Khalilzad, who later became ambassador to Iraq and ambassador to the United Nations under George W. Bush.[49]

Wohlstetter had a profound influence on them. "A key to understanding how Richard [Perle] and Paul [Wolfowitz] and I think is Albert," said James Woolsey, a staunch neocon hawk who met Wohlstetter in 1980 and later became head of the CIA. "He's had a major impact on us."[50] Wohlstetter's methodology, his applications of game theory and systems analysis to national security issues, became especially crucial to Senator Jackson, who used them as key elements in his hawkish Cold War policies.[51]

But Wohlstetter's influence was not merely a function of his ideas. In the national security world, being a member of Wohlstetter's entourage was a passport to the corridors of power. Indeed, in 1969, Wohlstetter was instrumental in sending two of his protégés, Richard Perle and Paul Wolfowitz, to Senator Jackson's office to work on a report about ballistic missile systems.[52] That summer, Wohlstetter also arranged for both Wolfowitz and Perle to work for the Committee to Maintain a Prudent Defense Policy, a Washington-based group cofounded by former secretary of state Dean Acheson and former secretary of the navy Paul Nitze. Wolfowitz and Perle remained close from that time on.[53] Later, in 1985, Wohlstetter introduced Wolfowitz and Perle to another acolyte, Ahmed Chalabi, who ultimately played a key role in helping them start the war in Iraq.[54]

As Wohlstetter's protégés got to know him, some of them felt the Strangelovian aspect had been overstated and was moderated by idealism and moral considerations. "I thought, well, maybe he was also associated with these sorts of cold-blooded systems analysts who kind of seemed to leave the moral piece of politics and strategy as though it wasn't part of the equation," Paul Wolfowitz said in an interview with Sam Tanenhaus for *Vanity Fair.* "It was terrifically gratifying to me as I got to know him better, to realize that there were intensely moral considerations in the way he approached these issues."[55]

"Albert believed he was put here on earth to be a man who would increase the security of the United States at the expense of those who threatened that security, and he was never going to be satisfied until there was nobody around at all who owned a slingshot that would allow them to be a potential David against the American Goliath," said Jude Wanniski, a one time Wohlstetter acolyte himself. "He basically believed that was the only way for a truly secure peace and that America was the only country that could get a secure peace for the world. And part of that means, if you look down the road and see a war with, say, China, twenty years off, go to war now."[56]

Not every bright analyst was the right fit for the Wohlstetter team. On one occasion, Wohlstetter and his wife unexpectedly dropped by the Pentagon offices of Colonel Patrick Lang, the director of the Defense Humint (Human Intelligence) Service in the Defense Intelligence Agency, and, not having announced the purpose of their visit, launched into a discussion of philosophy from Plato to Nietzsche. "They were both extremely Olympian," Lang said. "They acted as though they were God-like figures who had come to visit the earth. They weren't interested in telling me what they wanted. They just said that Paul [Wolfowitz] had wanted them to talk to me. So they sat there on my couch and we talked about how wonderful they were, their illustrious friends and associates, and did I understand world history and the classics."[57]

As they went on, Lang recalled, an unspoken question hung in the air. Later, he realized he was being tested. "After a while they became impatient with my responses and left, never to return," he said. "Clearly, I had failed the test."[58] That may have been because Lang actually had the temerity to challenge the Wohlstetters at various points in the conversation.

To join Team Wohlstetter, apparently, one had to embrace unquestioningly his worldviews, which eschewed old-fashioned intelligence as a basis for assessing the enemy's intentions and military capabilities in favor of elaborate statistical models, probabilistic reasoning, systems analysis, and game theory developed at RAND. As an analyst put it in the *Bulletin of the Atomic Scientists,* "This methodology exploited to the hilt the iron law of zero margin of error. . . . Even a small probability of vulnerability, or a potential future vulnerability, could be presented as a virtual state of national emergency."[59]

For Wohlstetter's disciples that meant that the Soviet Union was simply too evil to be treated as a rational actor. In other words, it didn't matter what weaponry the Soviets actually had, because they were so irrational. As one analyst summed it up, the essence behind Wohlstetter's policy was quite simple: "Expect and prepare for the worst case imaginable."[60]

It was a principle his acolytes would pursue for decades to come—with disastrous results.

CHAPTER FOUR

The Foreshadowing

To neoconservatives, Richard Nixon's landslide victory over George McGovern in 1972 was a casebook study of what was wrong with the defeatist, isolationist policies of the liberals who had captured the Democratic Party. If their wishes had been realized, Scoop Jackson, who had been a candidate for president, would have won the Democratic nomination. But with the Vietnam War grinding on interminably, hawkish Cold Warriors were in disfavor and antiwar forces gave it to George McGovern.

If the neoconservatives had been completely marginalized within the Democratic Party, the Republicans didn't have much more to offer them. That was because both Nixon and Gerald Ford, who succeeded him after the Watergate scandal, were still enthralled by another nemesis of the neoconservatives—Henry Kissinger. As a champion of realpolitik, Kissinger was scorned by neoconservatives who valued their ideology far more than his pragmatism. Yet in bureaucratic terms at least, Kissinger was at the apex of his power, simultaneously holding not one but two high-level posts, as secretary of state and national security adviser.

Meanwhile, the frostiness of the Cold War was giving way to an unlikely camaraderie between the two nuclear superpowers. There was one summit meeting after another, nearly a dozen bilateral commissions, and agreements with the Russians on everything from health care to strategic arms limitations. Meetings at the Kremlin between General Secretary Leonid Brezhnev and Kissinger were enlivened by bizarre frat house hijinks. According to official memoranda, the normally dour Brezhnev mischievously ran off with Kissinger's briefing books as if he were stealing them. Once a grim hard-liner who invaded Czechoslovakia, a villain straight out of central casting, Brezhnev now amused himself by mussing up the hair of National Security Council staffers and making one lame joke after another.[1]

If the neocons were unhappy with these developments, it was because this giddy new era of détente threatened to rob them of their favorite enemy—the Soviet Union. Not content to stay on the sidelines as Kissinger hogged the headlines, anti-McGovern Democrats—among them Henry Jackson; Irving Kristol; Norman Podhoretz; Midge Decter; Daniel Patrick Moynihan, then a Harvard academic and Nixon's domestic affairs adviser; Jeane Kirkpatrick, who, like Moynihan, would later become ambassador to the U.N.; Senator Hubert H. Humphrey, the former vice president and unsuccessful presidential candidate; Ben Wattenberg, a Lyndon Johnson speechwriter and Hubert Humphrey adviser;[2] and James Woolsey, general counsel to the U.S. Senate Committee on Armed Services, who later became director of the Central Intelligence Agency[3]—founded the Coalition for a Democratic Majority (CDM) to lobby for a tougher policy of "peace through strength" with the Soviet Union.

Through the summer and early fall of 1972, they met secretly in Washington at the old Federal City Club, under the guidance of Penn Kemble, yet another Trotskyist-turned-neocon. A former national chairman of the Young People's Socialist League, "at a tender age Penn had already learned, or perhaps designed, the big secret: that the Yipsels planned to take over the world in a blizzard of letterheads."[4] A lifelong socialist[5] and Scoop Jackson Democrat, Kemble and his colleagues launched a host of alphabet soup organizations that were essentially front groups for neocons and hard-line Cold Warriors.*

This was the beginning of a movement that wrote countless policy papers and scores of op-ed pieces in influential journals and newspapers throughout the country, mounted campaigns to discredit its adversaries, lobbied Congress, and bit by bit took over the entire foreign policy and national security apparatuses. Their first target was détente, the relaxing of tension between the two superpowers, which they viewed with alarm.[6] To that end, they assembled a group of hawkish anticommunist foreign-policy makers to put together intelligence showing that the "liberal" CIA had dangerously underestimated the Soviet threat. The group became known as Team B.

* * *

*They included the Coalition for a Democratic Majority (CDM), the Committee for a Free World (CFW), Committee to Maintain a Prudent Defense Policy (CMPDP); Friends of the Democratic Center in Central America (PRODEMCA), and, most important, the Committee on the Present Danger (CPD).

Team B's birth dated to a dinner party for a group of conservative strategic thinkers on June 4, 1974, in Santa Monica, California.[7] At the time, most of the nation was transfixed with the nationally televised Watergate hearings that would determine the fate of Richard Nixon's presidency. But as Anne Hessing Cahn relates in *Killing Detente: The Right Attacks the CIA,* this group had other things in mind. The host, James Digby, a senior researcher at RAND, had invited prominent journalists from *U.S. News & World Report, Time,* the *Wall Street Journal,* and the like over for Greek food and a 1964 Clos de Vougeot. But the most important personage present was the cohost, Albert Wohlstetter.

The next day, Digby's guests joined several dozen other like-minded intellectuals at a conference at the Beverly Hills Hotel called "Arms Competition and Strategic Doctrine" at which Wohlstetter held forth using specially declassified materials as ammunition to accuse the Pentagon of systematically underestimating Soviet military strength.[8] Not long afterward, Wohlstetter published his conclusions in *Foreign Policy,* the *Wall Street Journal,* and the *Strategic Review.*

The campaign had just begun. A few weeks later, in July, Paul Nitze, the brilliant Cold Warrior who had been one of the key figures behind U.S.-Soviet policy, used Wohlstetter's work, in testimony before the House Armed Services Subcommittee on Arms Control, as a basis to assert that Henry Kissinger and the CIA had dangerously underestimated the Soviet Union's strength and its intentions. Soon, one Cold Warrior after another—many of them nascent neoconservatives—began attacking the moderate policies of the détente and the CIA's intelligence.[9] This was the beginning of a thirty-year fight against the national security apparatus in which the neocons mastered the art of manipulating intelligence in order to implement hard-line, militarist policies.

The neocons weren't the only ones who were unhappy with the Ford-Kissinger policies. In the immediate aftermath of the Watergate scandal, and Nixon's resignation in August 1974, President Ford had been widely hailed as the nation's healer. But as the scandal receded from the headlines, a growing feud between Defense Secretary James Schlesinger and Kissinger gave the impression that the White House was at sea.[10] The fall of Saigon in April 1975 left indelible images of humiliation on the American consciousness. By late 1975, Ford's approval rating had fallen to 47 percent.

But the situation had also created a power vacuum ripe with irresistible opportunities for the two ambitious young White House officials. A handsome, lantern-jawed collegiate wrestler at Princeton, and later a navy fighter pilot in the fifties,[11] Donald Rumsfeld had served four terms in Congress before joining the Nixon administration*[12] as director of the Office of Equal Opportunity. To Henry Kissinger, Rumsfeld was "a special Washington phenomenon: the skilled full-time politician-bureaucrat in whom ambition, ability, and substance fuse seamlessly."[13] Nixon was more succinct, and paid Rumsfeld the highest compliment he knew. "He's a ruthless little bastard," Nixon said.†[14]

Immediately after Gerald Ford was sworn in, Rumsfeld had been appointed his chief of staff and brought on board his protégé, Dick Cheney, a thirty-three-year-old Yale dropout from Wyoming, as his deputy. At the time, aside from being Rumsfeld's devoted courtier, Cheney's greatest asset was his readiness to do routine scut work no one

*Under Nixon, Rumsfeld drew up what became known as "Rumsfeld's Rules" for serving in the White House, some of which might later be seen with considerable irony in light of his tenure in the George W. Bush administration:

Don't become, or let the President or White House personnel become, one President. Don't forget it and don't be seen by others as not understanding that fact.

Don't take the job, or stay in it, unless you have an understanding that you are free to tell him what you think, on any subject, "with the bark off"—and have the freedom—in practice—to do it.

Learn quickly how to say "I don't know." If used when appropriate, it will be often.

If you foul up, tell the President and others fast, and correct it.

In our free society, leadership is by consent, not command. To lead, a President must, by word and deed, persuade. Personal contact and experience are necessary ingredients in the decision-making process, if he is to be successful in persuasion and, therefore, leadership.

Where possible, preserve the President's options—he will very likely need them.

Know that it is easier to get into something than it is to get out of it.

Don't become, or let the President or White House personnel become, obsessed or paranoid about the Press, the Congress, the other Party, opponents, or leaks. Understand and accept the inevitable and inexorable interaction among our institutions. Put your head down, do your job as best you can, and let the "picking" (and there will be some) roll off.

Don't speak ill of another member of the Administration. In discussions with the President, scrupulously try to give fair and balanced assessments.

Never say "the White House wants." Buildings don't want.

†Rumsfeld's Machiavellian self-aggrandizement continued unabated during the Reagan era. According to Victor Gold's *Invasion of the Party Snatchers,* when asked if Rumsfeld would make a good running mate for the Gipper, Reagan aide Lyn Nofziger replied, "Rummy would be fine. But we'll have to hire a food taster for Reagan."

else wanted to do. Cheney was the guy who made sure the plumbing in the White House bathrooms got fixed. Cheney was the guy who made sure the right kind of salt shaker was available for White House table settings.[15] But as the 1976 presidential election season approached, he and Rumsfeld started gunning for bigger game. Like the neocons, their target was Kissinger and his policy of détente. But ultimately, Rumsfeld had his eye on the White House.

According to James Mann's *Rise of the Vulcans,* the maneuvering began when Rumsfeld heard about a minor conflict between a low-level Kissinger aide and White House press secretary Ron Nessen—and immediately told the president about it. "Rumsfeld was using the incident to drag me into a behind-the-scenes struggle to curb the power of Kissinger," Nessen later wrote.[16] Likewise, when he discovered that Kissinger secretly recorded all his phone calls, Rumsfeld went directly to Ford with the revelation. Soon, Kissinger struck back angrily. "Don't listen to [Rumsfeld], Mr. President," Kissinger snapped at Ford.[17] "He's running for president in 1980."

At the same time, Rumsfeld was taking on Vice President Nelson Rockefeller. With Rumsfeld, a hotheaded young hard-liner on the way up, and Rockefeller, head of the dying liberal wing of the Republican Party, a certain amount of discord was to be expected. But, according to one White House aide, the two men "loathed each other far beyond" the parameters of normal professional rivalries.[18] William Seidman, chief of Ford's Economic Policy Board, compared Rumsfeld's machinations against Kissinger and Rockefeller to those of the Wizard of Oz. "He thought he was invisible behind the curtain as he worked the levers," said Seidman, "but in reality everyone could see what he was doing."[19]

Meanwhile, Cheney helped keep Rockefeller in his place. "In the Ford Administration, we had major problems in managing the vice president," Cheney later said. "I was the SOB, and on a number of occasions got involved in shouting matches with the vice president."[20]

Finally, in what became known as the Halloween Massacre of 1975, Rumsfeld and Cheney helped orchestrate an extraordinary White House coup that toppled four giants who for years had stridden confidently across the world stage: Vice President Nelson Rockefeller, Secretary of Defense James Schlesinger, CIA director William Colby, and Secretary of State Henry Kissinger. Exactly how it happened is a matter of considerable dispute. In *Rise of the Vulcans,* James Mann makes the case that Rumsfeld's role has been greatly exaggerated and that the primary motivating force was really President Ford's strong resentment toward Secretary of Defense James Schlesinger.[21]

Nevertheless, when the dust settled, the big winners were Rumsfeld and Cheney. Schlesinger and Colby had been fired. Rockefeller had been dumped from the 1976 presidential ticket, thereby eliminating a key Rumsfeld adversary and potentially opening the vice presidential slot for Rumsfeld himself. Kissinger was ousted as national security adviser and, since Rockefeller, his chief patron, had also been wounded, actually considered resigning, before deciding to stay on as secretary of state.[22] Meanwhile, Rumsfeld, forty-three, was appointed secretary of defense, and Cheney, thirty-five, ascended from the bowels of the White House to become chief of staff. They were the youngest men in history to take their respective jobs.

For the coup de grâce, Rumsfeld, who nursed presidential dreams of his own, also went after one of his chief rivals—George H.W. Bush. It was not the first time Rumsfeld had Bush in his crosshairs. When Nixon resigned in 1974 and Gerald Ford became president, the Republican National Committee overwhelmingly recommended that Ford give Bush the vice presidency. Rumsfeld, however, who was then Ford's chief of staff, had convinced the president to pick Rockefeller instead.[23]

Now that Rockefeller had been thrown off the Republican ticket for 1976, Rumsfeld and Bush, who then headed the U.S. liaison office to the People's Republic of China,* were rivals again. With his spectacular résumé, Bush was again a natural choice to become Ford's running mate. But Rumsfeld was determined to thwart him.[†24]

The previous year Senator Frank Church (D-Id.) had led a Senate investigation of illegal intelligence gathering by the CIA. In the wake of its reports about the Agency's scandalous abuses of power, including attempts to assassinate foreign leaders, affiliation with the CIA had become political suicide for a politician with White House ambitions. According to Victor Gold, a former speechwriter for Bush and author of *Invasion of the Party Snatchers,* it was precisely because of the job's political liabilities that Rumsfeld persuaded President Ford to make

*Bush was effectively ambassador, but at the time, the United States had not yet extended diplomatic recognition to China.

†Ultimately, Bob Dole beat out both Bush and Rumsfeld to become Ford's running mate in 1976. The Ford-Dole ticket, however, lost to Jimmy Carter and Walter Mondale.

Bush director of the CIA. When Bush took the job, Rumsfeld said, that would "sink the sonofabitch for good."*[25]

During the confirmation hearings, Bush was publicly asked to promise that he wouldn't be Ford's running mate in 1976. "Bush thought it was a total violation of his constitutional rights," said Pete Teeley, who served as press secretary to Bush during his vice presidency.[26] "He was not happy. But he was always a team player and he did as he was asked."

Nevertheless, Bush had a good idea as to who was working against him behind the scenes. "Bush thought Rumsfeld planted that question," a Bush friend said in Andrew Cockburn's *Rumsfeld: His Rise, Fall, and Catastrophic Legacy.* "George Bush is the most polite of men. . . . I've only heard him speak with bitterness about two people."[27] One of them was Donald Rumsfeld.

Now that he was running the Pentagon, Rumsfeld, who had become a hard-line Cold Warrior during his tenure as ambassador to NATO, gave one speech after another attacking Kissinger's policies of détente.[28] In early 1976, when Kissinger was safely away in Moscow, he made his move. By orchestrating a secret rump meeting of the National Security Council in Kissinger's absence, he ultimately persuaded President Ford to shelve the Strategic Arms Limitation Talks, the series of ongoing negotiations between the United States and the Soviets to limit nuclear weapons.[29] "Rumsfeld won that very intense, intense political battle . . ." said Mel Goodman, head of the CIA's Office of Soviet Affairs at the time. "Now, as part of that battle, Rumsfeld and others, people such as Paul Wolfowitz, wanted to get into the CIA."[30]

Their objective, which prefigured the run-up to the Iraq War more than twenty-five years later, was to hijack the national security apparatus and find intelligence to support a much harsher view of the Soviet Union and its plans to fight and win a nuclear war.

When George H.W. Bush became CIA director in early 1976, one of

*In his autobiography, *Looking Forward,* George H.W. Bush wrote about the incident:

> White House chief of staff Donald Rumsfeld had a reputation as a skillful political in-fighter. It was inevitable that he'd be singled out in any rumor having to do with engineering my move to the CIA.
>
> In a meeting in his office, Rumsfeld vehemently denied the rumor.
>
> I accepted his word.

the first key decisions he faced was whether or not to allow neoconservative outsiders to pursue Wohlstetter's allegations. Saying yes meant giving the okay to a new team of intelligence analysts—Team B—to reexamine highly classified data and give an alternative assessment that might call into question the integrity of the CIA. Within the Agency, one faction warned that the deck would be stacked against them because Team B was dominated by hard-line ideologues who already subscribed to Wohlstetter's line. "Most of us were opposed to it because we saw it as an ideological, political foray, not an intellectual exercise," said Howard Stoertz, the national intelligence officer in charge of the national intelligence estimate (NIE) on the Soviets. "We knew the people who were pleading for it."[31]

Bush's predecessor, William Colby, had rejected the idea for precisely those reasons. But in a stunning break with CIA protocol, Bush followed the advice of William Hyland, the newly appointed deputy national security adviser. According to Hyland, the CIA "had been getting too much flak for being too peacenik and detentish. . . . I encouraged [Bush] to undertake the experiment, largely because I thought a new director [of the CIA] ought to be receptive to new views."[32]

It was unprecedented for the CIA to give access to so much highly classified data to outside critics. To ensure that it wouldn't jeopardize the integrity of the Agency, the entire enterprise was to be conducted in secret.

Now Team B went to work testing the Agency's analyses of the Soviet nuclear threat. Its leader, Harvard Sovietologist Richard Pipes, fit right in with the new cadre of hard-liners. As Pipes saw it, the Soviets were determined to fight—and win—a nuclear war.[33] Senator Jackson was so enamored of Pipes that he put him up at the Hay-Adams Hotel in Washington and had him shepherded around town by Jackson's young adviser, Richard Perle.*[34]

Not long after he got to Washington, Pipes went to Paul Nitze's office in Arlington, Virginia. "I had seen him on TV and thought he was so good, I asked him to be on Team B," Pipes said.[35] At Richard Perle's suggestion, Pipes also hired Paul Wolfowitz, the young protégé of Wohlstetter who had worked with Perle and Scoop Jackson and at the Arms Control and Disarmament Agency. One would be hard put to

*Asked about his role in Team B, Richard Perle replied by e-mail: "Sorry, but I want nothing to do with *Vanity Fair*." (The author is a contributing editor at *Vanity Fair*.)

find a more fitting intellectual heir to Wohlstetter than Wolfowitz, a trained mathematician-turned–Cold Warrior whose own father, Jacob Wolfowitz, had actually taught math to Wohlstetter at Columbia University. Having lost a number of relatives in Poland to the Nazis, Wolfowitz began to apply the lesson of the Holocaust to the Soviets and argued for the deployment of tactical nuclear weapons in Europe.[36]

According to Anne Cahn, who also worked at the Arms Control and Disarmament Agency, though in the Carter administration, Pipes's group had an "incestuous closeness"[37] because several of them had worked closely together and shared an "apoplectic animosity toward the Soviet Union."*[38] Officially, Wohlstetter himself was not a member, but his influence was pervasive: the panel's starting point was the work Wohlstetter published the previous year; his protégé Richard Perle had discovered Team B leader Richard Pipes; and throughout the process Wohlstetter remained in close contact with his former student Paul Wolfowitz and other panel members who were his colleagues.

On November 5, 1976, after months of examining highly classified documents, Team B finally debated its CIA adversaries, Team A, head to head in what was billed as "an experiment in competitive-threat assessment."[39] When it was over, by all accounts, the Agency was humiliated—largely because its young, unseasoned officers couldn't measure up against giants in the field such as Paul Nitze. "People like Nitze ate us for lunch," a member of Team A said. "It was like putting Walt Whitman High versus the Redskins. I watched poor GS-13s and -14s [middle-level analysts] subjected to ridicule by Pipes and Nitze. They were browbeating the poor analysts."[40]

But it was also because they didn't realize the extent to which ideology could override the factual basis of their intelligence-based policies. "If I had appreciated the adversarial nature, I could have wheeled up different guns," said Howard Stoertz.[41]

Stoertz had initially advised CIA director George Bush against participating in the enterprise, but even he did not fully appreciate that Team B was playing a different game than the CIA—very different. Most CIA

*Team B actually consisted of several panels, of which the one that was most frequently cited was the Strategic Objectives Panel. Its members included Seymour Weiss, the ambassador to the Bahamas; air force general John Vogt; USC professor William Van Cleave, who was known for making jokes about dropping bombs on civilians; Foy Kohler, former ambassador to the Soviet Union; retired air force general Jasper Welch; and Thomas Wolfe, a retired air force colonel working at the RAND Corporation.

officers who participate in the preparation of the National Intelligence Estimates (NIE) are well aware that what they do is not immune from politics.[42] After all, NIEs are called "estimates" because they contain a degree of uncertainty. They are the product of ongoing debates about policy, methodology, and political conflicts, but they are still based on facts. Dozens of high-level analysts spend months trying to create an intelligence product that is the gold standard in what is now a $40 billion-a-year industry—state-of-the-art intelligence on which the White House and Congress base their most vital national security decisions.

But Team B's objective was completely different. It attacked the CIA precisely because it relied too heavily on "hard data."[43] In other words, facts didn't matter. Pipes himself had told Congress several years earlier that whether or not the Soviets actually had the weaponry was irrelevant. "What mattered were not the capabilities of weapons but the psychology and political mentality of the people wielding them," he testified.[44] What the Soviets really wanted was "victory in the global conflict." Because Team B had already concluded that the Soviets wanted to wipe out America, getting an accurate assessment of Soviet strength was irrelevant. Its purpose was to politicize intelligence so the United States could prepare for an imminent global nuclear war. That meant creating a report showing a Soviet Union hell-bent on world domination.

Team B, for example, had no factual evidence to back up its assertion that the Soviets had a top secret nonacoustic antisubmarine system. Nevertheless, it concluded that the Soviets had probably "deployed some operational nonacoustic systems and will deploy more in the next few years."[45] The absence of evidence, it reasoned, merely proved how secretive the Soviets were!

Similarly, Team B issued dire warnings about the strategic capabilities of the Soviet Union's Backfire bomber, which, it turned out, had neither the range nor the refueling capability necessary for a round-trip mission—not to mention the fact that the Soviets produced less than half as many as Team B predicted.[46] Likewise, because it insisted on using a flawed methodology that had been rejected by the CIA, Team B exaggerated how effectively the Soviets could strike U.S. missile silos.[47] And even though Team B knew "that the Soviet Union was in severe decline," Anne Cahn noted,[48] it concluded that the Soviet threat was imminent, and would continue growing because it was such a wealthy country with "a large and expanding Gross National Product."[49]

"All of it was fantasy . . ." said Cahn, "if you go through most of Team

B's specific allegations about weapons systems, and you just examine them one by one, they were all wrong."[50] The CIA assailed the report as "complete fiction."[51] CIA director George H.W. Bush said Team B's approach "lends itself to manipulation for purposes other than estimative accuracy."[52] His successor, Stansfield Turner, a Democrat, came to the same conclusion. Ray Cline, a former deputy director of the CIA, asserted it had subverted the process of making the NIE by employing "a kangaroo court of outside critics all picked from one point of view."[53] And Kissinger said its only purpose was to subvert détente and sabotage a new arms limitation treaty.[54]

To Richard Pipes, however, such factual discrepancies were trivial. "It was General Vogt who overestimated the capacities of the Backfire bomber," he said. "But that is just detail. The important thing was reading the Soviet mind-set. We were saying they don't want war, but if they do have war, they will resort immediately to nuclear weapons."[55]

Likewise, Secretary of Defense Donald Rumsfeld, prefiguring his role in the Iraq War thirty years later, also decided the facts were unimportant and sided with Team B.[56] "No doubt exists about the capabilities of the Soviet armed forces," Rumsfeld said, after seeing Team B's report. "The Soviet Union has been busy. They've been busy in terms of their level of effort; they've been busy in terms of the actual weapons they're producing; they've been busy in terms of expanding production rates; they've been busy in terms of expanding their institutional capability to produce additional weapons at additional rates."[57]

Finally, even though the entire exercise was to have been conducted in utmost secrecy, Team B succeeded in launching a massive campaign to inflame fears of the red menace in both the general population and throughout the policy community—thanks to strategically placed leaks to the *Boston Globe* and later the *New York Times.* "[Richard] Pipes was jubilant," *Times* reporter David Binder said after meeting with him. "They had triumphed; they had poked holes at the agency's analyses."[58]

This was "an opportunity to even up some scores with the CIA," said Team B member General John Vogt. Borrowing a popular catchphrase of the era—from the TV comedy *Laugh-In,* no less—he made it clear how he really felt about the Agency: "Sock it to them!"[59]

So what had all this accomplished? At least one young neocon tied to the enterprise, Paul Wolfowitz, had won over powerful allies—namely former California governor and presidential candidate Ronald Reagan and Donald Rumsfeld[60]—and had cemented his relationship with

Richard Perle, another up-and-coming neocon. Rumsfeld and Cheney*[61] had emerged as powerful players at the highest levels of the executive branch and had started consolidating power in an executive branch that had come under assault because of Watergate. Finally, a band of Cold Warriors and neocon ideologues had successfully insinuated themselves in the nation's multibillion-dollar intelligence apparatus and had managed to politicize intelligence in an effort to implement new foreign policies.

In the seventies, however, all this maneuvering had its limits. Thanks to Watergate, Jimmy Carter beat Ford in the 1976 general election, and the neocons were completely frozen out. "We got one unbelievably minor job," said Scoop Jackson staffer Elliott Abrams. "It was a special-negotiator position. Not for Polynesia. Not Macronesia. But Micronesia!"[62]

But in the long run, that didn't matter. After all, they were true believers who were in it for the long haul. Whatever one thought of the neocons, they were resolute, committed to a lifelong struggle. Richard Perle saw the ongoing political battles in Churchillian terms, articulating his approach with a quote from the great English statesman that he kept framed in his study:

> *Never give in,*
> *Never give in*
> *Never, never, never, never*
> *In nothing great or small,*
> *Large or petty—*
> *Never give in*[63]

* * *

As it happened, the neocons were soon able to achieve power and taste the fruits of victory thanks to the ascendancy of Ronald Reagan and the Committee on the Present Danger (CPD), the most important front group the hard-liners put together in the seventies. When it was revived in 1976, its membership included an amalgam of Cold War stalwarts, ranging from William J. Casey, the gruff and disheveled but brilliant spy-

*A number of analysts, including Stefan Halper and Jonathan Clarke in *America Alone: The Neoconservatives and the Global Order*, describe Cheney and Rumsfeld as nationalists rather than neoconservatives even though the two men share the neoconservatives' views of American exceptionalism and unilateral military action and they were crucial to the implementation of neocon policies in the Bush administration.

master, to blue-blooded icons of the establishment such as Paul Nitze, and new stars of the neoconservative movement such as Jeane Kirkpatrick and Richard Perle.

When Ronald Reagan was elected in 1980, no fewer than thirty-three CPD members won prominent appointments in Reagan's first administration—twenty of them in national security.*[64] These were not trivial positions. One member of the committee was none other than Ronald Reagan.

Ultimately, in the CPD, one could see the emerging fault lines in the Republican Party, the ideological divide that separated hard-line neocons and Cold Warriors from the more moderate, pragmatic realists—i.e., practitioners of realpolitik such as Henry Kissinger and Brent Scowcroft, George H.W. Bush and James Baker. All of the latter were conspicuously missing from the CPD's roll call.

When they took power during the Reagan administration, the hard-liners succeeded in getting much of the country to believe that the United States faced grave dangers from Soviet WMDs—even though the Soviet Union was on the verge of disintegration. According to Melvin Goodman, who resigned from the CIA in 1990 because of the politicization of intelligence on the Soviet Union, the exaggerated estimates of the Soviet Union's military strength "meant that the policy community was completely surprised by the Soviet collapse and missed numerous negotiating opportunities with Moscow."[65] An extended study by the General Accounting Office concluded that in the 1980s, military officials exaggerated the threat posed by Soviet weapons and defenses, as well as U.S. vulnerabilities to that threat—all in order to get Congress to fund the largest defense buildup in the nation's history.[66]

But, most important, this was just the beginning. If the neoconservatives were to truly take power, a handful of small front groups and a

*Some of the most prominent members of the CPD who had high-level positions in the Reagan administration included: Ronald Reagan, president; Kenneth L. Adelman, U.S. deputy representative to the United Nations; Richard V. Allen, assistant to the president for National Security Affairs; William J. Casey, director of the Central Intelligence Agency; John B. Connally, member of the President's Foreign Intelligence Advisory Board; Jeane J. Kirkpatrick, U.S. representative to the United Nations; John F. Lehman, secretary of the navy; Michael Novak, representative on the Human Rights Commission of the Economic and Social Council of the United Nations; Richard Perle, assistant secretary of defense for International Security Policy; Richard Pipes, staff of the National Security Council; Eugene V. Rostow, director of Arms Control and Disarmament Agency; George P. Schultz, secretary of state.

couple dozen intellectuals would not suffice. What was needed, and was soon under way, was nothing less than a massive effort by right-wing billionaire philanthropists to completely reframe the entire national debate—about tax policy, the economy, welfare, the judiciary, foreign policy, national security, and so on—an elaborate institutional infrastructure that could sustain itself for generations to come.

"When I was young, the people who spoke of 'the movement' and who used 'radical' as an affirmative word were progressive," Bob Kuttner, cofounder of the liberal magazine *The American Prospect,* wrote after attending a national conference of conservative foundations in 2002. "The movement, at first, referred to the civil-rights movement; by the mid-1960s, it referred to a generalized movement for social justice. 'Movement people' boycotted nonunion grapes, worked on voter registration, opposed the war in Vietnam. Today, one hears the phrase 'movement conservatism.' The Right's think tanks and philanthropists alike understand that the enterprise is—above all—political."[67]

Consequently, the New Right started new think tanks and lobbying groups, and funded existing ones with ideologues of various stripes taking key positions as "scholars" and "experts" at the Hudson Institute,*[68] Freedom House,†[69] the Cato Institute, the Manhattan Institute, and others. The American Enterprise Institute‡[70] in particular became a

*Key neocons at the Hudson Institute include Robert Bork, Francis Fukuyama, Norman Podhoretz, Ben Wattenberg, and Meyrav Wurmser.

†Key policy makers at Freedom House have included James Woolsey, Kenneth Adelman, Jeane Kirkpatrick, Donald Rumsfeld, and Paul Wolfowitz.

‡Among the high-profile scholars and fellows to have taken positions at the AEI: John Bolton, the former ambassador to the U.N.; Lynne Cheney, former chair of the National Endowment for the Humanities and the wife of the vice president; David Frum, former speechwriter for George W. Bush and the man who penned the phrase "axis of evil"; Reuel Marc Gerecht, the director of the Project for a New American Century's Middle East initiative; Newt Gingrich, the former speaker of the house; Frederick Kagan, author of AEI's rival study to the Iraq Study Group report; Jeane Kirkpatrick, the former ambassador to the U.N.; Irving Kristol, arguably the founder of neoconservatism; Michael Ledeen, author, a founder of the Jewish Institute for National Security Affairs, and a Reagan administration operative involved in the Iran-contra scandal; Joshua Muravchik, a scholar at the Washington Institute for Near East Policy; Charles Murray, coauthor of *The Bell Curve*; Michael Novak, a Roman Catholic author who, among other accomplishments, introduced Donald Rumsfeld to Bernard Lonergan's categories of the "known, known unknown, and unknown unknown;" Norman Ornstein, a frequent contributor to the *Washington Post* who wrote part of the McCain-Feingold campaign finance bill; Richard Perle, Reagan official, member of the Defense Policy Board Advisory Committee, and a

home base for neocons. Its budget soared from $1 million in 1970 to more than $10 million in 1980,[71] and more than $25 million in 2006.[72] By 1985, the AEI had 176 people on staff and 90 adjunct scholars,[73] who constituted the kernel of a veritable shadow government.

With the help of Australian billionaire media magnate Rupert Murdoch;* Sun Myung Moon, the self-described Savior, Messiah, Returning Lord, and True Parent of Korea's Unification Church, and founder of the *Washington Times*; and Richard Mellon Scaife, the right-wing billionaire newspaper publisher, the New Right began building up and seizing editorial power at scores of media operations including the *New York Post*, Fox News, the *Washington Times*, the *American Spectator*, and the *Wall Street Journal* editorial page. In 1995, Murdoch launched *The Weekly Standard*, with Irving Kristol's son Bill at the helm as editor. Neocon columnists such as Charles Krauthammer and David Brooks won positions at the *Washington Post* and *New York Times* respectively.

The John M. Olin Foundation provided funding toward the research and writing of Allan Bloom's *The Closing of the American Mind*, David Brock's *The Real Anita Hill*, and Samuel Huntington's *The Clash of Civilizations*. Right-wing billionaire Richard Mellon Scaife contributed hundreds of millions of dollars† to the Heritage Foundation, the Hoover

---

vocal advocate of the war in Iraq; Danielle Pletka, a strong supporter of Iraqi exile Ahmed Chalabi and an advocate of the use of torture; Gary Schmitt, executive director of the Project for a New American Century, which was an advocate of regime change in Iraq; Fred Thompson, former U.S. senator, actor, and presidential candidate; Ben Wattenberg, moderator of various PBS television series; David Wurmser, one of the authors of the 1996 paper *A Clean Break*, calling for remaking the Middle East, part of Douglas Feith's secret intelligence unit in the Pentagon, and Middle East adviser to Vice President Dick Cheney; and John Yoo, legal adviser to President George W. Bush, author of parts of the Patriot Act and of controversial memos asserting the legality of torture, creating a narrow definition of habeas corpus and denying the rights of the Geneva Convention to "enemy combatants." Yoo is also the author of the Yoo Doctrine, or unitary executive theory, that holds that the president's war powers place him above the law, and who is said to have authored papers asserting that the president has the right to allow the National Security Agency to engage in warrantless surveillance of American citizens.

*A partial list of Murdoch's holdings include Twentieth Century Fox and its many subsidiaries in the movie business, Fox Television and its many subsidiaries, Fox Movie Channel, Fox News Channel, the *Weekly Standard*, *The Times* (London), *The Sunday Times* (London), *The Sun*, *News of the World*, multiple media holdings in Australia, the *New York Post*, HarperCollins publishers, and the *Wall Street Journal*.

†According to a 1999 *Washington Post* article, Scaife and his family's charities contributed "about $340 million to conservative causes and institutions—$620 million in current [1999] dollars, adjusted for inflation."

Institution on War, Revolution and Peace, the Center for Strategic and International Studies, the AEI, the Hudson Institute, the *American Spectator*, Accuracy in Media, and many other right-wing recipients.[74]

Meanwhile, neocon "scholars" such as William Kristol, Richard Perle, and a political operative named Michael Ledeen regularly provided fodder for right-wing talk radio hosts from Rush Limbaugh to Ollie North, G. Gordon Liddy to Michael Reagan—and became talking heads in the mainstream media as well. Increasingly, neocon ideologues played a key role in shaping the discourse on foreign policy. The so-called liberal press—the *New York Times*, the *Washington Post*—offered up its most precious editorial real estate to hundreds and hundreds of neocon op-ed pieces.

Noting that just four right-wing think tanks—the Cato Institute, the Heritage Foundation, the Manhattan Institute, and the American Enterprise—put up about $70 million a year between them, Roger Hertog, the wealthy part-owner of the once liberal *New Republic*, explained to a group of right-wing philanthropists he was addressing, "You get huge leverage for your dollars."[75]

He was right. After all, they were changing the course not just of American politics, but of the world. What they needed, of course, was a president.

CHAPTER FIVE

Into the Fray

Nestled in the Blue Ridge Mountains in Lynchburg, Virginia, a Southern town of 65,000 people, Liberty University is a private, coeducational, undergraduate and graduate institution founded by Jerry Falwell in 1971. It is situated deep in the heart of the Bible Belt, just twenty miles away from the Appomattox Court House where Confederate general Robert E. Lee surrendered to General Ulysses S. Grant, ending the Civil War. With 75 buildings on its 4,400-acre campus, an impressive football stadium, and a 9,000-seat domed basketball arena, Liberty offers the same amenities found in many other modern American universities. But as the world's largest Christian university, it differs from its secular counterparts in several important respects. Its state-of-the-art athletic facilities, with five basketball courts and an intercollegiate swimming pool, are named not after its sports idols or rich alumni, but after Tim and Beverly LaHaye, the apocalypse-obsessed coauthor of the Left Behind series, and his wife, both leading figures of the Christian Right. Where other universities show off their high-tech science facilities, Liberty takes pride in its Center for Creation Studies, where it teaches intelligent design and creationism. Finally, where Ivy League universities boast about their libraries and fine-arts museums, Liberty has the Jerry Falwell Museum, its homage to Liberty's chancellor. Filled with artifacts from Falwell's early life, the Falwell Museum displays portraits of the great heroes of evangelism from Puritan preacher Jonathan Edwards to Billy Sunday, the pro baseball star who became a fire-and-brimstone Prohibitionist, relics from the radio days of the *Old Time Gospel Hour* with Jerry Falwell, not to mention a photo gallery of Falwell himself meeting with the leading lights of evangelical culture and politics, among them Mel Gibson, Benjamin Netanyahu, Oliver North, Ronald Reagan, and both George H.W. and George W. Bush.

Perhaps the most striking edifice on campus is the Carter Glass Mansion, former home of the late U.S. senator from Virginia and a handsome manor house that now serves as Liberty's main administrative office. From its entrance, a magnificent view of the Blue Ridge Mountains is marred only by a newly constructed Wal-Mart.

Inside is the office of the late Jerry Falwell, who, as Liberty's chancellor, realized his dream of inculcating twenty thousand Christian evangelical students each year with the philosophy that "knowledge has validity only when viewed in the light of Biblical truth"[1] so that they can "serve the Lord in aggressive, fundamental, soul-winning Baptist churches."[2]

Serene, self-confident, and self-assured as ever,* clad in a dark suit and red paisley tie, Falwell answered questions in a May 2005 interview with the disarming candor that enabled him to build personal friendships with even his fiercest ideological foes, from Reverend Jesse Jackson to *Hustler* magazine pornography king Larry Flynt.

At seventy-two, Falwell looked back at the extraordinary success of the Christian Right with surprise. "There is no question in my mind that the greatest influence in America today is the New Testament Church," he said. "[But] when I started the Moral Majority, I was not sure at all that we could really make a difference. I just knew we had to try."[3]

Falwell was a product of the battles between the modern secular world and the world of faith and fundamentalism in the most deeply personal way. His father, Carey Falwell, before making his fortune as the owner of most of the retail gas stations in Virginia,[4] was a raffish, hard-drinking Prohibition bootlegger, who had won notoriety for killing his own brother in a heated argument.[5] In addition, the elder Falwell was an agnostic,[6] in sharp contrast to his wife, Helen, a devout evangelical who tuned in regularly to Charles E. Fuller's Old Fashioned Revival Hour radio broadcasts. "She listened to the Gospel Hour and would evangelize her twin sons and nobody would cut her off," Falwell said.

In 1952, Falwell, then an eighteen-year-old college sophomore, made his commitment to Christ and decided to join the ministry.[7] Four years later, in 1956, he founded the Thomas Road Baptist Church in Lynchburg, Virginia, and was soon broadcasting the *Old Time Gospel Hour* on both radio and TV.[8]

Even though evangelicals were still in retreat, many believed the secularist onslaught called for a forceful response. And nothing provoked a counterattack more than two sets of rulings by the U.S. Supreme

*Falwell died in May 2007, two years after the interview took place.

Court. The first took place in 1962 when the Court outlawed school prayer in public schools and, a year later, ruled that sanctioned organized Bible reading in public schools in the United States was unconstitutional. "[I]n my opinion . . . the Supreme Court . . . is wrong," Billy Graham told a UPI reporter.[9] "Eighty percent of the American people want Bible reading and prayer in the schools. Why should a majority be so severely penalized . . . ?"

Then, on January 22, 1973, in one of the most controversial and politically significant decisions in its history, *Roe v. Wade,* the U.S. Supreme Court ruled that most laws against abortion violate the constitutional right to privacy. More than any single event, the decision launched the Christian Right as a populist right-wing religious movement, and reshaped American politics for decades to come. This was the ultimate litmus test of the Christian Right. From this point on, one was either pro-choice or pro-life. "I don't think that the average liberal has ever understood, or will ever understand the depth of that issue in igniting evangelical Christians to participation in the public policy process," said Richard Land, director of the Southern Baptist Convention. "When we went to a baby being killed every 20 seconds, three babies a minute, 180 babies an hour within six months after *Roe v. Wade,* it ignited the evangelical movement to get involved in the public policy process in a way that nothing else ever has."[10]

Not long afterward, Falwell got a call from Francis Schaeffer, an electrifying Presbyterian evangelist and author who is probably the most important religious figure that secular America has never heard of. Widely regarded as the leading evangelical theologian of the twentieth century, Schaeffer, who died in 1984, was to the Christian Right what Marx was to Marxism, what Freud was to psychoanalysis. "There is no question in my mind that without Francis Schaeffer the religious right would not exist today," said Falwell. "He was the prophet of the modern-day values movement."[11]

A member of the resolutely fundamentalist Bible Presbyterian Church, Schaeffer won renown as a theologian for transforming the way in which the debate between modernism and fundamentalism was framed.[12] In the late forties, he and his wife, Edith, moved to Switzerland as missionaries and in 1955 founded L'Abri ("The Shelter" in French), a gorgeous mountain commune in the Swiss Alps.[13] Soon, according to *Christianity Today,* their daughters began bringing home fellow students from the University of Lausanne to enjoy meals and discussions about religion and philosophy.[14] Thanks to the Schaeffers' hospitality and an extraordinary intellectual discourse, L'Abri continued to grow in the late fifties, sixties, and seventies.

Schaeffer and Edith presided, he as the revered but totally approachable and accessible sage, she as a vital part of an intellectual team. The Schaeffers and their followers were adepts in the philosophy of Hegel, Kierkegaard and Heidegger, Camus and Sartre.[15] They discussed the films of Fellini and Bergman, the music of the Beatles, Bob Dylan, Cream, and Jimi Hendrix.[16] L'Abri attracted eclectic seekers from all over the world—Hindus, Buddhists, Zoroastrians, European intellectuals, American Jesus freaks, hippie drop-outs, and drug users. Searching, alienated, and morally confused, Schaeffer's followers were far closer to the ethos of the sixties counterculture than the right-wing, fire-and-brimstone traditions of their evangelical cohorts. Even Timothy Leary, the "Tune in, turn on, and drop out" guru of LSD, spent time at L'Abri.[17]

But it was Schaeffer's scrupulously intellectual approach of confronting the fundamentalism-modernism conflict head-on that was most striking. "They saw that if they didn't get involved in activism, they would be overrun by secularists," said John Berger, an editor at Cambridge University Press who visited L'Abri in the seventies and was active in the antiwar movement at the time. "They were very rigorous intellectually. They knew Western political thought. Schaeffer and his followers had a much larger worldview than American leftists at the time. These people were real scholars, not narrow-minded followers of a cause. They were there to learn. And Schaeffer was not interested in creating a cult. He wanted to start a movement with informed people."[18]

As one acolyte put it, "Schaeffer showed me that Christians didn't have to be dumb."[19] Added another person who attended his lectures, "There was one thing we felt we understood quite clearly after listening to Schaeffer at considerable length: the very real and immediate threat posed by secular humanism."[20]

By the early sixties, Schaeffer had become an international cult figure, and in 1965, he began taking his message back to America. Barely five feet tall, wearing an Alpine hiking outfit when he lectured—knickers, knee socks, and walking shoes—and, in later years, a white goatee to go with his long, flowing white hair, he cut an anomalous figure among straitlaced evangelical pastors, both in terms of his appearance and his eclectic intellectual pursuits. As John Fischer described him in *Christianity Today,* Schaeffer was hard to listen to, often emitting a grating, high-pitched scream that at first "sounded something like Elmer Fudd on speed. . . . After we had studied Kant, Hegel, Sartre, and Camus, the voice sounded more like an existential shriek. If Edvard Munch's *The Scream* had a voice, it would have sounded like Francis Schaeffer. [He] . . . understood the existential cry of humanity trapped in a prison of its own making. He was the closest thing to a 'man of sorrows' I have seen."[21]

Far from being seduced by the counterculture of the sixties, Schaeffer used his proximity to it to develop a critique of secular humanism in more than twenty books, several of which became key texts for leaders of the Christian Right over the next generation. In *How Should We Then Live? The Rise and Decline of Western Thought and Culture,* Schaeffer took a sweeping look at man's progress from the Middle Ages to the Atomic Age, and portrayed the Renaissance, secular humanism, and the entire post-Enlightenment rational modern world, as part of an ungodly, unholy ethos through which man replaces God with reason and with man-made things. In Schaeffer's view, it was but a short step from Darwin's theory of evolution to social Darwinism to Nazism.

In the late forties, Sayyid Qutb, the founder of the modern Islamic fundamentalist movement, had begun railing about how modernity, secularism, and materialism had polluted Islam. Similarly, to Schaeffer, the evils of modern culture—drug use, sexual permissiveness, pornography, abortion, and so on—were not merely accidents of history. Instead, they were the inevitable product of decadent ideas that had been flourishing in Europe for centuries, that had conquered the United States and replaced religion with a secular humanism in which man, not God, was central. Ultimately, he concluded that real Christians had a duty to sound the trumpet, so that "this generation may turn from that greatest of wickednesses, the placing of any created thing in the place of the Creator."[22]

In subsequent works, Schaeffer became more strident. In *Whatever Happened to the Human Race?,* coauthored with C. Everett Koop, who later became surgeon general of the United States under Ronald Reagan, he compared abortion to the Holocaust, asserting that *Roe v. Wade* had legalized the mass murder of millions of unborn children and it was evidence of "our seemingly unlimited capacity for evil."[23] For Christians, Schaeffer argued, fighting this secular evil was the ultimate test of moral values. Salvation depended on standing up against abortion. Calling for a massive social movement—a Christian political movement—he urged fundamentalists to do whatever was necessary to resist abortion, infanticide, euthanasia, and the eradication of human dignity resulting from the secularization of American culture.

Finally, Schaeffer believed, this massive social movement needed a manifesto. In *A Christian Manifesto,* his retort both to Karl Marx's

Communist Manifesto and to the more obscure *Humanist Manifestoes*,* he insisted that evangelicalism could no longer passively accommodate itself to the decadent values of a secular humanist world that murdered unborn babies. With his call to arms, he urged evangelicals to come out from behind the pulpit and into a full-scale cultural war with the secular world. Randall Terry, founder of the pro-life Operation Rescue, credits Schaeffer with prodding evangelicals to take political action against abortion. "You have to read Schaeffer's *Christian Manifesto* if you want to understand Operation Rescue," Terry said.[24]

Others who were deeply influenced by Schaeffer included James Dobson, the head of Focus on the Family, the powerful "traditional values" lobby; Christian broadcaster and presidential candidate Pat Robertson; former congressman and presidential candidate Jack Kemp; former aide to President Richard Nixon, Chuck Colson; syndicated columnist Cal Thomas; Florida secretary of state Katherine Harris; Tim LaHaye and his wife, Beverly, the founder of Concerned Women of America; Dorothy Walker Bush,[25] the mother of George H.W. and grandmother of George W. Bush; Marvin Olasky, a born-again Christian adviser to George W. Bush, and many, many more.[†26]

*Evangelical theologians frequently refer to the *Humanist Manifestoes* as if they are seminal works for secular humanists plotting against Christianity, when in fact many people are not familiar with them. The first *Humanist Manifesto* was written in 1933 by Raymond Bragg and spoke about humanism as a religious movement without a deity that potentially could challenge more conventional religions. In 1973, *Humanist Manifesto II* was written by Paul Kurtz and Edwin H. Wilson, noting that in light of World War II and the Holocaust, the first *Humanist Manifesto* seemed too optimistic. Its most frequently quoted line reads, "No deity will save us; we must save ourselves." Finally, in 2003, a shorter version of the manifesto, *Humanist Manifesto III*, was published by the American Humanist Association and stressed six "humanist ideals" such as empiricism, concluding, "Knowledge of the world is derived by observation, experimentation, and rational analysis."

Whatever one thinks of the manifestoes, it seems fair to say that evangelicals have overstated their influence on the secular world. At this writing, the *Humanist Manifestoes* were ranked 658,750 on the Amazon bestseller list.

†Not all of Schaeffer's acolytes saw him as someone who was declaring war on the secular world. In April 2007, John Fischer argued in *Christianity Today* that Schaeffer "took a strong stand against abortion and euthanasia and even called for serious measures, including political intervention. . . . But to conclude that this invocation to war was Schaeffer's crowning achievement is to truncate the man and his work. . . . Schaeffer's work is ultimately not a call to arms, but a call to care. Those who have taken up arms and claimed him as their champion have gotten only part of his message. . . . Schaeffer never meant for Christians to take a combative stance in society without first experiencing empathy for the human predicament that brought us to this place."

But no one was more indebted to Schaeffer than Jerry Falwell, who devoured everything he wrote. "I had several mentors in my life, but Francis Schaeffer was head and shoulders above the rest," Falwell said.[27] As it happens, Schaeffer had been keeping an eye on the rising young preacher in Lynchburg, and one day Falwell finally got a phone call from his hero. "Dr. Schaeffer had watched one of my TV shows and he called me to compliment me on the clear gospel presentation I had made," Falwell recalled.

More important, Schaeffer gave Falwell the go-ahead to tear down the wall that had kept politics out of the pulpit. "I was in search of a scriptural way that I as pastor of a very large church could address the moral and social issues facing American culture," Falwell said.[28] "Dr. Schaeffer shattered that world of isolation for me, telling me that while I was preaching a very clear gospel message, I was avoiding fifty percent of my ministry. He began teaching me that I had a responsibility to confront the culture where it was failing morally and socially. If it hadn't been for Francis Schaeffer, I would have been a pastor in Lynchburg, Virginia, but otherwise never heard of. He was the one who pushed me out of the ring and told me to put on the gloves."[29]

Not long afterward, Falwell got a call from Tim LaHaye, then a pastor in San Diego. At the time, LaHaye had just founded Californians for Biblical Morality, a coalition of extreme right-wing pastors who fought against gay rights,[30] and even sought to ban Dungeons and Dragons, a fantasy game with heroes and monsters, in public facilities on the grounds that it was witchcraft and a religion of the occult. When Falwell visited, he was impressed that LaHaye had organized conservative California pastors to confront the state government on moral and social issues. "I wondered why we couldn't do it on a national basis," said Falwell.

Courtly, genteel, and soft-spoken, LaHaye is the antithesis of the fire-and-brimstone preacher one might expect from a purveyor of End Times prophecies. Entranced by Rapturist theology since childhood,[31] LaHaye had graduated from ultraconservative Bob Jones University in Greenville, South Carolina, and began preaching in nearby Pumpkintown at a salary of $15 a week.[32] Later, he moved to San Diego, and in the sixties and seventies served as pastor at Scott Memorial Baptist Church, transforming it from a congregation of 275 into one with 3,000 members.[33]

Along the way, LaHaye avidly read Francis Schaeffer.[34] In *Battle for the Mind,* his 1980 homage to Schaeffer, LaHaye depicts America as a Bible-based country under siege by an elite group of secular humanists conspiring to destroy Christianity. He asserts that secular humanism is "the world's greatest evil and the most deceptive of all religious philosophies."[35] It is characterized by its "particular hatred toward Christianity,"[36] and it has been turning our godly nation into one that promotes Darwinism and the mass murder of the unborn, promiscuity, the homosexual agenda, and more.[37]

LaHaye's political work largely escaped the notice of the secular world, but according to Falwell, he has done more to set the agenda for evangelicalism in the United States than any other person in America.[38] Presbyterian televangelist D. James Kennedy hailed *Battle for the Mind* as "one of the most important books of our time," and Falwell said that every Christian must follow its tenets if we are to save America from becoming "another Sodom and Gomorrah."[39]

Central to LaHaye's work is the notion that secular humanism is an organized religion that is consciously trying to annihilate Christianity. "LaHaye writes as if there's a humanist brain trust sitting around reading John Dewey [the American philosopher and educational reformer], trying to figure out ways to destroy Christianity," said Chip Berlet, coauthor of *Right-Wing Populism in America* and founder of Political Research Associates, a think tank that studies right-wing political movements.[40] Of course, "humanism" is an abstract term, and while tens of millions of Americans might accurately be called secular humanists, very few think of themselves as members of a humanist movement. But to LaHaye that only proves that humanists are so devious they can control public debate by using the mass media, Hollywood, the government, and academia to brainwash unsuspecting Christians, turning our once godly nation into one that promotes Darwinism and the mass murder of the unborn, promiscuity and the homosexual agenda, and more.

In *Battle for the Mind,* LaHaye spells out his political goals clearly and precisely. As he sees it, the word *secular* is not merely a morally neutral term that means "worldly." It means "ungodly," and in his view, there are godly people, who are on the road to Rapture, and then there is the rest of the world, which is either complicit with the Antichrist, or, worse, actively assisting him. As a result, LaHaye argues, good evangelicals should no longer think of humanists merely as harmless citizens who just happen not to attend church. "We must remove all humanists from public office," he writes, "and replace them with promoral political leaders."[41]

At the time, America had elected Jimmy Carter as its first born-again evangelical Christian president. But as a supporter of the Equal Rights

Amendment and *Roe v. Wade,* Carter failed the most basic litmus tests of the evangelical movement. "He said that the Equal Rights Amendment was good for the family," LaHaye recalled disdainfully. "Well, I knew when he said that he was out to lunch."[42]

Meanwhile, a new generation of political strategists also believed that Christians should play a bigger role in the political arena and cast a watchful eye at their secularist foes, observing carefully how they had amassed power. Just by chance, in 1969, Paul Weyrich, a twenty-seven-year-old right-wing Catholic who was then press secretary to Senator Gordon Allott (R-Colo.), was invited to a strategy session for a civil rights coalition by mistake. Seizing this rare opportunity to be a fly on the wall behind enemy lines, the conservative Weyrich was astonished to encounter representatives from virtually every power base in the liberal political world, working together as if they were parts of a smoothly oiled machine. As Jason DeParle put it in the *New York Times,* "There he beheld the Liberal Behemoth in all its steely glory."[43]

There were congressional aides and policy wonks. Liberal think tankers from the Brookings Institution presented policy options. Liberal lawyers from the liberal ACLU cited the legal complexities. Liberal columnist Carl Rowan for the liberal *Washington Post* agreed to write a column. Liberal church groups planned demonstrations. And liberal lobbyists prepared to pressure Congress.

"There before me were all the different liberal groups, inside and outside Congress, the journalistic heavies, and it was a magnificent show," Weyrich said. "They orchestrated this particular bit of legislation in a very impressive way, each group playing its role—producing a study in time for the debate, drafting an amendment, planting stories. I saw how easily it could be done, with planning and determination, and decided to try it myself."[44]

"They put together a battle plan, right then and there. I was absolutely mesmerized. From that day on, I became absolutely insufferable," Weyrich said. "Knowing how the liberal side operated, it was my responsibility to set up something to replicate this on the right."[45]

Weyrich resolved that movement conservatives had to have it all: "think tank, lobby, legal arm, means of communication, political action—you know, the whole nine yards." Acolytes referred to him as "the Lenin of social conservatism—a revolutionary with a rare talent for organization."[46] Awed by the Brookings Institution, the think tank that was the embodiment of the liberal power structure, Weyrich secured a

$250,000 gift from conservative brewery mogul Joseph Coors and, with his friend Ed Feulner, established the Heritage Foundation in 1973.[47] "We are different from previous generations of conservatives," Weyrich said. "We are no longer working to preserve the status quo. We are radicals, working to overturn the present power structure of the country."[48]

Conservative activist Morton Blackwell joined the fray. "I believed that the theologically conservative Christians were the largest tract of virgin timber on the political landscape, and there was enormous potential for activism," said Blackwell, the executive director of the College Republican National Committee, which helped train aggressive young political operatives such as Lee Atwater, Karl Rove, Ralph Reed, Grover Norquist, and Jack Abramoff. "A decision was made to target a prominent skilled religious communicator to see if he would start moving people into political participation." The first issue they focused on was abortion.[49]

Consequently, in 1978, Jerry Falwell[50] gave his first sermon on "unborn babies, who, by the hundreds of thousands, are being murdered." When Francis Schaeffer's film *Whatever Happened to the Human Race?* was distributed to thousands of church groups the response was extraordinary. By this time, a Gallup poll showed that one out of three Americans said they'd had a "born-again" experience. Half the country believed in biblical inerrancy. Eighty percent saw Jesus as divine. And there were thirteen hundred evangelical radio and TV stations with a total audience of 130 million.[51]

Falwell looked ahead to the 1980 presidential election. "Jimmy Carter was making everyone mad," said Falwell. "He had gotten the evangelical support two to one in 1976. This guy had betrayed everything we stood for." By this time, Falwell's television ministry was booming. He had nearly eight million people on his mailing lists. As many as twenty thousand people attended his appearances. But at the time, only 55 percent of evangelicals in America were registered voters. Worse, as Falwell saw it, most of them voted the wrong way. "I asked, 'How can I, without violating my ministry, without breaking the law, put this group together?' They will listen to me if I have the facts."[52]

Falwell called together twenty or twenty-five top political operatives to his Thomas Road Baptist Church in Lynchburg. "I invited my friends from Washington with whom I had been talking about this thing," he said. There was Tim LaHaye, Paul Weyrich, and Presbyterian televangelist D. James Kennedy. There was Morton Blackwell, direct-mail guru Richard Viguerie, and Senator Jesse Helms. "We all made a commitment to God that day that for the first time in our lives we were going to get involved in the political process," said Tim LaHaye. "So I prayed to God, 'Dear God, we have got to get this man out of the White House and get someone in here who will be aggressive about bringing back traditional moral values.'"[53]

After the meeting had been under way for some time, Paul Weyrich spoke up. "Dr. Falwell," he said, "I believe there is a moral majority out there ready to be organized so that they see what they are battling."

Weyrich continued, but Falwell interrupted. "Go back to what you said earlier," he said. He was trying to recapture a "eureka" moment. Weyrich didn't understand what he was getting at, however, so Falwell patiently backtracked. "You started saying that out there was . . . What did you say?"[54]

"I had to think," said Weyrich, recalling the historic incident. He racked his brain for a moment, and finally, he spoke. "I said, 'Out there is a moral majority . . ."

"That's it!" said Falwell. "That's what I'm going to call the organization: The Moral Majority."[55]

In 1980, Falwell and his associates started going from state to state, meeting Catholics, old-line Protestants, and evangelicals, putting together chapters everywhere. Soon the Moral Majority had lined up 72,000 pastors in what it called "a pro-life, pro-traditional family" coalition. They called upon Christians to do three things: get saved, get baptized, and get registered to vote.

Although they strongly approved of Ronald Reagan as the Republican nominee, evangelical leaders were far less happy about the man Reagan picked to be his running mate—and made their feelings known at the Republican National Convention in July 1980 in Detroit. With a police escort blaring its sirens and flashing red lights, Paul Weyrich sped toward a last-minute meeting with Reagan, joining Jerry Falwell; Howard Phillips, the chairman of the Conservative Caucus; Phyllis Schlafly, the leader of Stop ERA; and other leading evangelicals who were attempting to talk Reagan out of putting George H.W. Bush on the ticket because they did not view him as a legitimate social conservative.[56]

The evangelicals did not seem to realize it at the time, but they had an ally who was also trying to blackball Bush—Donald Rumsfeld. "In 1980, when Reagan had the nomination in hand and it was clear that Bush was the choice of the convention to be [the vice presidential nominee], there was a real effort by Rumsfeld to get on the ticket," recalled

Pete Teeley, a Bush aide who accompanied him to the Detroit convention.

As Teeley walked across the hotel lobby, he saw a gaggle of reporters surrounding someone giving them a briefing.

"Bill [a Rumsfeld aide] is telling us Rumsfeld's strategy to get on the ticket," a reporter shouted to Teeley. "What's yours?"

"Thirty-six primaries and fourteen caucuses," Teeley replied, implicitly pointing out that Rumsfeld had not even entered the primaries.[57]

In the interests of party unity, Bush got the nomination anyway. Pragmatic political concerns had prevailed over ideology and theology. But both Rumsfeld and the evangelicals were now clearly aligned against the moderate wing of the Republican Party. One of their shared enemies was George H.W. Bush.

In the 1980 presidential election, evangelical voters ensured Jimmy Carter's defeat by going two to one for Ronald Reagan. It was a stunning victory for the right. Twelve Democratic senators—including liberal icons George McGovern, Frank Church, Birch Bayh, and Gaylord Nelson—went down in defeat, the biggest swing since 1958. Even with Bush on the ticket, evangelicals had managed to get most of their concerns written into the Republican platform. For the first time, those values would be taken seriously in the Oval Office. Christian broadcaster and *700 Club* host Pat Robertson chortled, "We have enough votes to run this country!"[58]

Even though it now had a friend in the White House, the Christian Right went full steam ahead building a new movement. In 1981, James Dobson founded the Family Research Council to be the political arm of *Focus on the Family,* his daily radio show with an audience of four million people.[59] Trained as a child psychologist, Dobson "dispensed stern but fatherly advice over the radio to help confused and alienated Middle Americans overcome personal crises," as the *Nation*'s Max Blumenthal put it, and in the process incorporated them into the Christian Right's political machine.[60] Creating a "family values" empire that ultimately consisted of more than seventy different ministries, ten magazines, and thirteen hundred employees, Dobson took on gays, abortion, the use of condoms, and the teaching of evolution. He later attacked the Girl Scouts for being "agents of humanism and radical feminism," the cartoon character SpongeBob SquarePants, and even the concept of tolerance itself, terming it "a watchword of those who reject the concepts of right and wrong . . . kind of a desensitization to evil of all varieties."[61] These were the family values that reinforced an emerging populist narrative of average people in the heartland doing battle against the sophisticated elites on the East and West Coasts. They were further promoted through Christian schools that were rapidly proliferating throughout the country—according to Falwell, growing at a rate of one every seven hours.[62]

Tim LaHaye and his wife, Beverly, had already founded Concerned Women for America[63] in 1979 "to bring biblical principles into public policy,"[64] and oppose the more liberal National Organization for Women (NOW). In 1981, LaHaye resigned as pastor to devote himself full-time to building the Christian Right. He began by meeting with moneyed ultraconservatives including two right-wing billionaires from Texas, Nelson Bunker Hunt and T. Cullen Davis.*[65]

With Hunt, Davis, and others putting up the money, LaHaye founded the Council for National Policy (CNP) in 1981 as a low-profile but powerful coalition of oil billionaires, fundamentalist preachers, and right-wing tacticians to come up with a coherent and disciplined strategy for the New Right. Though its membership list is highly secretive, the CNP rolls reportedly included powerful evangelicals like Falwell and Pat Robertson; top right-wing political strategists Richard Viguerie, Ralph Reed, and Paul Weyrich; and several senators and congressmen.

Still in its infancy as a political movement, the Moral Majority had registered 8.5 million voters in five years. They were powerful enough to swing presidential and senatorial elections. President Reagan appeared on the *Old Time Gospel Hour* with Falwell, who gave the benediction at the 1984 Republican National Convention. The CNP had access to the highest powers in the land.

Paul Weyrich's dream of replicating the infrastructure of the liberal establishment was being realized—with a vengeance. When it came to think tanks, the right now had Weyrich's Heritage Foundation as well as his Free Congress Research and Education Foundation, with a related organizing arm called Coalitions for America.[66] The American Enterprise Institute had been around since the forties, but was rising to prominence with the ascendancy of the New Right. There was the libertarian Cato Institute, foundations like the Bradley, Olin, and Scaife

*Cullen Davis is best known for having been indicted for and subsequently acquitted of the murder of his wife's lover and his stepdaughter, after which he became a born-again Christian.

trusts; "pro-family" groups like Dobson's Family Research Council, and the Christian Coalition.

Reagan appointed James Watt, a prominent Pentecostal, as secretary of the interior. His secretary of defense, Caspar Weinberger, espoused his belief in the Book of Revelation.*[67] As surgeon general, the nation's top medical post, Reagan appointed C. Everett Koop, who had collaborated with Francis Schaeffer on the seminal antiabortion books and films of the Christian Right.

And when President Reagan himself spoke before the National Association of Evangelicals in 1983, he assured the hundreds of clergymen in the audience that when it came to the ongoing war between faith and reason, religion and the secular world, he was on their side, even though it "puts us in opposition to . . . many who have turned to a modern-day secularism." Then, using deliberately apocalyptic rhetoric, Reagan famously reframed the Cold War in terms of the Manichaean spiritual struggle for "salvation" between the United States and the "evil empire" of the Soviet Union, which he called "the focus of evil in the modern world."[68]

Far as the Christian Right had come, however, there were still limits to its power. In 1981, Reagan appointed Sandra Day O'Connor to the Supreme Court, rather than an aggressively antiabortion jurist. Then, in 1982, Reagan halfheartedly honored his promise to support a constitutional amendment allowing school prayer, refraining from putting enough political muscle behind it to bulldoze it through the Senate. Worse, by the mid-eighties, fundamentalists found that one of their most trusted allies, Surgeon General C. Everett Koop, had betrayed them. As the nation's top doctor during the early era of the AIDS epidemic, Koop released a special report on AIDS in 1986 that advocated sex education, including teaching about homosexual behavior, as an essential tool in stopping the spread of the disease,[69] and even promoted the use of condoms for sex outside of marriage as a means of protection. To fundamentalists at Richard Viguerie's *Conservative Digest,* Koop was effectively "proposing instructions in buggery for schoolchildren."[70] Conservative author Phyllis Schlafly said she would "rather see her children infected with sexually transmitted disease than . . . know there was such a thing as a condom."[71] For all the money they had raised, the mil-

*Weinberger had a Jewish paternal grandfather, but was brought up as an Episcopalian.

lions of voters they had organized, the think tanks they had started, for all the paths to power they had charted, the Christian Right had found that its most sought-after goals were the most elusive.

Worse, in 1988, as Reagan's second term pulled to a close, waiting in the wings as heir apparent was Vice President George H.W. Bush, who was anathema to the new Christian Right both culturally and politically. Bush had been schooled at Andover and Yale—two of the great citadels of secular humanism in America. As a member of Skull and Bones, the elite secret society at Yale, as former ambassador to the United Nations, and later as director of the Central Intelligence Agency and the Trilateral Commission, Bush had key roles in no fewer than four institutions that, according to acolytes of End Times theology, were linked to the Antichrist and shadowy notions of One World Government. If one had deliberately set out to deeply alienate the God-fearing New South at the heart of the new Republican Party, one could scarcely have done better than to craft the patrician résumé of Vice President Bush.

Born-again Americans had played a crucial role in the 1976 Carter-Ford election when they constituted 26 percent of the American people. By the time of the 1988 presidential election, Gallup polls showed that figure had risen to 39 percent.[72] Winning that constituency was absolutely vital to any Republican who hoped to be president—and there was only one person in the world who could possibly help George H.W. Bush do it.

CHAPTER SIX

The Prodigal Son

Conventional wisdom has it that George W. Bush became a "born-again" Christian in the summer of 1985, after extended private talks with Reverend Billy Graham. As recounted by Bush himself in *A Charge to Keep: My Journey to the White House,* a ghostwritten autobiography prepared for the 2000 presidential campaign, one evening at Walker's Point, the Bush compound in Kennebunkport, Maine, Graham, spiritual confidant to Lyndon Johnson, Richard Nixon, and Ronald Reagan and a close friend of the Bush family, sat down by the fireplace and gave a talk.[1] "I don't remember the exact words," Bush wrote. "It was more the power of his example. The Lord was so clearly reflected in his gentle and loving demeanor."

The next morning, Bush and Graham went for a walk along the rugged Maine shore, past the Boony Wild Pool where Bush had skinny-dipped as a child. "I knew I was in the presence of a great man . . ." Bush wrote. "He was like a magnet; I felt drawn to seek something different. He didn't lecture or admonish; he shared warmth and concern. Billy Graham didn't make you feel guilty; he made you feel loved."

"Over the course of that weekend, Reverend Graham planted a mustard seed in my soul, a seed that grew over the next year," he continued.* "He led me to the path, and I began walking."[2]

There's just one problem with Bush's account of his conversion experience: it's not true. For one thing, when Billy Graham was asked about the episode by NBC's Brian Williams, he declined to corroborate Bush's account. "I've heard others say that [I converted Bush], and people have written it, but I cannot say that," Graham said. "I was with him and I used

*Bush is referring to the biblical parable of the mustard seed in the gospels of Luke, Mark, Matthew, and Thomas, in which Jesus compares the Kingdom of Heaven to the tiny mustard seed that ultimately grows into a huge mustard plant.

to teach the Bible at Kennebunkport to the Bush family when he was a younger man, but I never feel that I in any way turned his life around."[3]

Even if one doesn't accept Graham's candid response, there's another good reason to believe that the account in Bush's book is fiction. Mickey Herskowitz, a sportswriter for the *Houston Chronicle* who became close friends with the Bush family and was originally contracted to ghostwrite *A Charge to Keep,** recalled interviewing Bush about it when he was doing research for the book.[4] "I remember asking him about the famous meeting at Kennebunkport with the Reverend Billy Graham. . . ." Herskowitz said. "And you know what? He couldn't remember a single word that passed between them."

Herskowitz was so stunned by Bush's memory lapse that he began prompting him. "It was so unlikely he wouldn't remember anything Billy Graham said, especially because that was a defining moment in his life. So I asked, 'Well, Governor, would he have said something like, "Have you gotten right with God?"'"

According to Herskowitz, Bush was visibly taken aback and bristled at the suggestion. "No," Bush replied. "Billy Graham isn't going to ask you a question like that."[5]

Herskowitz met with Bush about twenty times for the project and submitted about ten chapters before Bush's staff, working under director of communications Karen Hughes, took control of it.[6] But when Herskowitz finally read *A Charge to Keep* he was stunned by its contents. "Anyone who is writing a memoir of George Bush for campaign purposes knew you had to have some glimpse of what passed between Bush and Billy Graham," he said.[7] But Hughes and her team had changed a key part. "It had Graham asking Bush, 'George, are you right with God?'"

In other words, Herskowitz's question to Bush was now coming out of Billy Graham's mouth. "Karen Hughes picked it off the tape," said Herskowitz.

There is yet another reason why the episode in Maine could not possibly have been the first time George Bush gave his soul to Christ. That's because Bush had already been born again more than a year earlier, in April 1984—thanks to an evangelical preacher named Arthur Blessitt.

*Herskowitz was paid for his work, and the interview tapes were given to the Bush campaign. Ultimately, the book was published as *A Charge to Keep* by George W. Bush and Mickey Herskowitz, but Bush communications director Karen Hughes oversaw completion of the project and is given credit as coauthor.

Whereas Billy Graham was a distinguished public figure whose fame grew out of frequent visits to the Oval Office over several decades, Arthur Blessitt had a very different background. His evangelicalism was rooted in the Jesus movement of the sixties counterculture. To the extent he was famous it was because he had preached at concerts with the Rolling Stones, Janis Joplin, the Jefferson Airplane, and others, and had run a "Jesus coffeehouse" called His Place on Hollywood's Sunset Strip during that turbulent decade.[8] His flock consisted of bikers, druggies, hippies, and two Mafia hit men.[9] The most celebrated ritual at Blessitt's coffeehouse was the "toilet baptism," a rite in which hippies announced they were giving up pot and LSD for Jesus,[10] flushed the controlled substances down the toilet, and proclaimed they were "high on the Lord."

In 1969, however, Blessitt was evicted from his coffeehouse and, in protest, chained himself to a cross in Hollywood and fasted for the next twenty-eight days.[11] Over the next fifteen years, "The Minister of Sunset Strip," as he was known, transformed himself into "The Man who Carried the Cross Around the World" by lugging a twelve-foot-long cross for Jesus through sixty countries all over the world, on what would become, according to the *Guinness Book of World Records,* the longest walk in human history.* Blessitt delivered countless lost souls to Jesus. He went to Jerusalem. He prayed on Mount Sinai. He crossed the Iron Curtain. Finally, in 1984, he came to Midland, Texas, to preach for six nights at the Chaparral Center before thousands of Texans night after night on a "Mission of Love and Joy."[12] He did not know it, but he was about to bring George W. Bush to Jesus.

Thirty-seven years old when Blessitt came to Midland, Bush had yet to make much of a name for himself and still struggled with the giant shadow cast by his father. The pattern had begun early, when Bush was playing sports in school. "His father had been the captain of the baseball team and star first baseman at Yale," said Mickey Herskowitz. "He had met Babe Ruth at home plate at the stadium at Yale to accept the manuscript of the Babe's autobiography. Dad was a star, a scholar, the leader of the team and the captain. And George never got much beyond Little League. He wanted to be a catcher, but one of his coaches said he had an unfortunate flaw—he blinked every time the guy swung the bat."[13]

*By November 2005, Blessitt had carried the cross more than 37,000 miles in 305 nations on all seven continents.

Whatever he did, his meager achievements were dwarfed by his father's spectacular résumé.

When he was in his twenties, his alcohol-fueled clashes with his father disturbed his parents so much that they asked friends to rein in their unruly son. In the spring of 1972, the elder Bush, then ambassador to the United Nations, called Jimmy Allison, an old friend from Midland, Texas, who was a political consultant and the owner of the *Midland Reporter-Telegram,* to ask if George W. could work on a Senate campaign Allison was running in Alabama for Winton "Red" Blount. "Georgie was raising a lot of hell in Houston, getting in trouble and embarrassing the family, and they just really wanted to get him out of Houston and under Jimmy's wing," Allison's widow, Linda, told *Salon*'s Mary Jacoby.*[14] "[The Bushes] wanted someone they trusted to keep an eye on him."

When the younger Bush got to Alabama, however, he continued drinking, according to Allison, often ambling into work at midday, boasting about how much he'd drunk the night before. One night at a party, she saw George W. urinating on a car in the parking lot.[15] He reportedly shouted obscenities at police officers, and trashed a home he rented, leaving behind broken furniture he refused to pay for. "He was just a rich kid who had no respect for other people's possessions," a member of the family who rented the house told the *Birmingham News.*[16]

When Bush returned to Washington for Christmas that year, he got drunk with his sixteen-year-old brother Marvin, ran over the neighbor's garbage cans, and found himself standing unsteadily in the doorway at home, confronting his father. "I hear you're looking for me," he said. "You wanna go mano a mano right here?"[17]

The elder George Bush didn't say a word. "He just looked at him over his glasses that had slid down the end of his nose," Barbara Bush told a friend of the family. "And he just looked until [George W.] walked away. Everything he needed to communicate was in that glance."

When young George went off to Harvard Business School in 1974, the differences between him and his father became more clearly defined.

*Bush maintained he was fulfilling his obligations to the Alabama National Guard during the year in question—despite the absence of documentary evidence or witnesses to substantiate his assertion. According to *Salon,* Linda Allison, asked if she'd ever seen Bush in a uniform, replied, "Good Lord, no. I had no idea that the National Guard was involved in his life in any way."

Where the older Bush embodied a genial and patrician preppy ethos, the son embraced the iconography of Texas as if determined to eradicate the last vestiges of East Coast elitism in his veins. At Harvard, his classmates "were drinking Chivas Regal, [but] he was drinking Wild Turkey," April Foley, who dated Bush briefly, told the *Washington Post.* "They were smoking Benson and Hedges and he's dipping Copenhagen, and while they were going to the opera he was listening to [country-and-western singer] Johnny Rodriguez over and over and over and over."[18]

After graduation, rather than join his classmates in the glittering canyons of Wall Street, Bush struck out for Midland's arid landscape of oil rigs and pump jacks, mesquite trees and horned lizards—where he fit right in.[19] But it was still unclear what he was doing with his life. A 1978 attempt to run for Congress was a disaster. Various stabs at making it in the oil industry—with companies named Arbusto Energy, Spectrum 7, and Harken Energy—failed.* Even after marrying Laura Welch in 1977 and becoming the father of twins four years later, Bush's reputation was that of an aging frat boy who worshipped what he called the four B's—beer, bourbon, and B&B. Family members still wondered what he was going to be when he grew up.[20]

Meanwhile, oil-rich Midland was going through its own spiritual crisis. When the price of oil soared in the seventies and early eighties, Midland had become a heady boomtown minting a new generation of hard-driving Texas oil barons. Its population exploded from 70,000 in 1980 to 92,000 just three years later.[21] There were shimmering skyscrapers, Lear jets, and Rolls-Royce dealerships.

But in the eighties, as oil plummeted from $40 a barrel to $8, Midland's boom gave way to unemployment lines, repo signs, and bankruptcies. In 1983, the First National Bank of Midland collapsed.[22] "Fear set in. . . ." said Midland evangelical Mark Leaverton. "Marriages broke up. People started having pretty serious emotional problems. . . . It was a scary time for all of us. . . . People started asking questions."[23]

By the time Arthur Blessitt came to Midland, several of Bush's friends had become born-again Christians, including two Midland oilmen named Don Poage and Jim Sale.[24] After preaching one night, Blessitt went over to Sale's house with Poage and a few other followers. Before Blessitt left, Poage asked if they could pray together. Blessitt anointed

*For an extended account of the unusual financial machinations at Arbusto, Spectrum 7, and Harken, see the author's *House of Bush, House of Saud: The Secret Relationship Between the World's Two Most Powerful Dynasties,* pp. 113–27.

him with Mazola oil because the Sales had no olive oil in their kitchen.[25] "I got down on the floor with him and a group of people," Poage said in the 2004 documentary, *With God on Our Side: George W. Bush and the Rise of the Religious Right.* "We prayed a very powerful prayer for me. And . . . I felt big white lightning bolts coming out of my shoulders and even though I was on my knees, I felt like I was about three feet off the ground."[26]

Baptized as an Episcopalian in Connecticut, Bush had been a regular churchgoer his entire life, but for the most part he had just been going through the motions. As Stephen Mansfield reported in *The Faith of George W. Bush,* when a Midland pastor asked his congregation what a "prophet" was, Bush replied, "That's when revenues exceed expenditures." Obvious quips were more important to Bush than spiritual quest.[27] But when Bush heard about Poage's encounter with Blessitt, he was so interested that a meeting was arranged.[28]

So, on the afternoon of April 3, 1984, Blessitt and Sale went to the coffeeshop in the local Holiday Inn. Bush had already arrived,[29] and got straight to the point. "I didn't bring up the subject of Jesus," Blessitt recalled.[30] "He did. That's his personality."

"Arthur," Bush said, "I did not feel comfortable attending the meeting, but I want to talk to you about how to know Jesus Christ and how to follow Him."

Stunned by Bush's directness, Blessitt silently prayed, "Oh Jesus put your words in my mouth and lead him to understand and be saved."

Then he picked up the Bible and leaned forward. "What is your relationship with Jesus?" Blessitt asked.[31]

"I'm not sure," Bush replied.

"Let me ask you this question. If you died this moment do you have the assurance you would go to heaven?"

"No," Bush said.

"Then let me explain to you how you can have that assurance and know for sure that you are saved."

"I like that."

Blessitt then quoted several verses on sin and salvation—from Matthew, Romans, Mark, and John. "The call of Jesus is for us to repent and believe!" he explained. "The choice is like this. Would you rather live with Jesus in your life or live without Him?"

"With Him," Bush replied.

"Had you rather spend eternity with Jesus or without Him?"

"With Jesus," said Bush.

Blessitt told Bush that Jesus wanted to write his name in the Book of Life, and extended his hand. "I want to pray with you now," he said.

"I'd like that," Bush replied. He joined hands with Sale and Blessitt. Then, Blessitt prayed a variation on the Sinner's Prayer aloud, one phrase at a time, with Bush repeating after him:

> Dear God, I believe in you and I need you in my life. Have mercy on me as a sinner. Lord Jesus as best as I know how, I want to follow you. Cleanse me from my sins and come into my life as my Savior and Lord. I believe You lived without sin, died on the cross for my sins and arose again on the third day and have now ascended unto the Father. I love you Lord, take control of my life. I believe you hear my prayer. I welcome the Holy Spirit of God to lead me in Your way. I forgive everyone and ask You to fill me with Your Holy Spirit and give me love for all people. Lead me to care for the needs of others. Make my home in Heaven and write my name in Your book in Heaven. I accept the Lord Jesus Christ as my Savior and desire to be a true believer in and follower of Jesus. Thank you God for hearing my prayer. In Jesus' name I pray.[32]

The three men smiled. "It was a happy and glorious time," said Blessitt. He explained to Bush exactly what had just happened. "Jesus has come to live within your heart," he told Bush. "Your sins are forgiven. . . . You are saved. . . . You have received eternal life . . .You are now the Child of God. . . . The Holy Spirit abides within you. . . . You have become a new person."[33]

Jim Sale was present during the entire discourse. "You can never tell what goes on in a man's heart and soul," he said.[34] "But the question was asked and answered." George W. Bush had invited Christ into his life. "Why God chose to move in our president's heart at that time, I don't know," Sale said. "I'm just glad he did."[35]

"A good and powerful day," Blessitt wrote in his diary. "Led Vice President Bush's son to Jesus today. George Bush Jr.! This is great! Glory to God."[36]

Bush was hardly alone with his newly found religiosity. Between 1976 and 1998, the number of Americans who defined themselves as born-again evangelicals went from 34 to 47 percent.[37] The United States was defying the widely held assumption that societies become more secular as they become more modern.[38] Astoundingly, postwar baby boomers, the generation of which Bush was a member but which was dominated by the sixties counterculture, were more likely than any other demographic group to be born-again Christians.[39] Moreover, throughout the seventies and eighties, evangelicalism, which had been largely a middle- and lower-middle class phenomenon, became increasingly acceptable among the more affluent social classes. Wealthy oil men like Bush and his friends in Midland were evangelical Christians.

None of which would have happened unless the evangelical church was able to fill real or perceived needs for tens of millions of Americans. As Robert Putnam documents in *Bowling Alone: The Collapse and Revival of American Community,* his account of the precipitous decline of community in postwar America, the social glue that held the United States together was disintegrating. In the generation after World War II, by any number of measurements—whether it was participation in bowling leagues or labor unions, the Knights of Columbus or Kiwanis clubs, the League of Women Voters, 4-H Clubs, or the Veterans of Foreign Wars—civic engagement in the United States went into free fall.[40] For tens of millions of Americans who lived in a world of bland suburban and exurban outposts and cookie-cutter shopping malls filled with fast-food outlets and faceless chain stores, the old-fashioned, homespun, small-town *Leave It to Beaver* American culture had given way to the anonymity of the mass market.

In a landscape ravaged by such alienation and isolation, the evangelical church increasingly provided both grand spectacles and a desperately needed source of community. "People are not looking for a friendly church as much as they are looking for friends," wrote Rick Warren, pastor of Saddleback megachurch in Orange County, California, in *The Purpose-Driven Church.*[41] "The average church member knows 67 people in the congregation. . . . A member does not have to know everyone in the church in order to feel like it's their church, but he or she does have to know some people."

"The modern megachurch* is like little towns in the Midwest that our parents and grandparents lived in two generations ago," explained Ted Haggard, president of the National Association of Evangelicals until he was exposed in a gay prostitution scandal in 2006, in the HBO documentary *Friends of God.* "It's about family."[42]

Millions of long-haired youths attended services each Sunday because

*Megachurches are generally considered to be those churches that have at least two thousand weekly worshippers.

church officials created an elaborate, fully developed evangelical counterculture that utilized all the marketing tools of the modern world they abhorred. In effect, they were acting like MTV marketing execs for Jesus, creating Jesus Nation, a parallel universe that insulated believers from the temptations of sinful, ungodly, secular America. Whereas Ray Charles had turned gospel into "the devil's music," Christian rockers reversed the phenomenon and brought evangelical lyrics to secular tunes. As rock festivals swept the country, the first religious Woodstock, aka "Godstock," brought nearly eighty thousand people to the Cotton Bowl in Dallas in 1972 for celebratory singing and sermons.[43] That same year, Larry Norman's "Why Should the Devil Have All the Good Music" became the first anthem for Christian rock. In the eighties, Amy Grant's Christian pop won Grammy Awards and hit number one on the *Billboard* pop charts. And for hairspray heavy metal addicts in the eighties, Stryper (Salvation Through Redemption Yielding Peace) went platinum with its megahit "To Hell with the Devil."

There were Christian movie nights for the family and Christian "poker runs" for dad, Christian summer camps for the kids, and Christian comic books, movies, and records for everyone. Instead of Disney World, evangelicals took the kids to Jim and Tammy Faye Bakker's Heritage PTL (Praise the Lord), a 2,300-acre Christian theme park in Fort Mill, South Carolina, that reportedly drew up to six million visitors a year in its heyday.[44] Likewise, the Holy Land Experience in Orlando, Florida, featured a replica of Herod's Temple and a re-creation of the Jerusalem street where Jesus walked toward his crucifixion.

To ensure that the next generation would not be polluted by secularist ideas, millions of fundamentalists homeschooled their children, and then sent them on to one of more than a hundred Christian colleges or universities* where they were taught that scientific phenomena that were taken for granted by the secular world, such as evolution or dinosaurs, were lies. "No one in our family read newspapers," said one former evangelical, a young interior decorator who left her church in Yuba City, California, and eventually moved to New York. "Growing up, our only source of information was the pastor. We believed in what God had told him to say because we were children, and he was our shepherd, and he had been chosen by God."[45]

*According to the Christian Council for Colleges and Universities, more than 900 institutions in the United States are religiously affiliated, but only 102 are "intentionally Christ-centered" educational institutions.

There were Christian singles groups, Christian marriage counseling, and ministries for singles, seniors, and the divorced—and countless Christian radio talk shows. Tim and Beverly LaHaye coauthored a bestselling marriage manual for Christians, *The Act of Marriage,* full of clinical tips on how to break the hymen, and other sage advice: "[C]unnilingus and fellatio have in recent years been given unwarranted publicity . . . [but] the majority of couples do not regularly use it as a substitute for the beautiful and conventional interaction designed by our Creator to be an intimate expression of love."[46]

For married evangelicals with more adventurous tastes, there were Christian "intimacy products" such as flavored Christian condoms ("Seven tasty flavors to choose from: Banana, chocolate, cola, grape, mint, strawberry and vanilla"),[47] edible Christian underwear, and Christian "happy penis" massage cream.[48] An online intimacy store for Christians called Book 22* even offered sex aids such as the "Screaming O" Christian cock ring, a disposable vibrating ring for the penis designed to supply "dynamic stimulation to the clitoris," and the "Double Humm Dinger," with dual vibrating bullets designed to provide forty minutes of continual vibration.

This profusion of evangelical consumer products had no small impact. As another former evangelical who moved to New York put it, "The marketing of moral Christian ideas sustains that culture. Going to a Christian store reinforces your identity as a Christian." In other words, the New Christian Right had created its own subculture—a subculture with its own myths, its own language, its own heroes and villains and its own nonnegotiable political issues. In the sixties and early seventies, millions of people who listened to Bob Dylan and smoked marijuana were overwhelmingly likely to be against the Vietnam War and Richard Nixon. Likewise, the strength of one's "walk with Jesus" became an equally reliable gauge through which evangelical Christians distinguished friend from foes of the pro-life movement or marriage-protection amendment. "The church played exactly the same role as the counterculture in the sixties," the lapsed evangelical said. "Everything that happened in the church environment was a pale carbon copy of secular culture. The pastor was the equivalent of a pop star or a TV star." When it came to politics, he added, that meant that "you vote for that which reinforces your belief system rather than that which will help you

*The twenty-second book of the Bible is the Song of Solomon, the most erotic book of the Bible.

economically. How you will appear in the eyes of the God you believe in—that's your anchor."

All of which boded ill for any candidate who was uncomfortable with evangelicals but who had his eye on the White House, including Vice President George H.W. Bush.

To help deal with the issue in the 1988 presidential campaign, Doug Wead, an adviser to the vice president who acted as a liaison to the religious right, met with the candidate's eldest son in Corpus Christi, Texas, in March 1987. Wead, an Assemblies of God evangelist who had been a motivational speaker for Amway,*[49] had been writing a series of memos for Vice President Bush on building a relationship with the Christian Right. Unbeknownst to Wead, the vice president had forwarded them to his son.

For young George, there was a certain irony in his father's request for help. After all, his brother Jeb, though seven years younger, was the favorite son—and designated heir to the family's political dynasty. "[His parents] felt [George W.] was too abrupt, and too unforgiving to be a politician," Mickey Herskowitz said. "There was a side to him that was truly charming, and there was a side that was just a blank slate. [George H.W.] said that Jeb would be the one in the family who would choose politics. Jeb, he felt, had the political gifts."[50]

Seasoned political operatives who had seen Jeb work for his father's 1980 presidential campaign, and later in the 1988 campaign, agreed. "I remember thinking how terrific he was, how charming he was, how earnest he came across, how honest—all these characteristics that as a political operative you look for in a candidate," said Ron Kaufman, a Bush campaign aide. "There wasn't a doubt in my mind that he had what it takes."[51]

By contrast, whatever George W. did paled in comparison to his father's accomplishments. Bush senior had become a millionaire after he cofounded Zapata Petroleum, but when George W. went into oil, even

*A cultlike marketing company closely allied with Pentecostal and charismatic sects, Amway was sued in 1997 by Procter & Gamble, its gigantic corporate rival, which accused Amway of spreading the urban legend that P&G's old "moon and stars" trademark represented the "666" identified in the Book of Revelation as the number of the beast. For many years, these false rumors have been spread by mass faxes, voice mails, and, more recently, e-mails, asserting that the head of P&G appeared on one talk show or another—Phil Donahue, Sally Jesse Raphael—proclaiming his company's ties to Satan.

after getting help from relatives, family friends, and his father's political cronies for investments, generous contracts, and magnanimous drilling concessions, three successive oil companies failed. "I'm all name and no money," he moaned.[52]

Even when he gave up alcohol, in 1986, and became a God-fearing family man, his father's success still loomed over him. "He loved and adored his dad, but hated and resented being the other George Bush and having to live up to that résumé," said Mickey Herskowitz. "The first time I ever brought up the relationship and what it was to live up to that name and the record his father had achieved over the decades, George W. said, 'I don't want to go there. All my life, all I've heard is, "What has the boy done on his own?" ' "[53]

Meanwhile, as the 1988 presidential campaign got under way, George H.W. Bush enjoyed the overwhelming advantages of being an incumbent vice president serving under a popular president. In the first major battle of the primary season, the Iowa caucus in February, Bush faced Senator Bob Dole and Representative Jack Kemp among others. Dole was widely seen as Bush's most serious rival, but as it turned out he had another powerful adversary as well. The Christian Right had fielded its own candidate—televangelist Pat Robertson.

Like Bush, Pat Robertson was the son of a U.S. senator—Willis Robertson,*[54] a Democrat from Virginia—not to mention the great-great-grandson of President Benjamin Harrison. And, like Bush, he had gone to Yale—having graduated from Yale Law School in 1955. But that's where the resemblance ended. Bush, of course, was an oil man; Robertson made his fortune by buying up FM radio stations at bargain-basement prices in the sixties, when they were considered worthless technology, and selling them later at huge profits. By the eighties, Robertson had his own TV network, the Christian Broadcasting Network, and hosted his own daily show, *The 700 Club,* which mixed "traditional values" lifestyle advice with right-wing politics and Pentecostal

*A Dixiecrat segregationist, Senator Willis Robertson was one of nineteen senators who signed the Southern Manifesto in 1956 condemning the U.S. Supreme Court's milestone *Brown v. Board of Education* decision, which resulted in public school desegregation. When President Lyndon Johnson sent his wife, Lady Bird Johnson, to barnstorm the South for the Civil Rights Act and the Voting Rights Act, Senator Robertson refused to meet her. In retaliation, President Johnson worked to unseat Robertson in 1966 and succeeded. But Robertson went on to become one of the architects of Richard Nixon's highly successful Southern Strategy.

faith healing, enabling him to announce that viewers were being cured of cancer, liver problems, and, on one occasion, hemorrhoids.[55] Over time, his empire grew to include the Christian Coalition, Regent University (formerly CBN University), the Family Channel, several radio and television conglomerates, and the American Center for Law and Justice, an ultraconservative public interest law firm.[56]

Robertson espoused a charismatic theology* unusual among his fellow Southern Baptists. Moreover, he saw fundamentalists as an aggrieved people whose forefathers had founded America as a Christian nation, only to have it stolen away by liberals whom he compared to Nazis during the Holocaust. "Just like what Nazi Germany did to the Jews, so liberal America is now doing to the evangelical Christians," Robertson said. "It's no different. It is the same thing. . . . It is the Democratic Congress, the liberal-based media, and the homosexuals who want to destroy the Christians."[57]

Far from being just another evangelical preacher, Robertson believed it was time for fundamentalists to redress the grievances imposed on them by the secular elites. "The Constitution of the United States, for instance, is a marvelous document for self-government by the Christian people," he said. "But the minute you turn the document into the hands of non-Christian people and atheistic people they can use it to destroy the very foundation of our society. And that's what's been happening."[58]

He claimed that the "separation of church and state in the Constitution . . . is a lie of the Left." He believed that only Christians and Jews should be allowed to hold public office.[59] He decried feminism as "a socialist, anti-family political movement that encourages women to leave their husbands, kill their children, practice witchcraft, destroy capitalism, and become lesbians."[60]

At times, as Michael Lind noted in the *New York Review of Books*, Robertson's theories were rooted more in the world of far-right populism than in fundamentalist theology.[61] Later, in *The New World Order*, for example, Robertson put forth an elaborate conspiracy theory about a satanic plot involving a secret society called the Illuminati, the Freemasons, and wealthy Jewish bankers including the Rothschild family. He even asserted that Jimmy Carter and Vice President George Bush,

*Though it is most typical of Pentecostal Christians, the charismatic movement is not confined to any single denomination. It refers to a wide range of manifestations of the supernatural "gifts of the Holy Spirit," including speaking in tongues (glossolalia), faith healing, miracles, and prophesying.

among others, were "unwittingly carrying out the mission . . . of a tightly knit cabal whose goal is . . . a new order for the human race under the domination of Lucifer."[62]

Such baroque hypotheses aside, Robertson was a savvy political organizer who began making low-profile visits to Iowa more than two years before its presidential caucus to recruit what he and his aides called "an invisible army."[63] When Jeanne Pugh, a journalist with the *St. Petersburg Times,* tried to attend a Robertson meeting, however, she was told that it was not open to the press—a highly unusual practice for a publicity-hungry political candidate. A photographer showed up, but he, too, was banned. "[It] soon became evident that the secret meetings had a definite objective: the recruitment of that invisible army," Pugh wrote.[64]

Using organizations largely unheard of by the liberal elite, Robertson put together a stunningly effective under-the-radar operation. After contacting various people who attended the meetings, Pugh determined that nearly all of them were associated with Concerned Women for America, the enormous anti-feminist organization headed by Beverly LaHaye.[65] Meanwhile, Tim LaHaye, who now headed the American Coalition for Traditional Values (ACTV), had recruited "Bible believing" pastors and hundreds of like-minded campaign workers in each of the nation's 435 congressional districts, registering millions of evangelicals in the process.

As the Iowa campaign drew to a close in late January, Robertson supporters were instructed to tell pollsters that they might *not* attend or vote in a caucus. Since the pollsters included only likely voters, Robertson fared poorly in the precaucus survey—thereby positioning himself to win a huge PR bounce for beating the expectations game.[66] Meanwhile, billboards sprouted up around Iowa depicting a fetus in a womb accompanied by the words "Support those who will protect us: Iowa caucuses, February 8."[67]

By late January, pundits widely predicted that the Iowa results would leave Vice President Bush with just one challenger, Kansas senator Bob Dole.[68] The only real question seemed to be whether Jack Kemp would survive as a serious challenger.

But when the caucus results came in on the night of February 8, 1988, network newscasters were agog at the results. Bob Dole soundly thrashed Bush 37 to 19 percent, but that wasn't the big story. The real news was that George H.W. Bush, an incumbent vice president under the most popular Republican president in decades, had come in third behind Pat Robertson. As the results were tallied, Peter Jennings and

David Brinkley, the coanchors for ABC News, were at a loss for words. How could a Bible Belt fundamentalist who engaged in the supernatural practices of backwoods religious sects have beaten a scion of the Eastern establishment like Bush?

Even more remarkable, Robertson, who had finished ahead of Bush in Michigan, the first GOP contest, continued to do so in Minnesota and South Dakota in late February.[69] "Robertson Burns Bush," "Robertson Steals Show" read the headlines. Moderate, Old Guard Republicans were aghast. "The notion of a fundamentalist preacher playing a major role in national politics is abhorrent to large numbers of people, including a lot of mainline Christians," Republican consultant John D. Deardourff told *BusinessWeek*.[70]

But slowly the truth sank in. The loyal, disciplined cadres working for Robertson constituted a powerful right-wing Christian political movement. "Robertson has perhaps the most powerful political machine in America," said Bob Dole's campaign chairman Bill Brock.[71]

"[They are] well organized at the local level, loosely connected to other cells and to the movement's independent leaders, highly motivated and intent on spreading their message," explained David Harrell in a biography of Robertson. "Charismatics [are] masters of grassroots organization. No group in American history has been better situated to permeate the nation's political system than charismatics."[72]

Neil Bush, the vice president's middle son, had scorned Robertson supporters as "cockroaches" issuing "from the baseboards of the Biblebelt."[73] Bush campaign aides called them "the aliens,"[74] and both Vice President Bush[75] himself and his adviser Lee Atwater[76] shared those sentiments. In the 1980 primaries, Bush had derided Reagan's "Voodoo economics," but, accommodating as ever, when he joined the team he eagerly embraced the same "supply side" policies he had once assailed. Now, in light of Robertson's ascendancy, Vice President Bush had to embrace the Christian Right.

Enter George W. Bush. His challenge—teaching his father how to win over evangelicals—was not an easy task. Try as he might, when evangelists asked him trick questions designed to reveal whether or not he was really one of the flock, the elder Bush almost always stumbled. Once, the *Washington Post* reported, Bush expressed dismay that a good Muslim would be barred from heaven simply because he hadn't accepted Christ. "I could see the color draining from [Jerry] Falwell's face as he said to himself, 'I can't believe I've been out there promoting this guy,'" said a source who attended the meeting.[77]

But when George W. came in, Doug Wead, the campaign's liaison to the religious right, said, "It was such a relief. I could see right away with this guy I wasn't going to have to write twenty-page memos explaining what born-again means." He impressed Wead with his ability to respond to theologically coded questions.[78] According to Wead, if asked what argument he would give to gain entry to Heaven, he would say "I know we're all sinners, but I've accepted Jesus Christ as my personal savior."[79]

Early on in the campaign, Wead saw young Bush in action with a First Assembly of God pastor in Iowa. "Junior knocked his socks off," said Wead. "In five minutes he was blown away. It's the way he talks, the humility in his voice. I've been to a thousand meetings with candidates and evangelicals and never have I seen them hit it off so instantly."[80]

"George W. Bush is an authentic evangelical," he adds. "And evangelicals can know that in an instant. He can go into a room and within five minutes they will know he is the real deal."[81]

Bush and Wead approached the evangelicals as the core of a new political base. "We had a memo produced called 'Targets,' and it was several hundred pages," says Wead, "and it listed a thousand evangelical leaders from a cross section of the evangelical world." The list included Tim and Beverly LaHaye, D. James Kennedy, Jimmy Swaggart, and countless others. "If there was a professor from Wheaton [the evangelical college in Illinois] who was a fan of C. S. Lewis [a favorite novelist of evangelicals], the Bush campaign would go out and buy an old *Time* magazine with C. S. Lewis on the cover and we'd get the VP to autograph it, and send it to the pastor."[82]

George W. was intoxicated by the heady taste of presidential politics. "There wasn't much he believed in," says a friend of the family. "It didn't really matter to him what his father's positions were on issues. He enjoyed the power he had as the gatekeeper. He enjoyed turning down people who wanted to speak to his father."

For the first time, his father depended on him, not vice versa. "I remember one meeting where we thoroughly prepped the vice president, and he had been in many sessions already," said Wead. "He was very good, but we were with a group of evangelicals. They were really tough. They started peeling the onion back so fast that I thought, 'Uh-oh.' Finally, the vice president said, 'Hey, fellas, you need to talk to my son. He's a real born-again Christian.' So the vice president was admitting, acknowledging that 'I'm trying to learn this stuff; my son's already there.'"[83]

Throughout the rest of the campaign, Wead and young Bush relentlessly pursued one evangelical leader after another, including Jim Bakker, the televangelist and host of *The PTL Club* television program who became embroiled in a highly publicized sex scandal that year and a subsequent accounting scandal that led to his imprisonment. According to Don Hardister, Bakker's bodyguard, Wead met frequently with Bakker to secure his endorsement of Bush, and even discussed the possibility of Bakker becoming a White House aide. (Wead has denied that the conversation took place.) "Bakker would have been the new Billy Graham," Hardister told the *Washington Post*, "and he would have gotten [a new White House position] without any question."[84] An oversized autographed picture of Bush was displayed prominently in the lobby at Heritage USA, the PTL theme park.

As a result of such machinations, Vice President Bush beat Robertson handily in such Southern evangelical strongholds as South Carolina, Georgia, Kentucky, Texas, Virginia, and ten other states. He not only won the nomination, but when the 1988 general election came in November, he won 81 percent of the evangelical vote. He lost the Jewish vote by a huge margin. He lost the Hispanic vote. He lost the Catholic vote. Yet Bush beat Democratic nominee Michael Dukakis in a landslide. "So if you can win in a landslide and the only identifiable constituency is the evangelical Christian—whew!" said Wead.[85]

For their first night at the White House, George H.W. Bush and Barbara invited more than two dozen family members to sleep over.[86] The five days of inaugural festivities were far more lavish than even Reagan's. This was the Bush family's moment of glory. It was also a noteworthy point from which to assess the Bush dynasty. How had it gotten this far? Where was it going?

Pundits who observed the inauguration were quick to point out that the new president had made peace with Jerry Falwell[87] and Pat Robertson.[88] But there were also signs that George H.W. Bush's alliance with them was a matter of expediency more than conviction. "Ronald Reagan talked a lot about God," noted Billy Graham, who gave the invocation. Asked why George Bush didn't talk more about God, Graham explained, "Well, you see, he's an Episcopalian."[89]

In contrast to the evangelicals, the neoconservatives, who had such a powerful presence in the Reagan administration, were virtually nowhere to be seen. With the Beach Boys playing at the Lincoln Memorial, a mariachi band at the Kennedy Center, and country music everywhere, the *New York Times* noted, the populist cultural fare was distinctly out of sync with "the sober elitism" they favored.[90]

But few seemed to be aware of what was the most important development of all. The prodigal son had redeemed himself at last. At a party sponsored by the huge Texas oil industry law firm, Baker Botts, young George and his wife, Laura, held forth as guests of honor, with him working the crowd, according to the *Washington Post,* "as if he were the candidate, with a slap on the back and a 'thank you' for nearly anyone who approached."[91]

By delivering the Christian Right, George W. Bush had played a vital role in making his father president. To the old man, who thought of fundamentalists as the product of a genetic disorder, the alliance was merely pragmatic. But to the son, it was very real. For most of his life, George W. Bush had been a failure, but now he was positioned to play a key role in the future.

For his part, Pat Robertson had never expected the elder Bush to be an authentic voice for evangelicals. After the elections, he founded the Christian Coalition, the powerful political advocacy group comprised of fundamentalists, evangelicals, Catholics, Pentecostals, and other sects. During the week of President George H.W. Bush's inauguration, Pat Robertson ran into Ralph Reed, a young Republican activist, still in his twenties, who had interned for Jack Abramoff* at the College Republican National Committee. Robertson instantly recruited Reed to run the day-to-day operations of the Christian Coalition. "The religious conservative movement had always gotten it backwards," says Reed. "It always tried to leapfrog over the preliminary steps to political influence by, in one long bomb at the end of the game, trying to win the White House. By going twelve rounds with [Bush aides] Lee Atwater, and Charlie Black and Jim Baker and Ed Rollins, Pat Robertson got a Ph.D. in the political school of hard knocks. The days of kowtowing to Republican presidents had come to an end and it was now time to build a permanent separate grassroots structure."[92]

As Reed saw it, unorthodox tactics were called for. "I want to be invisible," he explained. "I do guerrilla warfare. I paint my face and

*Touted by *Time* magazine as "The Man Who Bought Washington," Abramoff, of course, was the most powerful lobbyist in the Republican congressional machine during the Bush-Cheney era. He won notoriety as the central figure in a lobbying scandal in which he and partner Michael Scanlon, a former aide to Tom DeLay, bilked Indian gambling interests out of an estimated $85 million by lobbying against their own clients in order to force them to pay for lobbying services.

travel at night. You don't know it's over until you're in a body bag. You don't know until election night."[93]

Just as labor unions, long in decline, once served the Democratic Party, evangelical churches could serve as Republican political clubhouses. Pastors became precinct captains for the Christian Right in a new Republican political machine.[94] "The mission of the Christian Coalition is simple," Pat Robertson explained, "to mobilize Christians one precinct at a time, one community at a time, one state at time, until we are at the head and not the tail, at the top and not the bottom of our political system. We will develop the ability to elect majorities in the U.S. Congress and the legislatures in at least thirty states as well as the city councils, the city school boards, and other local bodies. As the Bible says, if God be for us, who can be against us."[95]

"If we execute this, in the coming ten years, we will be the most powerful force in American politics," Ralph Reed promised.[96] To some of the more radical Christian leaders, that meant the wholesale "Christianizing" of America—that is, the fulfillment of Puritan dreams by transforming the United States into a Christian theocracy. "It is dominion we are after," said Presbyterian televangelist D. James Kennedy, pastor at Coral Ridge Presbyterian Church in Fort Lauderdale, Florida. "Not just equal time. It is dominion we are after. World conquest. That's what Christ has commissioned us to accomplish. And we must never settle for anything less."[97]

CHAPTER SEVEN

The Age of Unreason

For years, George H.W. and Barbara Bush had thought about whether or not their eldest son could become heir to the Bush political dynasty. Usually, the answer had been no.[1] But in the 1988 presidential campaign, after two decades of uncertainty, aimlessness, and failure, for the first time in his life, George W. Bush had accomplished a real, clearly defined mission. Thanks to his relationship with the Christian Right, he had made an invaluable contribution to his father's campaign.

After the election he sat down to chat with Doug Wead. "We were sitting in his office," Wead recalled, "and we were talking about who on the staff is going to go to transition—who's going to go to the inaugural, who's going to go to the White House. He sighed in the middle of that and said, "What's going to happen to me?"

Wead hadn't thought about it. "You want me to do a memo on what happens to presidential kids?" he asked.

"Yes."

After doing some research, Wead was staggered by the high rates of divorce, alcoholism, and early, tragic death. "If you're a presidential son named after your father, it's almost like a curse," he said. "I mean, John Adams Jr. dies an alcoholic at thirty-one, and William Henry Harrison at thirty-five, and Andrew Johnson Jr. at twenty-six. . . . Calvin Coolidge Jr. is only sixteen years old when he dies after an accident on the White House tennis courts."

It was clear to Wead why there was so much stress on presidential sons—particularly those named after their fathers. "People come up to them when they're little kids and say, 'Well, when are you going to run for president?'" Wead thought about what kind of president Bush would make. "He's so decisive. He's so adamant, so dogmatic—makes a decision, never looks back. . . . I thought either he'd be terrific, or he'd be terrible. But he'd sure make news. There would be sparks."[2]

Wead may have been unaware of it, but another important factor in the younger Bush's future was that prevailing political forces were shifting directions. The pragmatic realism that helped put George H.W. Bush in the White House was about to encounter resistance from the powerful pressures that would later aid his son.

On January 20, 1989, the elder George Bush officially moved into the White House. As former ambassador to the United Nations, liaison to the People's Republic of China, and head of the CIA, he brought a wealth of foreign policy experience to the Oval Office. With his longtime friend James Baker as secretary of state, Brent Scowcroft as national security adviser, and, later, Colin Powell as chairman of the Joint Chiefs of Staff, Bush assembled an experienced and powerful team of policy makers who were deeply wedded to the realist school of foreign policy and determined to oversee the last days of the Cold War.

Initially, Bush designated former senator John Tower, a conservative Republican, as secretary of defense. But after questions were raised about Tower's ties to the defense industry, his personal relationships, and drinking problems, Bush appointed Dick Cheney (R-Wyo.), house minority whip, to the Pentagon's highest post instead. Cheney was considerably more conservative, but he was seen as so genial and flexible that the *New York Times* praised his "vaunted pragmatism."[3] It was widely speculated that Brent Scowcroft, who had served with Cheney in the Ford administration, was a major proponent of his appointment.[4]

Because Cheney was liked by such Democrats as Sam Nunn, the chairman of the Senate Armed Services Committee, and Tennessee senator Al Gore,[5] his nomination sailed through Congress. But his cordial demeanor masked a ferocious partisanship and grandiose dreams of an extraordinarily powerful executive branch. Cheney, for example, saw the Watergate scandal as nothing more or less than a partisan political power struggle—and an attack on the executive branch of the government. "He claimed [the Watergate scandal] was just a political ploy by the president's enemies," said Bruce Bradley, a colleague of Cheney's at the time.[6] "Cheney saw politics as a game where you never stop pushing. He said the presidency was like one of those giant medicine balls. If you get a hold of it, what you do is, you keep pushing that ball and you never let the other team push back."*[7]

Bush, as vice president, and Baker, as chief of staff, had both played key roles in the Reagan administrations, but on no issue did they diverge more sharply from their predecessor than Israel. Rather than attempt to resolve the Israeli-Palestinian conflict, Reagan had refrained from pressuring Israel for most of his administration, though he had initiated a dialogue with the Palestinian Liberation Organization (PLO) that began just as he was leaving office. Now that Bush and Baker were in charge, they put the Israeli-Palestinian conflict on the front burner.

All presidents, of course, are subject to the larger forces of history. But when it came to the Israeli-Palestinian conflict, resolving the most intractable problem on earth also meant confronting biblical myths about what transpired between God and Abraham four thousand years ago, myths that spoke to the most elemental questions of human existence, salvation and redemption, the End of Days and Armageddon. Consequently, to understand the obstacles the Bush administration and, later, the Clinton administration encountered, to understand how the Christian Right came together with neoconservatives and the Israeli right and ultimately how they came to support George W. Bush and the Iraq War, a bit of history is in order.

The Israeli-Palestinian conflict, of course, dated to the very creation of Israel and was gravely exacerbated by the Six-Day War in 1967, when Israel, fearing an imminent invasion, launched a preemptive strike against Egypt.† Israel's astonishing triumph over not just Egypt, but Jordan and Syria as well, allowed it to occupy vast stretches of new territory, including the Palestinian West Bank, Syria's Golan Heights in the north, and, in the south, the Gaza Strip and the Sinai Peninsula up to the Suez Canal. Most of the world saw Israel's victory as giving it the necessary bargaining chips to negotiate for a lasting peace that would also allow the Palestinians to return to the West Bank as their legitimate home.[8]

*Likewise, in 1987, when Congress investigated the Reagan administration's Iran-contra scandal, Cheney was the crucial defender of the illegal U.S. foreign policy initiatives conducted through privately funded back channels. Thanks to his leadership, not a single Republican congressman signed the investigating committee's final report charging "secrecy, deception and disdain for law."

†Once the war started, Iraq, Saudi Arabia, Kuwait, and Algeria also contributed troops and arms to Arab forces.

But winning the peace wasn't that simple. By occupying Palestinian territory, Israel had ripped open a wound that went back four thousand years. As Orthodox Jews and Christian evangelicals saw it, Israel had won back its birthright, the Holy Land—and it should never give back a single inch of land to the Palestinians. That's what the Bible said, in Genesis 13:15, where God told Abraham: "I will give it to you and your descendants forever."* This divine pledge, known among evangelicals as the Abrahamic Covenant, embodied the ultimate impasse in the battle between modernism and fundamentalism. To those who accepted the Bible as the literal word of God, who saw the birth of Israel in 1948 as a prelude to the return of the Messiah, who saw the Six-Day War in 1967 as the further fulfillment of divine prophecy, God's gift was irrevocable and eternal.

As a result, the ongoing Israeli occupation of the West Bank and the Gaza Strip† was a fault line, a litmus test, an event that was transformative not just to the millions of Palestinians who lived in apartheidlike squalor under Israeli rule, but to Israel itself, to the entire Middle East, and, in different ways, to American liberals, American Jews, and Christian evangelicals. To most of the world, Israel had no right to occupy the Palestinians' homeland. The Palestinians had been devastated. About one-quarter of the population of the West Bank, at least 200,000 Palestinians, had been forced into exile, angering more than a billion Muslims and tens of thousands of Islamic terrorists. Israel had violated international law. But to the Israeli Right, to Orthodox Jews, to neoconservative policy makers—and to American fundamentalists—Israel had every right, divine or otherwise, to reclaim the Land of Israel—Eretz Yisra'el.‡

*Genesis, of course, is one of the first five books (i.e., the Pentateuch) in the Torah, the Hebrew bible. Orthodox Jews believe that its laws are divine and were translated by God to Moses, and are immutable. Therefore, it is not surprising that Orthodox Judaism and evangelical Christianity should hold similar beliefs with regard to Israel as the Promised Land.

†The West Bank consists of the biblical lands of Judea and Samaria on the west bank of the Jordan River. By 2006, its population had grown to include about 2.5 million Palestinians and 260,000 Israeli settlers and another 185,000 Israelis in East Jerusalem. The Gaza Strip is about 25 miles long and 4 to 7.5 miles wide, bordered by the Mediterranean Sea on the west, Israel to the north and east, and Egypt to the southwest, with a population of about 1.4 million Palestinians.

‡The term "Land of Israel" or Eretz Yisra'el, is derived from the Hebrew bible, and is most often interpreted as referring to part of the Abrahamic Covenant in Genesis, 15:18–21, in which God promises Abraham: "To your descendants I have given this land, from the river of Egypt as far as the great river the Euphrates." When the

* * *

To many American Jews, this last part of the equation—the idea that Christian evangelicals were among the strongest supporters of hard-line Israelis—was particularly vexing. After all, the notion that Bible-thumping fundamentalists were now best friends with the Jews was a bit hard to swallow, especially in light of their long history of anti-Semitism.[9] More than a century earlier, evangelical theologian Arno Gaebelein had written that "there is nothing so vile on earth as an apostate Jew."[10] In the twenties and thirties, evangelists from Billy Sunday[11] to Gerald Winrod[12] expressed affinities with the Ku Klux Klan, the Nazis, and Adolf Hitler.[13] And in more recent times, Billy Graham* shared anti-Semitic asides with Nixon during the Watergate scandal; Bailey Smith, president of the Southern Baptist Convention, avowed that "God Almighty does not hear the prayer of a Jew"[14]; and Pat Robertson churned out anti-Semitic tales about scheming Jewish bankers and the like in his book *The New World Order.*[†15]

Yet theologically, both American evangelicalism and Israel shared deep roots in history that could be traced back to a shared strain of

---

Israeli Right refers to the land of Israel, it is generally referring to the area between the Mediterranean Sea and the Jordan River, including the West Bank and Gaza.

*Richard M. Nixon's Watergate tapes reveal that in 1972, Nixon, Billy Graham, and White House Chief of Staff H. R. Haldeman had a conversation in the Oval Office in which the Jews were targets:

BILLY GRAHAM: "This [Jewish] stranglehold has got to be broken or the country's going down the drain."

RICHARD NIXON: "You believe that?"

GRAHAM: "Yes, sir."

NIXON: "Oh, boy. So do I. I can't ever say that but I believe it."

GRAHAM: "No, but if you get elected a second time, then we might be able to do something."

GRAHAM: "By the way, [*Time* magazine editor in chief] Hedley Donovan has invited me to have lunch with [the *Time* magazine] editors."

HALDEMAN: "You better take your Jewish beanie."

GRAHAM: "Is that right? I don't know any of them now. . . . A lot of Jews are great friends of mine. . . . They swarm around me and are friendly with me because they know that I'm friendly with Israel. But they don't know how I really feel about what they are doing to this country."

NIXON: "You must not let them know."

†At times, Falwell himself had smiled upon anti-Semitism among his parishioners. "I know a few of you here today don't like Jews, and I know why," he told them in 1979. "[The Jew] can make more money accidentally than you can on purpose . . . still the Jews were the apple of God's eye."

Christian Zionism that first flourished in sixteenth-century England, and has continued to wax and wane over the centuries. Without it, neither Israel nor the neocons would have existed at all—and America would have been very different indeed. Astoundingly, discussions of this powerful historic force have been virtually taboo in the United States.*

Among English Puritans, the first stirrings of Christian Zionism began around 1585, when Anglican clergyman Thomas Brightman, in *Apocalypsis Apocalyseos,* wrote that only after the Jews returned to Palestine would a series of prophetic events unfold that would lead to the return of Jesus.[16] This marked the beginning of a powerful philo-Semitic doctrine that posited that the Jews were God's Chosen People, and as such were key players in the messianic drama of the Second Coming of Christ.

Even though its proponents were excommunicated and burned at the stake, Christian Zionism survived in England,[17] and in 1621, a prominent member of Parliament named Sir Henry Finch proposed the Jewish settlement of Palestine in order to fulfill biblical prophecy and facilitate the Second Coming of Christ.†[18] More than two hundred years later, in 1839, Anthony Ashley Cooper, seventh Earl of Shaftesbury, a conservative evangelical member of Parliament, argued "[T]he Jews must be encouraged to return [to Palestine] in yet greater numbers and become once more the husbandman of Judea and Galilee."[19] Like many American evangelicals more than a century later, Shaftesbury, a follower of John Nelson Darby,[20] cited Genesis 12:3‡ as a rationale for his cause, arguing that God would favor England if England supported the return

*An August 2007 search of the LexisNexis database shows that the term *Christian Zionism* has never appeared in the *New York Times* in the three decades the *Times* has submitted articles to the database.

†In America, the Christian Zionist impulse that was so powerful among the English Puritans who settled the country did not die out entirely and was revived in the second half of the nineteenth century with the arrival of John Nelson Darby. In 1891, William E. Blackstone, a disciple of Darby, finally put together the first major Zionist lobbying effort in the United States by bringing together J. P. Morgan, John D. Rockefeller, Charles B. Scribner, and other financiers to persuade President Benjamin Harrison to support—unsuccessfully, as it turned out—the establishment of a Jewish state in Palestine. But that was not the end of it. In August 1897, Blackstone journeyed to Basel, Switzerland, where he told Austrian journalist Theodor Herzl, who later became the father of political Zionism, that if he lobbied for the establishment of a Jewish state in Palestine, the "Christian nations" of the West would come to his aid. Ultimately, that meant England.

‡"I will bless those who bless you and through you will all the nations of the earth be blessed."

of the Jews to the Holy Land. Ultimately, Shaftesbury successfully lobbied for the appointment of a British vice-consul in Jerusalem whose duties included "protection of the Jews"[21] and lived to see Parliament authorize an Anglican bishopric in Jerusalem in 1843.[22] In describing Jews as "a people with no country for a country of no people," Shaftesbury effectively coined the slogan later used in a slightly different form by early Zionist leader Theodor Herzl.[23]

By the time World War I broke out, Christian Zionism, sometimes known as Gentile Zionism, had become a vital movement in England, a biblically rooted messianic dream that was finally being transformed into political reality. One of the key figures behind its ascendancy was Chaim Weizmann, a young chemist* at Manchester University who met Charles Prestwick Scott, the powerful editor of the *Manchester Guardian* and one of the great figures in British journalism, in 1914.[24] By converting Scott to Zionism, Weizmann, who many years later became the first president of Israel, thereby secured the support of one of the most influential newspapers in the country. Soon the *Guardian*'s chief military correspondent was writing dispatches asserting that "the whole future of the British Empire as a Sea Empire" depended on Palestine's being inhabited by Jews.[25]

In addition, Scott introduced Weizmann to Prime Minister David Lloyd George. Like many Christian evangelicals in America today, David Lloyd George was a premillennial dispensationalist.[26] Having served as the British attorney for Zionist pioneer Theodor Herzl,†[27] Lloyd George

*Weizmann became famous for discovering how to use bacterial fermentation to produce sought-after substances such as acetone, which was used in cordite explosives that were essential to the Allied effort in World War I.

†As David Fromkin writes in his prophetically titled history of the era, *A Peace to End All Peace: The Fall of the Ottoman Empire and the Creation of the Modern Middle East*, Herzl was the Paris correspondent for a Viennese newspaper who paid little attention to his Jewish background until the anti-Semitism of the Dreyfus Affair persuaded him that the world's Jews desperately needed a homeland. By the beginning of the twentieth century, however, Herzl realized that the Ottoman empire was not going to make Palestine available, and began to look elsewhere. To help him accomplish his mission, Herzl hired a politically connected lawyer, David Lloyd George, and finally got the British government to offer an alternative homeland in British East Africa—i.e., Uganda. In 1903, Herzl even managed to get the World Zionist Congress to approve the idea of settling in Uganda as "a way station" on the road to the Promised Land. There was just one problem: no one really wanted to move to Uganda. Herzl died in 1904. With the end of World War I and the fall of the Ottoman empire, Palestine finally fell into British control and the Zionist dream began to be realized.

was already deeply involved in the Zionist project, and when he became prime minister in 1916, he took on C. P. Scott as his most trusted political adviser[28] and Arthur Balfour, another Christian Zionist and a former prime minister himself, as foreign secretary. By the end of World War I, Zionists held powerful positions in the British government and media, and the realization of the Zionist dream was a real possibility.

In May 1916, British diplomat Mark Sykes,*[29] also a Christian Zionist, and his French counterpart, François Georges-Picot, hammered out what became known as the Sykes-Picot Agreement in which they essentially divvied up the spoils of the disintegrating Ottoman empire into their respective spheres of influence. A year later, on November 2, 1917, Great Britain took the first historic step that would ultimately lead to the birth of Israel when Balfour wrote Lord Rothschild a short letter that became known as the Balfour Declaration of 1917, asserting that His Majesty's government viewed "with favor the establishment in Palestine of a national home for the Jewish people, and will use their best endeavours to facilitate the achievement of this object."

From its roots in sixteenth-century Puritan England to the ascendancy of Christian Zionists to top posts in the media and the government, the Christian Zionist–Jewish alliance in World War I England shared striking similarities with and, in many ways, prefigured the American evangelical–neocon alliance more than eighty-five years later that fought the land-for-peace settlement to the Israeli-Palestinian conflict. In addition, Britain's intentions, just like America's today, were not confined to fulfilling the messianic dreams of Christian Zionists or finding a secure homeland for the Jews. After all, this was the period of the Raj, the British rule of India. Zionist ideals meshed conveniently with Britain's imperial ambitions, the increasing importance of Middle East oil, and the idea of establishing a Jewish state as a crucial strategic land bridge to India, the "Jewel of the Crown."

In other words, these noble religious ideals all just happened to serve

*Even though Sykes died in 1919 of the Spanish flu, his greatest gift to mankind may yet come to pass. In 2007, his body was exhumed in hopes of extracting DNA samples of the Spanish flu virus that killed him and 50 million other people. Because the modern bird flu virus is thought to have a similar genetic structure, scientists hope the samples can be used to develop a vaccine to ward off the disease. Sykes's body was preserved in a lead-lined coffin, giving virologists reason to believe that genetic samples may have been preserved as well. Said his grandson, Christopher Simon Sykes: "It is rather fascinating [to think] that maybe even in his state as a corpse, he might be helping the world in some way."

British interests in what was known as "the Great Game," Britain's never-ending struggle throughout the nineteenth century and into the twentieth to maintain lines of transport, communication, and trade that were essential to its global empire. It was a kingdom so vast that Britain had to prop up one floundering Islamic fiefdom after another against various European foes, especially Russia. As George Curzon, the viceroy of India put it, "Turkestan, Afghanistan, Transcaspia, Persia—to many these names breathe only a sense of utter remoteness. . . . To me, I confess, they are the pieces on a chessboard upon which is being played out a game for dominion of the world."[30]

Just because the British rulers were Zionist did not mean they had much regard for those who would be most affected by the Zionist cause. In a letter about the proposal for the new Jewish state, former prime minister Herbert Henry Asquith wrote about Lloyd George's feelings toward Jews: "Curiously enough, the only other partisan of this proposal is Lloyd George, who, I need not say, does not care a damn for the Jews."*[31]

In addition, in the process of carving up the Middle East and assigning spheres of influence to various Western powers, the British conveniently overlooked another deeply concerned party—namely, the millions of Arabs who lived there. Indeed, as details of the partition of Palestine were hammered out at the Paris Peace Talks of 1919, Lord Balfour voiced his distress at the notion that Muslims should participate in the discussions. "I am quite unable to see why Heaven or any other Power should object to our telling the Moslem what he ought to think," he said.[32]

"In Palestine, we do not propose even to go through the form of consulting the wishes of the present inhabitants of the country. . . ." Balfour added in a speech that year. "Zionism, be it right or wrong, good or bad, is . . . of far profounder import than the desires and prejudices of 700,000 Arabs who now inhabit that ancient land."[33]

Balfour's sentiments foreshadowed the conflicts that would haunt the Middle East for generations to come, beginning with one wave of Jews after another immigrating to Palestine, usually accompanied by violent

*Not everyone agreed with Asquith's assessment. In *Two Studies in Virtue*, Christopher Sykes, whose father, Mark Sykes, drafted the Sykes-Picot Agreement that dismantled the Ottoman empire, takes issue with the former prime minister's characterization of Lloyd George as anti-Semitic: "The contrary would have been much closer to the truth. Lloyd George cared intensely for the Jews; his devotion to them was one of the consistent things in his bewilderingly various nature."

clashes with Arabs. With the rise of Nazism in the thirties and, later, the Holocaust, British authorities tried to restrict Jewish immigration, but the floodgates had opened. From 1922 to 1945, the Jewish population in the region increased from 83,790 to 608,230.[34]

With the end of World War II, revelations about the Holocaust fueled the Zionist impulse to create a Jewish nation in the land of Israel. Jewish immigration to Palestine increased, as did Arab-Jewish violence. The British decided to abandon the Palestine Mandate, the United Nations General Assembly partitioned the territory into two states, one Jewish, one Arab, and in 1948, the state of Israel was proclaimed. Egypt, Iraq, Jordan, Lebanon, and Syria immediately declared war on the infant state, and the ensuing 1948 Arab-Israeli War resulted in more than 700,000 Palestinian refugees and the creation of the most intractable conflict in contemporary history.

To Christian evangelicals in America, however, Israel's birth in 1948 was nothing less than the fulfillment of their millennial dreams. "I consider it the greatest event, from a prophetic standpoint, that has taken place . . . perhaps since 70 A.D. when Jerusalem was destroyed," said premillennialist Louis Talbot of the Bible Institute of Los Angeles.[35] The Institute's monthly magazine, the *King's Business,* predicted that Israel's birth would lead to "the appearance of the greatest Jew of all time, the Lord Jesus Christ."[36]

To most Americans in the modern era, however, the shared roots of Israelis and Christian Zionists were ancient history. "The Bible said Christians should be supportive of Israel," explained Rabbi Yechiel Eckstein, an Orthodox rabbi who founded the International Fellowship of Christians and Jews to harness support from evangelicals. "But were they willing to translate that into realpolitik?"[37]

The answer began to emerge in 1977 after Jimmy Carter became the first American president to assert that "Palestinians deserve a right to their homeland." It was Carter, of course, who oversaw the historic Camp David Accords in which Egyptian president Anwar Sadat and Israeli prime minister Menachem Begin agreed to Israel's withdrawal from the Sinai and its restoration to Egypt in return for normal diplomatic relations with Egypt and guarantees of freedom of passage through the Suez Canal.

Camp David was a historic and lasting step forward, and as such represented a strong case for the land-for-peace initiatives. But there was still no resolution to Israel's occupation of the West Bank and Gaza

Strip, and Carter soon reached an impasse with Begin because the latter was intent upon expanding Israeli settlements in the West Bank and flouting Carter's attempt to resolve the Israeli-Palestinian conflict with a two-state solution.

Israel had long been close to liberal Democrats in the United States, but now Prime Minister Begin sought new allies. A romantic nationalist and a serious biblical scholar who pointedly referred to the West Bank by its biblical names of Judea and Samaria, he reached out to American evangelicals who were finally emerging from their fifty-year-long political hibernation. "The prime minister said a person who has got the Bible in his home and reads it and believes it cannot be a bad person," recalled Yechiel Kadishai, who served as Begin's longtime aide and closest confidante. "He said the evangelicals have to know that we are rooted in this piece of land. There should be an understanding between us and them."[38]

One of the first people Begin sought out was Jerry Falwell, who had just begun to achieve national recognition and who saw the birth of Israel as the fulfillment of biblical prophecy. "It was obvious to me, beginning with the birth of the Israeli state, in 1948, and the Six-Day War, in 1967, that God was bringing his people back home," Falwell said in a 2005 interview. "So I came to believe that it was in America's best interest to be a friend of Israel. . . . If America blessed the Jew, Israel in particular, God would bless America."[39] On the wall behind him was a framed front page from the *Palestine Post,* dated May 16, 1948, with the headline "State of Israel Is Reborn."

"Long before I became a political activist," Falwell added, "I had been taught that the Abrahamic Covenant—Genesis 12 and Genesis 15—was still binding, where God told Abraham that I will bless them that bless you and curse them that curse you. I later came to believe that Israel is America's best friend in that part of the world."*

He and Begin got along famously. In 1980, Begin presented Falwell with the prestigious Jabotinsky† Award, making him the first gentile to receive it. He gave Falwell's ministry a private jet. Then, in June 1981, not long after President Ronald Reagan and Vice President George H.W. Bush took office, the prime minister called Falwell to give him advance notice about a historic event. "Tomorrow you're going to read

*Falwell is citing Genesis 12:3.

†Ze'ev (Vladimir) Jabotinsky, for whom the award is named, was a founder of the Revisionist Zionist movement in the twenties who sought to turn Palestine into a Jewish state based on the British model.

some strong things about what we are going to do," Begin told him. "But our safety is at stake. I wanted you, my good friend, to know what we are going to do."[40]

More specifically, Prime Minister Begin was concerned because the United States had supplied Israel with F-16s and other arms on the condition that they be used only for defensive purposes, and Israel was about to use them for a preemptive strike. Before the end of the conversation, however, Falwell assured Begin of his support. "Mr. Prime Minister," he said, "I want to congratulate you for a mission that made us very proud that we manufacture those F-16s."[41]

The next day, Israeli bombers knocked out the Osirak nuclear reactor in Iraq, thereby demolishing Saddam Hussein's nuclear weapons program. "Sure enough, they put one down the chimney," said Falwell.[42]

Inside the Reagan White House, however, the attack was not well received. After all, Israel had violated the stipulation that the F-16s were to be used only for defensive purposes. "Reagan went around the room and asked each of us our opinion on the Osirak raid," recalled Alexander Haig, then secretary of state.[43] "I remember [Vice President] Bush and then [chief of staff James] Baker making it very clear that they thought Israel needed to be punished."

But that didn't matter as much now. Because Prime Minister Begin had touched base with Falwell, Israel had plenty of political cover on the American right.

At the same time, even though the world appeared to be dominated by modernism, a fundamentalist backlash began taking place all over the globe. In 1979, the entire Western world had been stunned by the toppling of the shah of Iran, Reza Pahlavi, the autocratic but pro-West guardian of the Persian Gulf, who gave way to the Islamist revolution led by Ayatollah Ruhollah Khomeini. How was it possible that oil-rich Iran—the most powerful country in the Middle East, with access to all the temptations of the liberal, secular, Western world—had cast aside all those freedoms and all that wealth and reverted to Islamic dress, medieval ideals, and a fundamentalist Islamic revolution to transform Iran into a theocratic society ruled by Sharia, Islamic law, based on the Koran? Secular elites in energy-dependent America and Europe were staggered that atavistic faith had trumped the post-Enlightenment values of reason.

Meanwhile, the Age of Unreason was dawning not just with Islamist fundamentalists in the Middle East, but with Christian fundamentalists and Orthodox Jews as well. Both the United States, the citadel of post-Enlightenment Jeffersonian democracy, and Israel, the only democracy in the Middle East, had powerful political blocs that were deeply committed to implementing the rule of God rather than the rule of reason. In the United States, Jerry Falwell and the Moral Majority, Pat Robertson and the Christian Coalition, and thousands of pastors had begun leading tens of millions of people in a new but powerful political movement battling humanist values—fighting the evils of abortion, permissive sex, drug use, and "the homosexual agenda." Likewise, in Israel, hundreds of thousands of Orthodox Jews had settled in the biblical lands of Judea and Samaria and were fighting to win control of the Holy Land in clear violation of international law—because the Bible said it had been given to them by God.

In 1982, Falwell brought more than three dozen evangelical leaders from the Moral Majority to Israel to foster ties between it and American evangelicals.[44] He cultivated personal friendships and political alliances with Menachem Begin, Benjamin Netanyahu, and Ariel Sharon.[45] Israel gave multimillion-dollar grants to Falwell's Liberty University, enabling him to bring as many as three thousand students at a time from Liberty University to tour the Holy Land.[46]

Other evangelicals followed suit. Prime Minister Begin reached out to David Lewis, an Assemblies of God evangelist, to initiate package tours for American evangelicals.[47] The International Christian Embassy of Jerusalem, a Christian Zionist organization with no diplomatic standing, set up shop in the holy city with the intention of providing evangelical support to the Israeli government.[48] New travel agencies sponsored Bible Prophecy Tours.[49] Bible tourism soared.

Troubled by the fact that many American Jews looked askance at evangelical Christians, in 1983, Rabbi Eckstein[50] founded the Fellowship to bridge the gaps between the two groups and to support Israel. In 1984, he joined with Pat Robertson and Jerry Falwell in celebrating the first annual Day of Christian and Jewish Solidarity with Israel. Eckstein brought Congressman Tom DeLay (R-Tex.), a self-professed Christian Zionist who later became House majority leader; Ralph Reed; and other leaders of the Christian Right into the fold.[51] James Dobson of Focus on the Family; Gary Bauer, who later became president of Family Research Council; and key figures in the Moral Majority and the Christian Coalition joined. "They began getting their hands dirty," Eckstein said.[52]

Soon Eckstein found support for the alliance in the Oval Office, where even President Ronald Reagan expressed Christian Zionist sentiments. "You know," Reagan told Tom Dine, executive director of the powerful Israeli lobbying group, the American Israel Public Affairs Committee (AIPAC), "I turn back to your ancient prophets in the Old Testament and the signs foretelling Armageddon, and I find myself wondering if we're the generation that's going to see that come about."[53]

In March 1984, the Reagan White House brought together top Likud lobbyists from AIPAC, various State Department officials, and about 150 Christian fundamentalist leaders including Falwell, Pat Robertson, bestselling End Times novelist Hal Lindsey (*The Late Great Planet Earth*),* televangelists Jim and Tammy Faye Bakker, Jimmy Swaggart, and Tim and Beverly LaHaye.[54]

Since there are roughly ten times as many evangelicals in America as Jews, Israel welcomed the political support. "Israel's relationship with America can't be built only on the AIPAC and the 2.5 percent of the population in America who are Jews," said Benny Elon, leader of the right-wing National Union, former tourism minister, and a frequent guest of the Christian Coalition in the United States.†[55]

"When Israel enjoys support because we are the land of the Bible, why should we reject that?" asked Uzi Arad, former national security affairs adviser to Prime Minister Benjamin Netanyahu and head of the Institute for Policy and Strategy, a think tank in Herzilya, Israel.[56] "Whether it is because of expediency or because we are soul mates, each side offers the other something they want. And the evangelicals are a political force to be reckoned with in America."

American lobbyists for Israel, well aware that most American Jews were secular liberals, embraced the evangelicals, albeit somewhat reluctantly. "Sure, these guys give me the heebie jeebies," said Lanny Davis, head of research for AIPAC, in a moment of candor. "But until I see Jesus coming over the hill, I'm in favor of all the friends Israel can get."[57]

But other Jews, such as Gershom Gorenberg, a Jerusalem-based journalist, cautioned against embracing Christian fundamentalists who yearn for the Battle of Armageddon.[58] "This is incredibly dangerous to Israel," said Gorenberg, the author of *The End of Days: Fundamentalism and the Struggle for the Temple Mount*, a chronicle of messianic Christians

*Arguably the bestselling book of the entire decade, *The Late Great Planet Earth* was published in fifty-four languages, sold more than 35 million copies, and was the precursor in apocalyptic blockbuster fiction to LaHaye and Jenkin's Left Behind series.

†According to the 2006 American Jewish Yearbook population survey, there are 6.4 million Jews in the United States, approximately 2.1 percent of the total population.

and Jews and their struggle with Muslim fundamentalists over the Temple Mount. "They're not interested in the survival of the State of Israel. They are interested in the Rapture, in bringing to fruition a cosmic myth of the End Times, proving that they are right with one big bang. We are merely actors in their dreams."[59]

Few realized it, but one of the most profound consequences of the Israeli-Arab conflict was that it had created this unlikely and rarely examined but nonetheless historic alliance between evangelical Christians, the Israeli right, and neoconservative policy makers—an alliance that ultimately would have enormous power in domestic American politics. This was to be a relationship that stretched from the most sacred sites in the Holy Land to the deepest recesses of the Bible Belt; from the Orthodox Jewish settlements in Judea and Samaria to the world of NASCAR dads and megachurches; from the Knesset in Jerusalem to the Thomas Road Baptist Church in Lynchburg, Virginia; from powerful pro-Israeli lobbying groups such as the American Israel Public Affairs Committee to neoconservative think tanks such as the American Enterprise Institute to the White House.

So these were the potent forces with which George H.W. Bush and James Baker had to contend if they were to tackle the Israeli question. In mid-March 1989, just six weeks into the new administration, Secretary of State Baker asserted that Israel had to negotiate with the PLO.[60] Two weeks later, he urged Israel to withdraw troops from the Palestinian territories it occupied on the West Bank and the Gaza Strip.[61] Israel, he said, must "lay aside, once and for all, the unrealistic vision of the greater Israel."[62]

Israeli prime minister Yitzhak Shamir, a leader of the right-wing Likud Party, was livid. At an April 6 appearance before the American Enterprise Institute in Washington, Shamir told neoconservatives that he was rejecting the Bush administrations requests.[63] Israel would not end its occupation of the West Bank and Gaza. Shamir said that the policy of trading land for peace—that is, returning the Palestinian land Israel conquered in the 1967 war in return for political recognition and a promise of peace—"is a deception."* "If we leave," he asserted, "there will almost certainly be war."

Deteriorating relations with Israel were not the only problem the Bush White House faced in the Middle East. By early 1990 there were

*Shamir's written text called it a "hoax."

signs that Saddam Hussein, the brutal Iraqi dictator, was becoming a serious regional threat.[64] Three months later, those fears were realized when Iraqi troops invaded Kuwait, precipitating the Gulf War.

Bush 41's Middle East policy had been far from flawless—especially when it came to his generous treatment of Saddam. After all, the United States had supported the brutal Iraqi dictator for seven full years during the Reagan-Bush era after learning he had weapons of mass destruction. As the evidence of Saddam's ruthlessness mounted during the Iran-Iraq War, the United States publicly condemned Saddam, but secretly continued to fund him, provided him with detailed plans for battles, air strikes, and even allowed him to obtain pathogenic material that could potentially be used for biological warfare. In 1986, as vice president, Bush had even made a secret trip to the Middle East to convey strategic military intelligence to Saddam. And when he became president, Bush continued to give Saddam even more aid.* All of which had been done to help Iraq provide a counterweight to Iran's growing influence in the region. This disastrous policy resulted in countless deaths among the Iranians and Kurds, and led to Iraq's invasion of Kuwait, thereby demonstrating the limits of a policy of ruthless pragmatism.

Now that he was president, however, and the Iraqi dictator had shown his hand, Bush finally saw Saddam as the increasingly dangerous monster the United States helped to create. Working closely with National Security Adviser Brent Scowcroft, Secretary of State James Baker, and Chairman of the Joint Chiefs of Staff Colin Powell, Bush and his foreign policy team assiduously stroked both its allies and its adversaries. Bush had invited the somewhat difficult French president François Mitterrand out to Kennebunkport—to great effect. He cultivated obscure rulers of the Middle Eastern emirates, and put together a coalition of more than thirty countries, no fewer than eight of which were Arab.†

*For a more complete account of U.S. aid to Saddam in the Reagan-Bush era and George H.W. Bush and James Baker's roles in it, see the author's *House of Bush, House of Saud,* pp. 65–77.

†The U.S.-led coalition included: Argentina, Australia, Bahrain, Bangladesh, Belgium, Canada, Czechoslovakia, Denmark, Egypt, France, Greece, Italy, Kuwait, Morocco, Netherlands, New Zealand, Niger, Norway, Oman, Pakistan, Poland, Portugal, Qatar, Saudi Arabia, Senegal, South Korea, Spain, Syria, Turkey, United Arab Emirates, United Kingdom, and the United States of America. Germany and Japan provided financial assistance and military hardware. The United States asked Israel not to participate in the war.

In the end, Operation Desert Storm was widely seen as Bush's finest hour and greatest triumph. At a minimal cost—the total number of casualties for the entire coalition was fewer than four hundred killed*—it had accomplished its stated objective of pushing Saddam out of Kuwait. It also strengthened Israeli security, and shored up relations with both Arab and European allies—giving the United States unparalleled global prestige just as the Cold War was coming to an end.

In the postwar euphoria, Bush's approval rating hit an unprecedented 90 percent. But perhaps his greatest accomplishment of all was what he didn't do—namely, that he maintained the discipline to refrain from marching on to Baghdad and toppling Saddam. As Bush and Scowcroft explained in *A World Transformed,* they thereby avoided a costly occupation, with no exit strategy, that would have turned the United States into an occupying foreign power in a faraway Arab land that would have alienated both European and Arab allies and would have betrayed their U.N. mandate.

Not everyone was happy with that decision, however. That was especially true in the Pentagon, where, far from seeing the Gulf War as a great triumph, Dick Cheney's underlings, including then Under Secretary of Defense Paul Wolfowitz, Deputy Under Secretary of Defense for Strategy and Resources I. Lewis "Scooter" Libby, and Deputy Under Secretary of Defense for Policy Planning Zalmay Khalilzad, saw the failure to oust Saddam as a disastrous lost opportunity.[65] Interestingly, in what critics later termed Chickenhawk Groupthink,[66] the moderate, pragmatic, somewhat dovish policies implemented by men with genuinely stellar military records—George H.W. Bush, Brent Scowcroft, and Colin Powell—were under fire by men who had managed to avoid military service—Cheney, Wolfowitz, Libby, and Khalilzad.

In addition, in many ways, what was happening was a reprise of the Team B experiment in 1976, with familiar faces playing similar but more exalted roles. This time around, sixteen years later, in 1992, President George H.W. Bush had unwittingly allowed Secretary of Defense Dick Cheney to empower Wolfowitz and Khalilzad, with Wolfowitz, the third-highest official in the Pentagon, leading a small team of neoconser-

*There have been contradictory reports as to the exact number of deaths stemming from the war. Coalition deaths were widely reported to have totaled 378. But the Department of Defense reported that the United States alone had 147 battle-related deaths and 235 nonbattle-related deaths.

vatives whose job was to chart a new course for American power now that the Cold War was over.

Working with his former political science student at Yale Scooter Libby,[67] and Zalmay Khalilzad, Wolfowitz came up with a radical, visionary, and highly controversial forty-six-page classified document known as the Defense Planning Guidance (DPG). (The document is also sometimes referred to as the Defense Policy Guidance paper and as the Wolfowitz doctrine.) "Paul Wolfowitz believed then that it was a mistake to end the war," said Richard Perle.[68] "They underestimated the way in which Saddam was able to cling to power, and the means he would use to remain in power. That was the mistake." The boys from Team B were at it again.

Written just after the demise of the Soviet Union, in the triumphal glow of a fresh, new "unipolar" world, the DPG asserted that America's goal should be to "prevent the reemergence of a new rival" to U.S. supremacy as the lone global superpower. The document made it clear that one of the most vital regions in the world strategically was the Middle East because of its vast oil reserves: "In the Middle East and Southwest Asia, our overall objective is to remain the predominant outside power in the region and preserve U.S. and Western access to the region's oil."[69] To achieve America's new objectives, however, Wolfowitz and company decided it was necessary to rewrite the long-standing rules of engagement under which the United States might take military action, employing stratagems that violated the most sacred principles of American foreign policy.

Astoundingly, in light of the success of the coalition that had been assembled for the Gulf War, Wolfowitz's team declared that we "should expect future coalitions to be ad hoc assemblies, often not lasting beyond the crisis being confronted."[70] Translation: in the future, the United States, if it liked, would go it alone.[71]

As if unilateralism were not radical enough, the DPG also argued that the United States should be prepared to act preemptively as well—initiating military action when necessary, not just in response to an attack or an immediate threat.[72] The new policy paper insisted that the United States must have a wide latitude in deciding exactly what situations called for military action. "We will retain the preeminent responsibility for addressing selectively those wrongs which threaten not only our interests," it read, "but those of our allies or friends or which could seriously unsettle international relations."[73]

The published excerpts of the DPG did not mention Israel, but its pledge to maintain "regional stability" in the Middle East implicitly meant a commitment to Israel's defense.[74] Likewise, the DPG said that U.S. interests in the Middle East could be threatened by "[restricted] access to vital raw materials, primarily Persian Gulf oil; proliferation of weapons of mass destruction and ballistic missiles, threats to U.S. citizens from terrorism."[75] Moreover, deterring "potential competitors" from aspiring to a larger role meant "punishing" them before they could act.[76] In other words, the slightest false move by Iraq, or, for that matter, Iran, could lead to preemptive military action.

This time, sixteen years after Wolfowitz et al. had conjured up "alternative intelligence" to battle Bush and the CIA, the nascent neoconservative echo chamber had begun to acquire some of the characteristics of a cultlike sect with its reverence for certain modes of thinking and for various intellectual heroes. In joining Wolfowitz, Scooter Libby, then a forty-one-year-old Washington lawyer, seemed to have undergone an especially sweeping transformation. He had been a radical in college who wore Malcolm X T-shirts and organized Vietnam antiwar demonstrations. "He was such a good person," a girlfriend from that era told *The American Prospect.* "He still is."[77]

But since then, she added, "The Dark Side" had gotten to him. And what dark force was that?

"Paul Wolfowitz." As soon as he fell under Wolfowitz's sway, she said, Libby became an apprentice of sorts. "It was as if he had joined a secret society," she said. At the time, Libby was serving his fourth tour of duty with Wolfowitz, having worked with him at Yale, the State Department, and once before at the Pentagon.

Secretary of State James Baker told President Bush to watch out for the "kooks" working for Cheney.[78] When the documents' recommendations were leaked to the press in March 1992, they encountered such a hostile reaction that Bush ordered Cheney to have them rewritten.[79] Subsequent versions were more measured, but Bush still rejected them. As he and National Security Adviser Brent Scowcroft later famously wrote in *A World Transformed,* this call for preemptive and unilateral regime change in Iraq could trap the United States in a costly, futile quagmire overseeing age-old ethnic and religious conflicts.

Bush's neoconservative foes had drafted their grandiose vision of reordering the world apparently unaware that he would react so strongly against it. By the time excerpts from the Wolfowitz doctrine were leaked, the glory from Bush's Gulf War victory had already begun to fade. Thanks largely to the nation's economic woes, the president's approval ratings had plummeted from 90 to 40 percent.[80] The presidential primary season was under way, but it was still too early for the neoconservatives to know if the president would recover, win reelection—and presumably marginalize them for another four years. Likewise, they could not have known that one of Bush's Democratic adversaries, an obscure governor from Arkansas named Bill Clinton, would turn out to be one of the most skilled politicians of the century. Finally, they could not possibly have known that the man who would finally implement their plans was a businessman in Texas who had not yet entered politics, and was the son of the president who scorned them. His name was George W. Bush.

CHAPTER EIGHT

First Son

The most important factor behind George H.W. Bush's defeat in the 1992 presidential election was the recession that gripped the nation—or, as the Clinton campaign mantra put it, "It's the economy, stupid." The implications went beyond high unemployment and declining property values. Bush lost much of his electoral base because related budgetary woes forced him to violate his sacred pledge: "Read my lips, no new taxes."

Moreover, by this time, many evangelicals who had voted for Bush in 1988 had figured out that there wasn't an evangelical bone in his body. "I can't say there was a moment when I came to my boss and there was a disconnect, where [he said] 'That's evangelical, I don't understand.' It was the other way around," explained Doug Wead. "There was never a time in my life that I took one of these evangelical issues to my staff and there was a connect."[1]

A case in point was the ceremonial signing of the Hate Crimes Statistics Act to which President Bush, clueless as ever to the ways of the faithful, invited gays and lesbians to the White House without considering what fundamentalists would think about sodomites in the Oval Office.[*2] Ultimately, Bible Belt evangelicals had little in common with the patrician transplant from the leafy suburbs of Greenwich, Connecti-

*In response to the event, Richard Land, who was president of the Southern Baptist Convention's Ethics and Religious Liberty Commission, wrote Bush, "Large numbers of Southern Baptists want to know why you are giving such official recognition to a homosexual-lesbian lifestyle they find abhorrent." As an antigay lobbyist, Land was later appointed by George W. Bush to the U.S. Commission on International Religious Freedom and declared, "It is time . . . to stop the sexual paganization of America . . . and boldly proclaim what God's intention is for man and woman in the confines of Holy Matrimony as God designed it."

cut—especially when they were being wooed by the honeyed Southern accent of Bill Clinton.

In any case, on January 20, 1993, the political career of George H.W. Bush, still a vigorous sixty-eight, was over. Counting Prescott Bush's tenure in the U.S. Senate, the Bushes had given more than forty years to public service. The former president's two eldest sons, George W. and Jeb, were acutely aware that if the family political legacy were to pass on to a new generation, they had better get moving. In 1990, George W. had borrowed $500,000 to invest in and become a managing partner with the Texas Rangers baseball team—an investment that eventually reaped $15 million. Now that he had finally started to build a reputation for himself in Texas, he was considering running for the governorship of the state,[3] and Jeb, a Miami businessman, was mulling over the same job in Florida. In other words, two presidential sons were positioning themselves for a potential run for the presidency—and it was anybody's guess as to which of them might inherit their father's legacy.

Meanwhile, William Jefferson Clinton moved into the White House. With Saddam Hussein still in power in Iraq and two-thirds of the American people favoring the use of military force to topple him,[4] Clinton continued his predecessor's policy of containment, and, on his very first day in office, bombed Iraqi radar sites to enforce the "no-fly zones" in the northern and southern parts of the country.[5]

In April, less than three months later, former president Bush returned to Kuwait for a triumphal visit to the country whose sovereignty he restored during the Gulf War. Unbeknownst to him at the time, he was allegedly the target of an assassination attempt by Saddam Hussein. Whether or not the plot to bomb President Bush was real remains unclear,* but even if it wasn't, the alleged conspiracy had serious consequences.[6] After receiving reports about it from the FBI and the CIA, President Clinton ordered a missile strike on Iraqi intelligence headquarters in June that killed six civilians.[7] Even more important, to both George W. and Jeb Bush, Saddam Hussein was the man who tried to kill their dad.

Nor was Iraq the only issue in the Middle East on the front burner.

*In late 1993, in a *New Yorker* piece entitled "A Case Not Closed," Seymour Hersh had seven independent experts examine the forensic evidence. They concluded that the plot was bogus because the purported bomb parts were really mass-production items used for walkie-talkies and other devices.

Much to the consternation of the neocons, Israeli prime minister Yitzhak Rabin and Yasser Arafat, chairman of the Palestine Liberation Organization (PLO) were making progress in the Oslo Accords, the ongoing Israeli-Palestinian negotiations to trade the Israeli-occupied West Bank to the Palestinians in return for peace. In July 1993, when Itamar Rabinovich, Israel's ambassador to the United States under Rabin, addressed the America-Israel Public Affairs Committee (AIPAC) in Washington about the resulting territorial concessions Israel was going to make, however, he was greeted by stony silence from its largely Likudnik, neoconservative membership.

AIPAC vice president Harvey Friedman went further, and called Rabin's deputy foreign minister Yossi Beilin "a little slime ball" for his willingness to trade land for peace.[8] Friedman was asked to resign from AIPAC, but the message was clear: the neocons were not so much pro-Israel as they were pro-Likud. Trading land for peace was the official policy of both the Israeli government and the United States, but it was anathema to the neocons. One of them, Douglas Feith, the former Scoop Jackson aide who was also special counsel to Richard Perle in the Reagan administration, had even attached himself to Itamar Rabinovich's staff in the Israeli embassy. As legal consultant to the Israelis, Feith actively lobbied against the Oslo Accords by organizing symposia and lobbying congressmen. Ultimately, he lobbied so aggressively against Oslo—in opposition to both official Israeli and American policies at the time—that, as Rabinovich said, "We parted ways because he was agitating against the peace process."*[9]

Such opposition notwithstanding, on September 13, 1993, Bill Clinton presided at the White House as Rabin and Arafat met for the ceremonial signing of the Oslo Accords. Twenty-seven years after the Six-Day War, Israel had finally agreed to relinquish Palestinian territory. Before the ceremony began, President Clinton addressed the delicate question of whether Rabin and Arafat would engage in a symbolic handshake. Both men were tentative—especially Rabin—because the agreement was bound to incite extremists on both sides. "The whole world will be watching, and the handshake is what they will be looking for," Clinton told the Israeli prime minister.[10]

Clinton opened with introductory remarks saluting their determination to achieve "a peace of the brave." When the signing was completed, Clinton shook hands with Arafat and quickly executed a well-rehearsed

*Feith declined a request to be interviewed.

and elaborately choreographed "blocking maneuver."[11] To prevent Arafat from kissing him, the president turned away from him and shook hands with Prime Minister Rabin. Then he stepped back between the two leaders and spread his arms to bring them together. Arafat extended his hand, but Rabin was still hesitant. As President Clinton described it in his memoir, *My Life,* Rabin finally put forth his hand, and the crowd gasped. "All the world was cheering," wrote Clinton, "except the diehard protesters in the Middle East who were inciting violence, and demonstrators in front of the White House claiming we were endangering Israel's security."[12]

But that was precisely the problem. The diehard protesters were not just militant Islamist fundamentalists who felt Arafat was selling them out. There were Christian fundamentalists, neoconservative ideologues, and Orthodox Jews as well. Among the evangelicals, Ed McAteer, one of the founders of the Moral Majority with Jerry Falwell, asserted, "Every grain of sand between the Dead Sea, the Jordan River, and the Mediterranean Sea belongs to the Jews. This includes the West Bank and Gaza."[13]

But the most intense reaction of all came from the settlers in the occupied territories and the Orthodox Jews who believed land that was to be returned to the Palestinians had in fact been given to the Jews by God. About six thousand miles away from Washington, one of them, Yigal Amir, an Orthodox Jewish law student at Bar Ilan University, watched the television in horror as Rabin and Arafat shook hands, and told his father Shlomo Amir that Rabin's decision to forfeit Israeli land was forbidden by the Torah.[14] Then, according to *Murder in the Name of God: The Plot to Kill Yitzhak Rabin* by Michael Karpin and Ina Friedman, Yigal added that it might be necessary "to take down" the prime minister.

"Everything is in God's hands," Shlomo Amir responded.

"But this time," Yigal replied, "it's necessary to help Him."[15]

Meanwhile, the neoconservatives were no happier with Clinton than they had been with his predecessor. Among the hard-liners, only one, CIA director James Woolsey, wound up with a significant post—and even he left after less than two years on the job. Containing Saddam through occasional bomb strikes wasn't enough. The neocons wanted regime change in Iraq, and ultimately, they wanted to redraw the map of the entire Middle East.

One of the key figures behind their new strategy was yet another acolyte of Albert Wohlstetter's, Ahmed Chalabi. The scion of a prominent

Shia family, Chalabi had fled Iraq in 1958, when he was just thirteen and the Communist Party overthrew the royal family. He lived in Jordan, Lebanon, and Britain before studying at MIT. In 1969, he completed his doctoral studies in mathematics at the University of Chicago,[16] where he met Wohlstetter, who later introduced him to Richard Perle and Paul Wolfowitz, who had become dean of the School of Advanced International Studies (SAIS) at Johns Hopkins University.[17] Eight years later, Chalabi moved to Jordan to found the Petra Bank, and built it into the second largest commercial bank in the country.[18]

Given to elegant silk suits and Rolex watches, Chalabi took pleasure in seeing his children enjoy their equestrian pursuits with the Jordanian royal family, but he also nursed a secret ambition. "Ahmed wanted to avenge his father's ouster and the deprivation of his lands," Imad Khadduri, an Iraqi exile who was a schoolmate of Chalabi's, told Jane Mayer, who wrote the definitive profile of Chalabi in *The New Yorker.* "Now he's trying to fit in his father's shoes."[19]

He also seemed to be trying to re-create a history that went back to the days of the Balfour Declaration and the Sykes-Picot Agreement, when Britain's David Lloyd George had carved up the Middle East after World War I, and had empowered a Hashemite dynasty in Iraq by supporting Faisal I as king.* Not averse to working with colonial forces, Chalabi's father, Abdul Haydi Chalabi, served as president of the Iraqi Senate and was a member of the council of ministers to King Faisal II, who ruled until he was killed in the 1958 coup d'état.[20]

In 1989, Chalabi and his family fled Jordan for London just as multiple criminal charges were being drawn up against him. Three years later, a Jordanian military tribunal concluded that Chalabi was guilty of thirty-one charges of embezzlement, theft, forgery, currency speculation, and other crimes. He was sentenced in absentia to serve twenty-two years of hard labor and was ordered to pay back $230 million.[†21]

*The Hashemite dynasty refers to a powerful network of tribes in the Hejaz region of Arabia, bordering the Red Sea, which traces its roots back to the great-grandfather of Muhammad. Starting in 1908, with the demise of the Ottoman empire, Hussein bin Ali, the leader of the Hashemites, took one title after another, proclaiming himself sharif of Mecca, emir of Mecca, king of the Hejaz, and finally caliph of all Muslims, until 1924 when he was defeated by Abdul Aziz al Saud, his chief rival and later the first king of Saudi Arabia. Hussein's son Abdullah became king of Jordan, which remains a Hashemite monarchy. Another son, Faisal, became king of Iraq, which remained a Hashemite kingdom until a coup d'état in 1958.

†Chalabi claimed that he was framed because he was a fierce critic of Saddam Hussein, to whom the Jordanians were indebted for oil and economic reasons.

By the time of his conviction, however, Chalabi was safely ensconced in London, where he began putting together the Iraqi National Congress (INC), a loose grouping of exiles who hoped to topple Saddam. His charm and intelligence were such that his circle of supporters extended far beyond the usual suspects. "Chalabi is one of the smartest people I know," said Peter Galbraith, who served as ambassador to Croatia under Bill Clinton.[22]

But the neocons—the Scoop Jackson aides, the Team B alumni, the other Wohlstetter protégés and their associates—were the core of his support. A longtime acquaintance of Richard Perle,[23] Chalabi won favor throughout the nineties with other Wohlstetter disciples and neocons including Paul Wolfowitz, James Woolsey, Douglas Feith, and David Wurmser, a protégé of Perle.[24] He met with Princeton professor Bernard Lewis,* Zalmay Khalilzad, and General Wayne Downing, all of whom would later win influence in George W. Bush's administrations.[25] He went to conferences and symposia with them and their colleagues. He cultivated new friends at pro-Israeli think tanks such as the Jewish Institute for National Security Affairs (JINSA)† and the Washington Institute for Near East Policy (WINEP).‡[26] His appeal to the neocons could be explained in one word: Israel. He assured the neocons that the new Iraq he would build would have strong diplomatic and commercial ties with Israel.[27] He even told the *Jerusalem Post* that he would eventually restore the oil pipeline from Kirkuk, in oil-rich Kurdish Iraq, to Haifa, Israel, which had been inoperative since the forties.[28]

Such promises were so enticing that the neocons turned a blind eye to Chalabi's past. According to a report by John Dizard in *Salon,* in the eighties, Chalabi's Petra Bank had funded Amal, a Shia militia allied with

*A specialist in the history of Islam, Lewis coined the phrase "clash of civilizations," which was used as the title for Samuel Huntington's eponymous book. He became enormously influential among neocon policy makers, and was often cited by Vice President Cheney as an authoritative scholar. But critics of neoconservativism such as the late Palestinian-American scholar Edward Said often had harsh words for his work. According to Said, Lewis's views of the Middle East were so biased, his work should not be taken seriously. Said added: "He knows something about Turkey, I'm told, but he knows nothing about the Arab world."

†JINSA's boards have included at various times Elliott Abrams, Michael Ledeen, Richard Perle, James Woolsey, Dick Cheney, John Bolton, Zalmay Khalilzad, Douglas Feith, and Paul Wolfowitz.

‡At various times, WINEP's board has included Richard Perle, Paul Wolfowitz, Jeane Kirkpatrick, Robert McFarlane, Joshua Muravchik, James Woolsey, and Martin Peretz.

Iran in Lebanon, and Chalabi himself had maintained close ties to the Iranian regime throughout the nineties.[29]

To neocon apparatchiks, however, Chalabi was not a convicted criminal or someone with dubious ties to Iran and to Shi'ite militants, but a heroic Iraqi exile who was on his way to becoming the George Washington of his country. "He's a rare find," Max Singer, a trustee and cofounder of the Hudson Institute, told *The American Prospect.*[30] "He's deep in the Arab world and at the same time he is fundamentally a man of the West." Added Patrick Clawson, deputy director of WINEP, "He could be Iraq's national leader."[31]

After founding the Iraqi National Congress, Chalabi went to the CIA for help in pursuing their goal of overthrowing Saddam. Thinking Chalabi could provide useful intelligence, the Agency gave the INC millions of dollars to set up a "forgery shop" in Salahuddin, Kurdistan, where it created phony mock-ups of Iraqi newspapers filled with stories about Saddam's abuses. "It was something like a spy novel," said former CIA agent Robert Baer.[32] "It was a room where people were scanning Iraqi intelligence documents into computers, and doing disinformation. There was a whole wing of it that he did forgeries in." According to *The New Yorker,* the operation also forged a letter to Chalabi on bogus stationery that purported to be from President Clinton's National Security Council, and that asked Chalabi to help put together an assassination plot against Saddam.[33] Francis Brooke, a lobbyist for Chalabi, conceded that the INC had run a forgery operation, but he said Chalabi had nothing to do with the phony assassination letter. "That would be illegal," he told *The New Yorker.*

With the help of the Rendon Group, a CIA-funded PR firm, Chalabi took the phony intelligence back to INC's offices in London for use in a global media offensive to portray Saddam Hussein not just as a brutal dictator, but a genuine threat to world peace.[34] Chalabi was not only charming and accessible to reporters, but he offered them tantalizing if highly dubious scoops about Saddam as well. "When I visited him in London, he told me 'You can have anything you want,'" recalled Carlo Bonini, a reporter for the Italian daily *La Repubblica.*[35] "It was like a shopping mall for intelligence." Like other more rigorous reporters, however, Bonini found that nothing Chalabi offered could be independently verified. And in fact, many of his allegations turned out to be fabrications.

At about the same time, Chalabi was also reportedly cultivating his ties with Iran. "He was given safe houses and cars in [Kurdish-controlled] northern Iraq, and was letting them be used by agents from the Iranian Ministry of Intelligence and Security [Vevak], and the Iranian Revolutionary Guard Corps," a CIA case officer told *Salon.*[36]

In November 1993, Chalabi approached the Clinton administration with a plan for regime change in Iraq. According to Colonel Patrick Lang, a former intelligence officer for the Defense Intelligence Agency (DIA), End Game, as it was called, was meant to start with a revolt by Kurdish and Shi'ite insurgents that would supposedly trigger an insurrection by military commanders who would replace Saddam with a regime that was friendly to both the United States and Israel.[37] Initially, Washington approved. But Saddam learned about the operation before it took place. Chalabi proceeded anyway, and the military insurrection he had predicted never occurred. Instead, the Iraqi army fought the rebels and more than a hundred INC combatants were killed.[38] Ultimately, Ahmed Chalabi was persona non grata with both the Clinton administration and the CIA. But he was not yet out of the game for good.

Meanwhile, the Christian Right was anything but dead. Even though the Democrats had Clinton's victory to celebrate, more than a thousand Christian evangelical candidates sponsored or endorsed by Pat Robertson's Christian Coalition had run for office in 1992, and more than half had won.[39] By any measure, they were formidable foes, and the ascendancy of Bill Clinton had galvanized them. A Southern Baptist from Arkansas, "Slick Willie" could talk church. He spoke fluent Elvis. He knew their language. Yet by initiating policies that allowed gay men and lesbians into the armed forces, and by supporting stem cell research, he became their worst nightmare—one of their own who betrayed them.*[40]

The Clinton wars had begun, and with them one pseudoscandal after another. From Travelgate† and Whitewater,‡ to Vince Foster's sui-

*In 1993, the Southern Baptist Convention, of which Clinton was a member, first passed a resolution separating itself from his policies in support of homosexual and abortion rights. In June 1999, the SBC formally rebuked him for designating that month Gay and Lesbian Pride Month, and called for him to revoke his appointment of James Hormel, an openly gay diplomat, as ambassador to Luxembourg.

†The Travelgate controversy arose in May 1993 after the incoming Clinton administration discovered financial irregularities and chaotic bookkeeping in the White House Travel Office, fired seven employees, and began to reorganize it. To Republicans, however, this was merely a ruse to funnel business to Clinton cronies. After a long investigation, independent counsel Kenneth Starr exonerated President Clinton of any wrongdoing in the affair.

‡The Whitewater controversy concerned a failed real estate investment by Bill

cide,*[41] the attack dogs of the right—Rush Limbaugh, the *American Spectator,* the *Wall Street Journal*'s editorial page, the *New York Post,* and more—accused the Clintons of cronyism, fraud, conflicts of interest, and destroying evidence. With Foster's death in July 1993, Rush Limbaugh even suggested that Hillary may have had Vince Foster murdered.[42]

Meanwhile, courtesy of the Christian Right political network, doomsday videos such as *The Crash: The Coming Financial Collapse of America,* backed by Tim LaHaye's right-wing umbrella group, the Council for National Policy, went out to millions of evangelicals, asserting that the dollar would soon be worthless and the United States would be rife with riots, breadlines, and a Hitleresque strongman at the helm.[43] A Jerry Falwell video, *Bill Clinton's Circle of Power,* linked the Clintons to a series of unsolved murders, accidental deaths, and suicides—including Vince Foster's.[44] It was followed by *The Clinton Chronicles,* a litany of false charges tying Clinton's cronies to drug smuggling, money laundering, and the like.[45]

With the millennium and the Second Coming just around the corner, the fundamentalists, full of passionate intensity, were saying, in effect, that America had already fallen under the sway of the Antichrist, Bill Clinton. Spurious as these scandals were, they worked their way up the media food chain—in some cases, all the way to the front page of the *New York Times* and network news.

All of which fed the growing power of the religious right within the Republican Party. In 1992, Democrats controlled both legislative houses in twenty-five states, the Republicans eight. In 1994, Democrats had majorities in eighteen, the Republicans nineteen. To prepare for the midterm congressional elections in 1994, the Christian Coalition distributed 40 million copies of the "Family Values Voter's Guide" in more than 100,000 churches.[46] In November, Republicans won control of the House of Representatives for the first time in forty years,[47] and Newt Gingrich became Speaker of the House.

---

and Hillary Clinton in the Whitewater Development Corporation in Arkansas. Three separate investigations found no evidence that the Clintons had engaged in criminal activity.

*A longtime friend and colleague of the Clintons, Foster joined the administration as deputy White House counsel and was soon attacked by the *Wall Street Journal*'s right-wing editorial page in a series of articles as one of the Clinton's "legal cronies from Little Rock" who engaged in "legal corner-cutting that leads to trouble" and "carelessness about following the law." Foster reportedly exhibited a number of symptoms of depression. He committed suicide on July 20, 1993.

* * *

But Gingrich's ascent was not the most important one for the Republicans. On Tuesday, November 8, 1994, former president George H.W. Bush spent the evening watching election returns come in with Prince Bandar bin Sultan bin Abdul Aziz al Saud, the longtime Saudi ambassador to the United States.[48]

To Bush senior, the flamboyant Bandar wasn't just another foreign dignitary, his relationship with the Saudi prince was one of the deepest, most genuine friendships he made in all his years in Washington. The fact that Bandar was the multibillionaire scion to the royal House of Saud, the richest family in the world and owner of the largest oil reserves on the planet, also meant that their friendship had profound geopolitical implications. Regular lunch companions throughout the Reagan era, they went hunting and fishing together, played tennis together, and Bandar was known to surprise Barbara Bush by popping in, uninvited, at Kennebunkport, and whipping something up in the kitchen. When Bush senior lost the 1992 election, his Saudi friend had been devastated. "It was like I lost one of my family, dead," Bandar told *The New Yorker.*[49]

Now, two years later, even though Bush 41 was still out of politics, watching and listening to returns that evening was a deeply personal event. That was because Bush's two eldest sons, George W. and Jeb, were running for the governorships in Texas and Florida, respectively. The outcome could determine if the Bush family dynasty could be extended for another generation—and if so, which of the sons might lead it.

Thanks to his stake in the Texas Rangers baseball team, George W. had finally made a name for himself in Texas. On one occasion in the early nineties, at a meeting of Major League team owners in Denver, they were being herded around on an excruciatingly long bus drive when the bus driver got lost and began circling round and round interminably. According to an article by David Brooks in *The Weekly Standard,* Gene Autry, the aging cowboy singing star who owned the California Angels, "needed to relieve himself . . . [so] the bus pulled over . . . [and] an embarrassed Autry got out and urinated. It was an awkward moment for everybody."

When Autry got back on the bus, the first voice that rang out was that of George W. Bush. "Hey, Gene," he shouted, "you still got a great spray for a guy your age." The bus reportedly erupted with laughter.[50]

Skilled as he was at defusing such awkward moments, Bush showed little concern about the nuts and bolts of the business. "Uninterested in doing the things he was not good at, [Bush] delegated day-to-day management of the club and spent his time on climate control," Brooks wrote. "He was a constant presence in the ballpark, keeping everybody, from the ushers to the players, feeling good about the franchise. His ownership group was an ever-shifting stew of between a dozen and two dozen millionaires; he spent a lot of time keeping them happy. During games he sat in a box next to the dugout, not in the normal owner's box above. He ribbed the players, passed out autographed baseball cards of himself to fans, and shouted jokes to the managers."[51]

In other words, he was the master of ceremonies far more than he was the man who ran the team on an operational basis. "It's cruel to say this, but his character was still being shaped. . . ." said one longtime friend of both George Bushes. "There was so much that was being sculpted or molded. There wasn't anything final about his philosophy."

Living up to his father's expectations remained a sensitive issue. "I think that's what drove him to run for governor," said Mickey Herskowitz, the Houston sportswriter who was hired to ghostwrite George W. Bush's autobiography for the 2000 presidential campaign. "It appealed to him that his father had not been governor of Texas."[52]

As the 1994 elections approached, his younger brother Jeb clearly was winning the family oedipal sweepstakes. "[His father] never thought George would have a career in politics," said a friend of the family. "He was thinking about Jeb or maybe even Neil.* His parents always felt that Jeb had the gifts and the brightness."

Another obstacle for George W., however, was that he faced Ann Richards, a particularly formidable incumbent with an astoundingly high 70 percent approval rating.[53] A sassy good ole girl, Richards had made her name at the 1988 Democratic National Convention by deriding the elder Bush as a spoiled and privileged son of the Eastern aristocracy. "Poor George," she had famously drawled, "he can't help it. He was born with a silver foot in his mouth." As a result, this battle was personal. Victory would be particularly sweet, defeat bitter.

The races in Texas and Florida both promised to be nail-biters. Throughout the summer, the smart money said Jeb had a better shot at beating incumbent Lawton Chiles in Florida than George did of beating

*Neil Bush, the third son of George H.W. and Barbara Bush, was a member of the board of directors of Silverado Savings and Loans, which collapsed in the 1980s, costing taxpayers $1 billion. He was not indicted but a related civil action destroyed his political prospects.

Richards. But as election day approached, the odds switched.[54] One reason may have been a sub rosa whispering campaign engineered by Bush political strategist Karl Rove suggesting that Ann Richards was a lesbian.

When the results came in, Jeb lost narrowly to Lawton Chiles, but George W. beat Richards by nearly eight percentage points.[55] He had not only avenged his father's humiliation by Richards, he had also become First Son. "I have very mixed emotions," said Bush 41 after the results came in. "Proud father, is the way I would sum it all up."[56]

What Bush senior didn't say was that he knew at the deepest, most fundamental level that if George W. ever made it to the White House, he would be in way over his head.

CHAPTER NINE

The Righteous Assassin

Even as George W. Bush was celebrating his victory as governor, a series of events was taking place on the other side of the world that was seemingly unrelated but would ultimately affect his legacy enormously. Few people had more firsthand experience confronting the explosive issues at the heart of such matters than Yitzhak Fhantich, who spent twenty-eight years in Israeli intelligence, many of them as head of the Jewish Department of Shin Bet,* the Israeli counterintelligence and internal security service. At fifty-eight, with his blue-checked shirt half-buttoned, a bushy mustache, and a thick mop of brown hair, Fhantich is a compact but powerful physical presence. One warm day in the summer of 2005, he leaned back in his chair at a bustling Iraqi market in Jerusalem where the produce stalls were filled with fresh fruit, sipped a glass of Israeli merlot, and took a puff on a Marlboro. Behind him, elderly men drank coffee and played backgammon.

Now a private security and intelligence consultant, in the nineties, Fhantich had been in charge of investigating right-wing religious extremists who posed a threat to the most combustible place on earth—the Temple Mount, the massive 144,000-square-meter platform, 32 meters high, built by King Herod as a base for the biggest and most grandiose religious monument in the world, the shining white stone Temple of the Jews.

To Jews, the Temple Mount, also known as Mount Moriah, marks the holy of holies,† the sacred core of the Temple where Jews worshipped for

*Shin Bet, Israel's internal security service, is also known as Shabak.

†The term *holy of holies* is used by a number of religions, including the Ethiopian Orthodox Church, the Church of the Latter-Day Saints, and Hinduism, to refer to the sacred core of a sacred building. In Judaism, it refers to the inner sanctuary of the Temple of the Jews, which is to be entered only by the high priest on Yom Kippur.

centuries. Beneath it, Orthodox Jews believe, is the foundation stone of the entire world. The Mount is the disputed piece of land over which Cain slew Abel. It is where Abraham took his son, Isaac, when God asked him to sacrifice the boy. At its outer perimeter is the Western Wall, or Wailing Wall, where Jews worship today. And messianic Jews believe the Mount is where the Temple must be rebuilt for the coming Messiah.

To Christians, the Temple is where Jesus threw out the money changers. Its destruction by the Romans in A.D. 70 came to symbolize the birth of Christianity, when a new Temple of Jesus, eternal and divine, replaced the earthly Temple made and destroyed by men.

To both Christians and Jews, reclaiming the Temple Mount and rebuilding the Temple symbolizes the complete restoration of the nation of Israel. That includes winning back Judea and Samaria, also known as the West Bank. Nothing is more important to the messianic vision of Orthodox Jews, but to Christians it is equally important because it is a prerequisite to the Second Coming of Christ. Ultimately, for true believers, the stakes could not be higher—eternal salvation.

There was, however, one great obstacle to reclaiming the Temple Mount—namely the fact that the site is also sacred to Muslims. They refer to the Temple Mount as the Noble Sanctuary because they believe that the Mount's Dome of the Rock is the foundation stone from which Muhammad ascended to heaven nearly fourteen hundred years ago, making it the third-holiest site in Islam, behind Mecca and Medina.

And so, to all three religions, the Temple Mount is the gateway to Heaven, the locus of the immensely powerful forces of millennialism, salvation, and redemption. Because it is so holy to all three religions, it is a microcosm of the ongoing Middle East conflict. No single place on earth is more likely to be the fuse that will ignite an international catastrophe.

In contemporary times, the battle for the Temple Mount dates back to its capture by Israeli forces during the historic Six-Day War in June 1967. After they entered the Old City through the historic eastern entrance, the Lions' Gate, hundreds of Israeli soldiers stood in the hallowed Dome of the Rock. "We stood in front of the rock, which was in the midst of the holy of holies," recalled Gershom Solomon, a messianic Jew who now leads the Temple Mount Faithful and was a soldier in the Israeli army at the time. "And I will never forget that moment. I tell you, we did not believe."[1]

Soon the soldiers were joined at the Western Wall by the venerated Israeli defense minister who won the battle, Moshe Dayan, wearing his trademark eyepatch.[2] "We stayed with [Dayan] inside the Dome of the

Rock," Solomon said. "The Temple Mount is again in our hands! Again in our hands! And I cannot describe this moment in words. . . . I felt again the presence of God, like He was speaking to my heart, to all of Israel. . . . It was a great moment, the greatest in the history of Israel, because this was the place where this nation was born. This is the place where the main event of history took place! This is the place where the dialogue between Israel and the God of Israel took place!"[3]

From the beginning, Moshe Dayan had said, to the outrage of both Orthodox Jews and Christian fundamentalists, that Muslims would continue to control the Temple Mount, which was soon handed over to the Waqf, a Muslim administrative body. Since Dayan's decision, Muslim authorities have usually allowed non-Muslims to come to the Temple Mount, but they have been forbidden to move their lips in ways that suggest they are praying. As a result, the Temple Mount remains one of the most explosive tinderboxes on earth. Gershom Solomon, then a member of Menachem Begin's right-wing Herut Party, a precursor to Likud, immediately founded the Temple Mount Faithful and began demonstrating in hopes of taking it back. Others, including Jews who had immigrated to the Holy Land from Orthodox communities in the United States, began to consider taking more drastic actions.

Even though Jewish settlements on the West Bank were certain to provoke hundreds of thousands of West Bank Palestinians, religious Zionists launched their holy crusade to settle Judea and Samaria. "I tell you explicitly that the Torah forbids us to surrender even one inch of our liberated land," declared Rabbi Zvi Yehuda Kook, a leader of the settlement movement known as Gush Emunim (Block of Faithful). "There are no conquests here. And we are not occupying foreign lands. We are returning to our home, to the inheritance of our ancestors. There is no Arab land here, only the inheritance of our God."[4]

The most fanatical of these right-wing Zionists would stop at nothing in their zeal to push Palestinians out of these biblical lands. In 1971, Meir Kahane, founder of the militant Jewish Defense League (JDL), immigrated to Israel from Brooklyn, New York, started the radical right-wing Kach Party,* won a seat in the Knesset, and launched a racist campaign

*In 1980, Kahane was jailed for six months on suspicions that he had planned a provocative act of sabotage on the Temple Mount. He was assassinated in 1990 in New York by El Sayyid Nosair, who was later sentenced to life imprisonment as a coconspirator in the 1993 bombing of the World Trade Center.

to expel Arabs that was characterized by its fusion of genocidal invective and messianic fervor. "In two years time," Kahane predicted, "[the Arabs] . . . will come to me, bow to me, lick my feet, and I will be merciful and allow them to leave. Whoever does not will be slaughtered."[5]

The 1973 Yom Kippur War, which began with a sneak attack on Israel by Arab forces, only hardened the resolve of Orthodox settlers. Over the next thirty years more than 250,000 Jews settled 150 communities in Palestinian territories in the West Bank and Gaza—creating a formidable political obstacle to forces who wanted to trade land for peace. Resolving the Israeli-Palestinian conflict became impossible without dislocating tens of thousands of Jews.

Of course, thanks to the evangelicals, the Orthodox Jews were no longer alone. In 1977, a group called the Institute for Holy Land Studies took out full-page ads in newspapers throughout the United States to make that abundantly clear. "We affirm as evangelicals our belief in the promised land to the Jewish people. . . ." the text read. "We would view with grave concern any effort to carve out of the Jewish homeland another nation or political entity." It added, "The time has come for evangelical Christians to affirm their belief in biblical prophecy and Israel's divine right to the land."[6]

As head of Shin Bet's Jewish Department, Fhantich's job entailed investigating Orthodox Jews who might resort to terrorism to fulfill their messianic dreams. "The vast majority of settlers in the West Bank are positive people with sincere religious beliefs," he said. "They are good people. They love Israel. They are the kind of people you would let marry your daughter. I know the leader [of the Temple Mount Faithful], Gershom Solomon, will never break the law. But the problem is when someone else will use his organization, or one like his, to surprise me and blow it up.[7]

"If someone believes God told him to do something you cannot stop him. The mosques on the Temple Mount are like red for the bull. You have to be prepared minute by minute. These Christians, they believe what they are doing is sacred. Some of them are so naive they can be used. If something happens to the Temple Mount, I think these American evangelicals will applaud such an act. After all, religion is the most powerful gun in the world."

In hopes of preventing such violence, Fhantich and his team of twenty-five Shin Bet operatives kept an eye on Meir Kahane, whose ultra-right-wing Kach Party had put together a terrorist campaign that included bombings and assassinations of Arab leaders with the aim of expelling *all* Arabs from Israeli occupied territory. In 1984, Fhantich's team arrested twenty-six Jewish terrorists who were plotting to blow up the mosques on the Temple Mount to pave the way for the rebuilding of the Temple and the return of the Messiah.[8]

Then, in the mid-nineties, Fhantich opened a file on Yigal Amir, the law student at Bar Ilan University who was so outraged when Rabin shook hands with Arafat. "He was a typical religious type," said Fhantich. "He served in a combat tour, and then went to study law. We knew about him." Amir had been raised and educated in the customs of *haredi* Judaism,* the most theologically conservative form of Orthodox Judaism.[9]

At the time, it would have been difficult to find anyone in Israel with a more distinguished military record than Yitzhak Rabin. During Israel's War of Independence, he had commanded elite troops for the Haganah, the Jewish paramilitary organization. As chief of staff to the Israeli Defense Forces (IDF) under Prime Minister Levi Eshkol, Rabin had been the architect of Israel's victory in the 1967 Six-Day War—including the taking of the Temple Mount. Having won the sobriquet "Mr. Security," Rabin had emerged as an icon whose feats allowed his countrymen to put aside the horrific imagery of the Holocaust and replace it with that of the indomitable and invincible new Israeli.[10] "This was the peak of my life," Rabin later recalled. "For years I had secretly harbored the dream that I might play a role . . . in restoring the Western Wall to the Jewish people. . . . Now that dream had come true, and suddenly I wondered why I, of all men, should be so privileged."[11]

After being elected prime minister in 1974, Rabin took a hard-line stance against the Palestinian Liberation Organization (PLO), against relinquishing the Golan Heights, and against releasing Palestinian political prisoners.[12] But in 1992, when he was elected prime minister for the second time, Rabin abruptly became a dove and announced that "we do not intend to lose precious time" in terms of peacemaking.[13] He blocked funding to expand the settlements in the occupied territories. He released eight hundred Palestinian prisoners. Then he opened unofficial back channels to have exploratory talks with the PLO. According to Fhantich, Yigal Amir, who "was fanatically against any compromise whatsoever with the Arabs,"[14] was stunned by Rabin's about-face. What he was doing was heresy.

*About 12 percent of Israelis consider themselves *haredim.* More than 40 percent are secular.

* * *

Amir's militant views were not unusual among Orthodox Jews, especially the *haredim,** who trace their religious practices back to when God gave the Torah to Moses on Mount Sinai. Just as militant Islamic fundamentalists view Sharia, religious law based on the holy Koran, as higher than secular law, so do many militant Orthodox Jews, including the *haredim,* see *halachic* law—Jewish religious law—as higher than the laws of the state. To them, Rabin's move toward peace was in violation of *halachic* law that forbade turning Jews or Jewish property over to gentiles. Orthodox Jews in Israel were now at war with many of their secular counterparts.[15]

Rabin's adversaries on the right jumped into the fray—including Christian evangelicals. "I was ambassador for four years of the peace process, and the Christian fundamentalists were vehemently opposed to the peace process," said Itamar Rabinovich, who served as Israeli ambassador to the United States between 1993 and 1996. "They believed that the land belonged to Israel as a matter of divine right. So they immediately became part of a campaign by the Israeli right to undermine the peace process."[16]

During the first Knesset debate over Oslo, Likud chairman Benjamin Netanyahu characterized Oslo as a Munich-like act of appeasement that would pave the way for another Holocaust. "You are worse than [former British prime minister Neville] Chamberlain," Netanyahu shouted at Foreign Minister Shimon Peres.[17]

Yigal Amir and his Orthodox cohorts listened with glee. By this time, it was not uncommon for militant Orthodox Jews to harbor thoughts of violence in response to Oslo. Dr. Baruch Goldstein, an Israeli-American physician who was a protégé of Meir Kahane and a member of the Kach Party, had sunk into a period of deep depression since the 1990 assassination of his mentor. According to *Murder in the Name of God,* after the signing of Oslo, Goldstein resolved, "Only an act of Kiddush ha-Shem [self-sacrifice for the sanctification of God] could perhaps change history and return the messianic process to its course."[18]

At dawn on February 25, 1994, Goldstein went to the Cave of the Patriarchs in Hebron, the second most sacred site in Judaism after the Temple Mount because it is said to be the burial place of four great bib-

*The term *haredi* is sometimes interpreted to mean "one who trembles in awe of God."

lical couples—Adam and Eve, Abraham and Sarah, Isaac and Rebekah, Jacob and Leah. But like the Temple Mount, the Cave is also sacred to Muslims and is known as Ibrahimi Mosque (the Mosque of Abraham), because Abraham is also thought of as a great prophet in Islam. Wearing his army uniform, Goldstein appeared to be a reserve officer on active duty, carrying with him his assault rifle and four magazines of ammunition. Other guards assumed he was entering the Cave to pray in an adjoining chamber reserved for Jews. Once he arrived, however, Goldstein opened fire on scores of Palestinians who were praying there, killing 29 and wounding another 150. Survivors eventually overcame him and beat him to death.

Afterward, seventeen hundred Orthodox students took out a full-page ad in the *New York Times* condemning the carnage. But in the settlements themselves, a stunning ambivalence foreshadowed more violence. "Indeed, sympathy for Goldstein's desperation, if not for the massacre itself, was widespread among settlers," wrote Yossi Klein Halevi.[19] Added a spokesman for the settlers' rabbis committee: "We don't think there is anything to gain by condemning the action. . . . I don't recognize the Machpela [Cave of the Patriarchs] as a place for Muslims."[20]

Ongoing attempts to implement the Oslo Accords only fed the tensions. On July 1, 1994, PLO chairman Yasser Arafat went to Gaza as outlined by the peace process, and the next day, as many as 100,000 right-wing protesters[21] took over Zion Square in Jerusalem, chanting "Rabin is a homo!" and "Rabin is the son of a whore!"[22] Near a huge banner reading "Death to Arafat," Benjamin Netanyahu assailed the "blindness" of the Rabin government for allowing "the destruction of the Jewish state."[23] "Arafat, who is personally responsible for the murder of thousands of Jews and non-Jews, this war criminal, is being hoisted aloft by the Government of Israel. . . ." Netanyahu said. "What Arafat truly wants is not an Arab state beside Israel, but an Arab state in place of Israel."[24]

Increasingly, Oslo only incited extremists on both sides who were committed to undermining the agreement. In October 1994, twenty-two people were killed in a suicide bombing attack by Palestinian terrorists on a bus in Tel Aviv. It was just one of a series of murderous attacks on Israelis by car bombs or at bus stops that had been going on for months.

At the same time, on the Israeli side, the Orthodox right did everything it could to obstruct the accords. By January 1995, forty militant

Orthodox rabbis in the West Bank settlements, including Rabbi Eliezer Melamed, secretary of the Rabbinical Council of the Land of Israel, began deliberations to consider whether Rabin and his "evil government" were in violation of *halachic* law.[25] Orthodox rabbis began talking specifically about whether Rabin's actions called into play two such halachic edicts, a *din rodef* ("the duty to kill a Jew who imperils the life or property of another Jew") and a *din moser* ("the duty to eliminate a Jew who intends to turn another Jew into non-Jewish authorities"), which have been characterized as roughly the Orthodox Jewish equivalent of a fatwa.*[26] For the most part, such discussions took place in private with the militant rabbis being careful not to leave a paper trail.

Meanwhile, in early 1995, Yitzhak Fhantich personally informed Prime Minister Rabin of the danger he faced. "I told him, on the hit list, you're number one," Fhantich said.[27] Rabin responded with a silent shrug and a wave of the hand that seemed to say what will be, will be.

At about the same time, Yigal Amir, who had been taking part in the many anti-Rabin demonstrations, told friends, "Rabin has to be killed." But his incendiary remark was immediately followed by a smile. "They didn't know what I meant," he later explained. "No one thought I would kill Rabin. Even I didn't know I would kill Rabin."[28]

"It wasn't a matter of revenge or punishment, or anger, Heaven forbid, but what would stop [the Oslo process]. . . ." Amir continued. "If I took Rabin down, that's what would stop it."[29]

Fhantich was on a leave of absence for much of 1995, but his team continued surveillance of Amir. "We were watching him," Fhantich said. "He was very active, very public, organizing students on the weekends to encourage people to fight Oslo. But everything he did was legal. He was planning it all alone."[30]

Meanwhile, in the summer of 1995, the most violent threats toward Rabin came not from Israel, but from Orthodox Jews in the United States. In June, Rabbi Abraham Hecht, a leading figure in New York's rabbinical establishment, the head of the 540-member Rabbinical Alliance of America,[31] asserted that Jewish law permitted the assassination of Prime Minister Rabin because of his willingness to cede land to the Palestinians.[32]

In August, on a Voice of Israel radio broadcast, Rabbi Nachum Rabinovitch, an implacable foe of the Oslo Accords, declared that Rabin was

*A fatwa is a legal ruling made by an Islamic scholar who is an authority on Sharia (Islamic law).

a *moser* "who according to Maimonides* is liable to death." Rabinovitch quickly added, "I didn't say that it's permissible to harm him."[33] But Rabinovitch's pronouncement was taken seriously by many other Orthodox rabbis who repeated it again and again that fall to their students. "There were lots of rabbis who said *din rodef* applied to Rabin," Amir said.[34]

In fact, that same month, *Hashavua,* a *haredi* weekly, charged that Rabin and Peres "must be placed before a firing squad." In the face of such threats against the prime minister, key leaders of the Likud Party, including Knesset members Benny Begin, the son of the former prime minister, and Dan Meridor, both of whom were critics of Oslo, pointedly distanced themselves from the extremists.[35]

But Benjamin Netanyahu, far from withdrawing, aggressively sought their favor. One such episode occurred at the Knesset in October 1995 when Netanyahu stepped up to speak. "You, Mr. Prime Minister, made an appalling remark the other day," he said, addressing the phantom Rabin. "You said the Bible is not our land registry. I say: The Bible is our registry, our mandate, our proof of ownership. Only if we examine remarks you and members of your cabinet made can we fully understand the meaning of what you have done—an unprecedented deed in the history of the Jewish nation. It is true that in the past we were defeated, exiled and forced to cede parts of our homeland. . . . It has never happened, however, that Jews willingly, gleefully, gave away parts of their homeland, forsaking the recognition that we have a right to this land."[36]

To many people, Netanyahu's intended audience was not so much his fellow legislators in the Knesset as the tens of thousands of right-wing *haredim* who were listening in.[37] Later that night, the protesters gathered in Jerusalem's Zion Square where they had put up posters of Rabin in a Nazi SS uniform.[38] One banner read "Rabin, Arafat's Dog." Hoping to bring down Rabin's government, young men carried Benjamin Kahane, son of Meir Kahane, on their arms, shouting, "Death to Rabin! Nazis! *Judenrat!*"†[39]

Housing Minister Benjamin Ben-Eliezer found himself trapped in the frenzy of the crowd. "I've never experienced anything like it!" he said.

*Moses Maimonides was the twelfth-century rabbi and philosopher whose work forms a cornerstone of Orthodox Jewish thought.

†*Judenrat* is German for "Jewish council" and refers to the administrative bodies that the Nazis required Jews to form in the ghettoes to expedite the deportation of Jews for slave labor or extermination in the concentration camps during the Holocaust.

"I've fought in all [of Israel's wars] and seen death before my eyes. But never was I so close to death as I was tonight."

When Ben-Eliezer finally got to Netanyahu, he yelled, "You'd better restrain your people. Otherwise it will end in murder. They tried to kill me just now. . . . Your people are mad. If someone is murdered, the blood will be on your hands. . . . The settlers have gone crazy, and someone will be murdered here, if not today, then in another week or another month!"[40]

Then, with the crowd chanting "Bibi! Bibi! Bibi,"* Netanyahu was introduced as the next prime minister of Israel, and took the podium.

On October 6, 1995, the Knesset approved Oslo II, a complex agreement between Israel and the PLO about the future of the Gaza Strip and the West Bank, which had been signed a week earlier in Washington. To Yigal Amir, it was the last straw. In the previous eight months, he had made three somewhat halfhearted attempts to kill Rabin.[41] As he waited for another opportunity, he contented himself by reading *The Day of the Jackal*, Frederick Forsythe's novel about an attempt to assassinate Charles de Gaulle, who had been accused of treason by the right for withdrawing French troops from Algeria.

Meanwhile, one journalist after another warned about the risk of assassination.[42] One of them, Victor Cygielman, the Israel correspondent for the French weekly *Le Nouvel Observateur*, wrote about a statement from Shin Bet which warned that the poisonous atmosphere of violence had set the stage for the assassination of Rabin.[43] He also reported on a small group of militant religious fanatics led by Avigdor Eskin, a thirty-five-year-old businessman from the former Soviet Union who had been a disciple of Meir Kahane, who went to Rabin's house on Yom Kippur in early October and chanted the Pulsa da-Nura,† the mystical, kabbalistic curse of death. As Michael Karpin and Ina Friedman describe it in *Murder in the Name of God:*

> Unlike his infamous mentor [Kahane], Eskin was glib and polished, an accomplished demagogue who insisted that violence was abhorrent to him. Rocking back and forth on the sidewalk that day, he raised his eyes to the prime minister's house and solemnly intoned the words: "I deliver to you, the angels of wrath and ire, Yitzhak, the son of Rosa Rabin, that

*Bibi has been Netanyahu's nickname since childhood.

†Pulsa da-Nura is Aramaic for "lashes of fire."

you may smother him and the specter of him, and cast him into bed, and dry up his wealth, and plague his thought, and scatter his mind that he may be steadily diminished until he reaches death." As Eskin declaimed the Aramaic text, the men around him chimed in: "Put to death the cursed Yitzhak, son of Rosa Rabin, as quickly as possible because of his hatred for the Chosen People." For the finale, Eskin filled his lungs and shouted up at the building: "May you be damned, damned, damned!"[44]

Two days after Cygielman's story hit the stands, on November 4, more than 100,000 people rallied in Tel Aviv's Kings of Israel Square to show their support for Rabin and his peace policies in the face of the militant campaign against him. The rally, organized by the Labor and Meretz Parties and the advocacy group Peace Now, had a lighthearted spirit, in deliberate contrast to the confrontational tone of militant rightists, and consisted of songs and entertainment. At one point, the *Washington Post* reported, liberal activist Miri Aloni began singing her trademark number, "The Song of Peace," and passed the microphone to Rabin and Shimon Peres.[45]

In the past, Rabin, as Mr. Security, had mocked Aloni's anthem of peace.[46] In addition, given his limited vocal abilities, he had always been reticent about singing in public. But this time he joined in, singing horribly off-key. "We don't know how to sing," Shimon Peres told the crowd, somewhat apologetically. "But in the making of peace we won't be off-key."[47]

After the song, Rabin gave a short radio interview before leaving the stage. "People have doubts about their personal security," he said, "but they do not have doubts that the path of peace should be pursued. I think this rally gave voice to many of the people."[48]

Meanwhile, on the platform, a journalist from *Ha'aretz,* the liberal Israeli daily, approached Leah Rabin, the prime minister's wife, and asked whether her husband was wearing a bullet-proof vest. "Why all of a sudden a protective vest?" she asked.[49] "Have you gone crazy? What are we—in Africa?" But, the reporter persisted, what if a lunatic should try to assassinate Rabin? "Why all of a sudden protection?" she replied. "I don't understand the ideas you journalists have."

At that moment, Yigal Amir was sitting on a concrete flower planter in the parking lot. Shimon Peres came down from the stage area, accompanied by only one bodyguard. The guard looked at Amir suspiciously. "For God's sake," he whispered into his microphone. "What's that dark guy doing down there? Is he one of us?"[50]

As Peres walked through the parking lot, several supporters shouted words of encouragement to him. A few moments later, Rabin descended from the stage area, walked past Amir, and strode quickly toward his car. Amir, who was about five yards away, spotted him almost immediately and stood up. Just as Rabin reached out to grab the car door, Amir went for his gun and fired three shots. Two hollow-point bullets hit Rabin, severing major arteries in his chest and destroying his spinal cord.[51] Amir's third bullet hit Rabin's bodyguard in his left arm.

"It's nothing!" Amir shouted. "It's nothing! It's just a joke. . . . Blanks, blanks."

But Rabin slumped to the ground. The wounded bodyguard pushed him into the car, which rushed off to Ichilov Hospital a few blocks away. About an hour and a half later, at 11:00 p.m., Rabin died on the operating table.

Not long afterward, at a Tel Aviv police station, the police told Amir that Rabin was dead. "Do your work," Amir replied. "I've done mine." Then he turned to one of them and added, "Get some wine and cakes. Let's have a toast."

Later, someone went through Rabin's pockets and found a blood-stained piece of paper with the words to "The Song of Peace," the song the prime minister had reluctantly decided to sing. "Just sing a song of peace," went the lyrics. "Don't whisper a prayer, sing a song of peace loudly."[52]

On the night of Rabin's assassination, President Clinton appeared outside the White House and delivered a brief statement. "The world has lost one of its greatest men, a warrior for his nation's freedom and now a martyr for his nation's peace," he said. He recalled that Rabin and Arafat had been with him at the White House just a month earlier signing yet another agreement, and he quoted Rabin's words: "'We should not let the land flowing with milk and honey become a land flowing with blood and tears. Don't let it happen.' Now it falls to us, all those in Israel, throughout the Middle East and around the world who yearn for and love peace, to make sure it doesn't happen.

"Yitzhak Rabin was my partner and my friend. I admired him, and I loved him very much. Because words cannot express my true feelings, let me just say *shalom haver.* Good-bye, friend."[53]

The realists of the previous administration, George H.W. Bush and James Baker, expressed similarly warm sentiments. Former president George H.W. Bush declared, "He was a true peacemaker whose efforts and sacrifice will be remembered through the ages."[54] Added Baker, "Peace is going to be Yitzhak Rabin's legacy."[55]

But it wasn't that simple. In part because of his legacy as a great Israeli military commander, no one in Israel was, or could ever be, a more forceful figure than Rabin in promoting the peace process. As a result, his murder was a devastating blow to the Oslo principle, the principle of land for peace.

Perhaps saddest of all, millions and millions of American fundamentalists thought that when Rabin signed the Oslo Accords and offered to trade land for peace, it was not just a mistake, it was a sin. "They were going against the word of God," said Kay Arthur, a founder of Precept Ministries International. "You cannot go against the word of God. And I believe that God stopped it . . . by the things that happened."

She believed that God punished Rabin by assassinating him. "I think that God did not want that Oslo Accord to go through."[56]

Immediately after her husband's assassination, Leah Rabin lashed out at Netanyahu. "Mr. Netanyahu incited against my husband and led the savage demonstrations against him,"[57] she said. When Rabin's body was lying in state, she refused to shake hands with Netanyahu—relenting only after making it clear that she preferred shaking hands with PLO leader Yasser Arafat.[58]

She was not alone in her feelings. As the campaign to elect the next prime minister began in early 1996, even fellow Likud leaders thought Netanyahu didn't stand a chance.[59] At a demonstration after the assassination, the crowd chanted "Bibi's a murderer! Bibi's a murderer!"[60] Meanwhile, Netanyahu's opponent, Shimon Peres, basked in the halo of his martyred colleague. The Clinton administration tacitly endorsed Peres as Rabin's heir in hopes of continuing the implementation of the Oslo Accords.[61]

But Netanyahu had had his eye on the prize for years and brought an unusual set of skills to the table. Thanks to his polished oratory and mastery of American idioms—he attended high school in suburban Philadelphia and went to college at MIT—Netanyahu knew how to appeal to affluent, cosmopolitan, secular Jews in the United States and Israel. At the same time, Bibi struck chords from the Bible and the Holocaust that resonated deeply with hard-line neoconservatives, militant settlers in the West Bank who he hailed as "the New Pioneers," and fundamentalists with a Christian Zionist bent.

Embracing the latest American campaign techniques, Netanyahu hired Arthur Finkelstein, a top GOP political consultant who hammered home the same points repeatedly. One ad showed bombed-out buses juxtaposed with a photo of Peres shaking hands with Arafat, with a voice saying, "A dangerous combination for Israel."[62] The message was clear: Shimon Peres was a weakling who would do anything for peace.[63]

No Israeli was more adept at playing the American card. As one of the most powerful voices on the Israeli right, Netanyahu was effectively the über-neocon and, as such, made a point of promoting the Israeli-evangelical alliance. That meant wooing Christian fundamentalists who gathered to celebrate the Feast of Tabernacles in Jerusalem, and to whom he described the alliance between the Jews and Christian Zionists as "a partnership that has endured for more than a century and, if anything, is growing stronger."[64] Shortly afterward, he convened a newly created Israel Christian Advocacy Council with seventeen Christian fundamentalist leaders in the Holy Land to put forth a Likud-like policy statement backed by evangelical theology, rejecting all calls to abandon occupied parts of the West Bank, East Jerusalem, Gaza, and the Golan Heights.[65] Soon ads appeared in newspapers all over the United States citing Genesis, Leviticus, and Deuteronomy, asserting that "Israel's biblical claim to the land" was "an eternal covenant from God." Signatories included the Christian Broadcasting Network's Pat Robertson, Ralph Reed of the Christian Coalition, Jerry Falwell, and many others. The neocon–Likudnik– Christian Zionist message was now reaching millions and millions of people.

In the end, Palestinian terrorists came to Netanyahu's rescue. In February and March, one Hamas suicide bomb after another rocked Israel.[66] After an eight-day barrage killed 59 men, women, and children, Netanyahu's ads blamed Peres for failing to stop the bombings. On Election Day, May 29, the final tally gave victory to Netanyahu by 29,507 out of 3.1 million votes.[67] Clinton White House officials who had been involved in the peace process were not happy. "Our collective relief became a collective dread," wrote Dennis Ross, U.S. envoy to the Middle East, in *The Missing Peace: The Inside Story of the Fight for Middle East Peace.*[68]

But it was not just the peace process that was in jeopardy. Saddam was still in power. Iran was a growing threat. And in July 1996, just after taking office, newly elected Israeli prime minister Benjamin Netanyahu was on his way to the United States to speak before a joint session of Congress. Just as important, he would meet with Richard Perle and other principals of the neoconservative movement. His trip would ultimately lead to a new American foreign policy, based on a radical new vision of the Middle East—a vision that would change history, and much of the world.

CHAPTER TEN

Ripe for the Plucking

The neocons had come to power in Israel, but they had not been so fortunate in the United States. With Clinton still in the White House, they had to find other channels through which to pursue their policies. On July 8, 1996, about six weeks after his election as prime minister, Benjamin Netanyahu flew to Washington to see Richard Perle, one of several neoconservative analysts who had mapped out a new strategy for him in a policy paper entitled "A Clean Break: A New Strategy for Securing the Realm."[1] Commissioned by an Israeli-American think tank, the Institute for Advanced Strategic and Political Studies, "A Clean Break" was effectively an Israeli-centric update of the radical vision for a new Middle East that had first emerged in the 1992 "Defense Policy Guidance" paper. The term *clean break* referred to the notion of completely jettisoning the land-for-peace formula behind the Oslo Accords[2] and replacing it with "peace through strength," which, in turn, meant "reestablishing the principle of preemption."[3]

In other words, Richard Perle and his associates were working for the right-wing Likud Party in Israel in hopes of subverting official U.S. foreign policy. In addition to Perle, the authors of the paper included Meyrav Wurmser,* an Israeli-American who later became director of the Center for Middle East Policy at the Hudson Institute; her husband, David Wurmser, who later became head of Middle East policy for Vice President Dick Cheney; Douglas Feith, the future under secretary of defense for policy; and several academics.[4]

In "A Clean Break," the Likud and the neocons had come up with an

*Meyrav Wurmser was also cofounder, with Israeli colonel Yigal Carmon, of Memri, the Middle East Media Research Institute, which translates texts from Arab media and disseminates them to the West. Ibrahim Hooper of the Council on American-Islamic Relations told the *Washington Times*: "Memri's intent is to find the worst possible quotes from the Muslim world and disseminate them as widely as possible."

ideology that was effectively a secularized version of the theology of the Christian Right. As the Christian Zionists interpreted the Abrahamic Covenant and as the settlers in the West Bank would have it, the Jews were ordained by God to reclaim the biblical lands of Judea and Samaria in the West Bank. Likewise, in "A Clean Break," Likud and the neocons asserted Israel's claim to the West Bank went back two thousand years.[5] Just as the Christian Right argued that one did not compromise over Zion, the neocons demanded "the unconditional acceptance by Arabs of our rights, *especially* in their territorial dimension."[6]

Even the boundaries as defined in the Bible thousands of years ago were similar to those the neocons were drawing up in their grandiose plan to overhaul the Middle East. As defined in Genesis, the Promised Land extended far beyond Israel's borders: "To your descendants I have given this land, from the river of Egypt as far as the great river, the river Euphrates."[7] Depending upon one's interpretation,* the land referred to in the Abrahamic Covenant theoretically encompassed all or parts of Egypt, Syria, Lebanon, Jordan, Iraq, and even Saudi Arabia. In "A Clean Break," the authors asserted that by waging wars against Iraq, Syria, and Lebanon, Israel and the United States could reshape the "strategic environment" and balance of power in the region.†

Specifically, they called for overthrowing Saddam Hussein, which raised one rather obvious strategic question. Beginning in 1980, the brutal eight-year Iran-Iraq War cost over a million casualties and hundreds of billions of dollars. The war was disastrous to both countries, but, from the point of view of the West, it had the convenient consequence of paralyzing two powerful destabilizing forces in the Middle East. The United States did not want to see Iran's Islamic revolution spread throughout the region, potentially jeopardizing Western oil supplies, so it was happy to have Iraq as a counterweight. As a result, at various points during the Reagan-Bush era, it generously aided Saddam,

*Most scholars interpret "the river of Egypt" as a reference to the Nile, which, if taken literally, would mean the Abrahamic Covenant defines Israel as encompassing not just all of modern-day Israel, including Gaza, the West Bank, and the Golan Heights, but also Lebanon, Jordan, much of Iraq, Syria and Kuwait, and even some parts of Egypt, Turkey, and Saudi Arabia. In modern times, Israel has never sought such vast territory. When the Israeli right refers to Eretz Yisra'el, the land of Israel, it is generally referring to the area between the Mediterranean Sea and the Jordan River, including the West Bank and Gaza.

†Jordan and Egypt were already considered at peace with Israel. Later, some neoconservatives added Saudi Arabia to the list.

giving him billions of dollars—even after he had used chemical weapons. Likewise, the United States saw Saddam's Iraq as a threat to Israel and secretly armed Iran in part to keep Iraq in check. Even after their war ended in 1988, Iran and Iraq were still at loggerheads, much to the delight of the West.

But if the West were to overthrow Saddam, what would that do to the balance of power in the Middle East? What would happen to the Islamic Republic of Iran—a massive, oil-rich country with 70 million people, the size of the United Kingdom, France, Germany, and Spain combined—if its chief foe were eliminated? Shia Islam was the official state religion of Iran, and even though it was the majority religion in Iraq, it had been suppressed by Saddam. If he were eliminated, would a new Shia power base begin to dominate the Middle East?

According to "A Clean Break," these issues could best be dealt with by promoting the restoration of the Hashemites in the new Iraq after Saddam was overthrown. The Hashemite Kingdom of Jordan* was already pro-West, and if the neocons' bold new vision came to fruition, the newly empowered Hashemites in Iraq could influence Najaf, the holiest Shi'ite city in Iraq and the center of Shia political power, "to help Israel wean the south Lebanese Shia away from Hezbollah, Iran, and Syria."[8]

In other words, the neocons assumed the Hashemites would be able to persuade powerful Shi'ite leaders in Iraq to do the bidding of the United States and Israel. This assumption was critical to the strategic vision behind "A Clean Break." Yet beyond the bald assertion that the Shia "retain strong ties to the Hashemites," the factual basis behind the assumption was unexplained. Moreover, the paper did not address the fact that the Al Dawa Party ("the Call"),† one of the two main Shi'ite parties in Iraq, had a long history of ties to Islamist terrorism, and

*The country's official name is in fact the Hashemite Kingdom of Jordan.

†The Al Dawa Party was formed in the late fifties by a number of Shi'ite leaders including Mohammad Baqir al-Sadr, the uncle of Moqtada al-Sadr. Despite some differences in its vision of how an Islamic Republic should function, the party supported Ayatollah Khomeini and the Islamic Revolution in Iran, and received financial support from Iran in return. According to Juan Cole, a Middle East specialist at the University of Michigan, in the early eighties, parts of Dawa's Islamic Jihad coalesced into what became Hezbollah and were connected to attacks on the American embassies in Kuwait and Lebanon, including the 1983 attack that killed more than 240 American servicemen. In more recent years, however, Al Dawa has repudiated these attacks. Nuri al-Maliki, who became prime minister of Iraq in 2006, had been deputy leader of Dawa.

its spin-off, the Supreme Council for Islamic Revolution in Iraq (SCIRI) had close ties to Iran's Revolutionary Guards.[9]

Thus the neoconservative vision of democratizing the Middle East was born. "It was the beginning of thought," said Meyrav Wurmser. "It was the seeds of a new vision."[10]

Netanyahu certainly seemed to think so. Although he never implemented the policies put forth in the paper, two days after meeting with Perle, the prime minister addressed a joint session of Congress with a speech that borrowed from "A Clean Break" and called for the "democratization" of terrorist states in the Middle East and warned that war might be the only way such noble goals could be accomplished. But Netanyahu also made one significant addition to "A Clean Break." The paper's authors were concerned primarily with Syria and Saddam Hussein's Iraq, but Netanyahu saw a greater threat elsewhere. "The most dangerous of these regimes is Iran," he said.

Together, the "Defense Planning Guidance" and "A Clean Break" represented the fruits of a radical political movement led by a right-wing intellectual vanguard that over three decades had put together the capital, infrastructure, and political organization—think tanks, institutes, foundations, political journals, and popular media outlets—to finally implement their ideas. But there were still a few things missing—namely, support among the intellectual elite for these radical new policies, a popular electoral base so they could win power, and, finally, a candidate who would support and implement their policies.

After "A Clean Break," the neocon machine shifted into a higher gear. The drumbeat for war began. In the summer of 1996, William Kristol and Robert Kagan put forth the proposition in *Foreign Affairs* that now was the time for the United States to become a "benevolent global hegemony" and begin "a remoralization" of American foreign policy.[11] In the *Weekly Standard*, the same duo specified that such a policy meant that "Saddam Must Go."[12] Likewise, syndicated neoconservative columnists Charles Krauthammer[13] and A. M. Rosenthal[14] called for regime change in Iraq. In the *Washington Post*, Zalmay Khalilzad and Paul Wolfowitz joined the fray.[15] "A Clean Break" coauthor David Wurmser took the message to the *Wall Street Journal*[16] and wrote the book *Tyranny's Ally: America's Failure to Defeat Saddam Hussein*—in which he proposed that the United States redraw the map of the Middle East.[17]

In *Tyranny's Ally,* Wurmser, a protégé of Richard Perle and Ahmed Chalabi, asserted that this was not just about Iraq, but that Iran must also be "severed from its Shi'ite foundations. And this can be accomplished by promoting an Iraqi Shi'ite challenge."[18] He added that Saddam's tyranny "leaves Iran by default as the arbiter of Shi'ite politics in the region. The Iraqi Shi'ites, if liberated from this tyranny, can be expected to present a challenge to Iran's influence and revolution."[19]

In other words, if the United States overthrew Saddam and liberated the Shi'ites, far more than just greeting Americans with flowers, the Shi'ites would also take on the Islamist state of Iran. Democracy would spread throughout the region! Israel would be secure and the U.S. would have allies in the oil-rich states of Iran and Iraq!

But, again, the text contained no facts to back up Wurmser's assumptions. According to the acknowledgments, however, the key figures "who guided my understanding" included Ahmed Chalabi; Richard Perle, who wrote the introduction; and fellow Scoop Jackson alum Douglas Feith; among others. In other words, the neocon echo chamber had begun to rely on itself to reinforce its own myths. "They deluded themselves into thinking that these links operated only one way—with Najaf (the holiest Shia city in Iraq) undermining Qom (Najaf's counterpart in Iran)," said Vali Nasr, the Iranian-American scholar and author of *The Shia Revival.* "They assumed that it was Iraq that would influence Iran, not that Iran would influence Iraq."[20]

Whereas evangelicals had expressed their belief that Americans were a Chosen People whose mission was ordained by God, neocons now did so in secular terms, and they further proclaimed that the United States had a moral duty to *project* that greatness throughout the world—using American military power, if necessary. Calling for a "national greatness conservatism," William Kristol and David Brooks asserted, "Our nationalism is that of an exceptional nation founded on a universal principle, on what Lincoln called 'an abstract truth, applicable to all men and all times.' "[21]

Other neocons promoted the same doctrines using language that was more forthright but somewhat less refined and, for that matter, less sensitive to the horrors of the warfare they advocated, wars that would kill hundreds of thousands of people, maim countless more, and create literally millions of refugees. Writing in the *National Review,* for example, Jonah Goldberg saluted his colleague Michael Ledeen, a contributor to the same publication and an AEI fellow and neocon firebrand, with giddy imperial triumphalism, urging the implementation of what he called the Ledeen Doctrine: "Every ten years or so, the United States needs to pick up some small crappy little country and throw it against the wall, just to show the world we mean business."[22]

At the time, however, without a patron in the White House, and with no electoral base whatsoever, the neocons had gone as far as they could. The vast majority of Americans were completely unfamiliar with them and their ideology. As a result, the founding fathers of the movement—Irving Kristol and Norman Podhoretz—now argued that it was time to forge an alliance with the Christian Right even if it meant overlooking their undemocratic attitudes that were often expressed by various fundamentalists. "[C]onservatives and the Republican Party must embrace the religious if they are to survive," wrote Kristol in *Neoconservatism: The Autobiography of an Idea.* "Religious people always create problems since their ardor tends to outrun the limits of politics in a constitutional democracy. But, if the Republican Party is to survive, it must work on accommodating these people."[23]

Norman Podhoretz, Kristol's longtime collaborator, added that Jews should even turn a blind eye to Pat Robertson's anti-Semitism because such an alliance with fundamentalists was vital to Israel. "[I]n my view Robertson's support for Israel trumps the anti-Semitic pedigree of his ideas," he wrote.[24] Michael Ledeen, who worked with Richard Perle and David Wurmser at the American Enterprise Institute, began making scores of appearances on Pat Robertson's televised *700 Club,* to promote their strategy to overhaul the Middle East before millions of fundamentalist viewers.[25]

As Ledeen himself pointed out in *The First Duce: D'Annunzio at Fiume,* his paean to Italian fascist Gabriele d'Annunzio, the poet-warrior who was Benito Mussolini's mentor, religious imagery had the power to inspire the masses to embrace extraordinary nationalistic goals: "The radicalization of the masses in the twentieth century . . . could not have succeeded without the blending of the 'sacred' with the 'profane.' The timeless symbols that have always inspired men and women to risk their lives for higher ideals had necessarily to be transferred from a religious context into a secular liturgy if modern political leaders were to achieve the tremendous control over their followers' emotions."[26]

Meanwhile, in the twenty plus years since *Roe v. Wade,* the Christian Right had grown enormously. Megachurches—those with more than two thousand parishioners—had proliferated across the country, growing from just fifty churches in 1980 to nearly nine hundred a generation later—one of which, Saddleback Church in southern California, boasted eighty thousand members on its rolls.[27] Revenue at Falwell's Thomas

Road Baptist Church topped $200 million a year—and was projected to exceed $500 million by 2010.[28] Prestonwood Baptist Church in Plano, Texas, just outside of Dallas, drew seventy thousand people at $20 a ticket and more to its Christmas spectacle featuring its five-hundred-person choir. Its sports programs, with eight playing fields and six gyms, served sixteen thousand people and brought in countless young converts.[29]

Likewise, evangelicals used the latest Madison Avenue marketing techniques. There were cowboy churches for ranchers, country music churches for the C&W market, gospel and rhythm and blues churches for black evangelicals, motorcycle churches for bikers, and sandals and electric guitars for the long-haired Birkenstock crowd.[30] There was Christian miniature golf, a Christian Wrestling Federation, and ministries for evangelical skateboarders, NASCAR drivers,[31] and Harley-Davidson motorcycle owners.*[32]

In parts of the Bible Belt, no force dominated life as much as the evangelical church. Greenville, South Carolina, had more than 700 churches for its 56,000 inhabitants. Christian rock far outsold jazz and classical music combined. By 2000, there were at least fifteen annual Christian rock festivals, including the Cornerstone Festival; the Sonshine Festival; Spirit West Coast; Rock the Desert, at Universal Studios Florida; and Night of Joy at Walt Disney World.[33] As Ted Haggard put it, "Cool kids like to go to church if church is cool."[34]

Whereas secularists had Macy's and Bloomingdales, evangelicals had LifeWay Christian Stores, a division of the Southern Baptist Convention† that expanded throughout the South, becoming a chain of more than 120 stores offering Christian books, music, and apparel. There were Christian pencils, pro-life T-shirts ("Mommy, please let me live" and "Former Embryo"), and greeting cards and bookmarks with Bible verses. There was Christian furniture[35] ("Episcopalian style at Pentecostal prices!"), Christian trailer parks, Christian cutlery, and Christian spatulas. If your car broke down, there were Christian auto repair shops, and if you needed to refinance your home, there were Christian mortgages. For vacation, there were Christian Caribbean cruises. For obesity, there

*To reassure its biker-parishioners, on its website, the Heralds of the Cross Motorcycle Ministry advises its members, "How Can I Know for Sure That I Will Spend Eternity in Heaven?"

†The words *Southern Baptist Convention* refer to the denomination's name, not just its annual convention. It is the largest Protestant denomination in the United States, with more than 16 million members.

was a whole host of Christian weight-loss options—diet books such as *What Would Jesus Eat?*, the fatfree4jesus.org website, and Steve Reynolds's Bod4GOD program.[36]

This was an alternative world to secular America—and it was Republican. Throughout the South, billboards proclaimed "Evolution Is Science Fiction," "Darwin Is Dead, Jesus Is Alive," "Meet at My House Before the Game—(signed) God," and "One Nation Under Me—(signed) God." The Institute for Creation Science trained biochemists who did not believe in evolution.

Moreover, as the millennium approached, the apocalyptic themes in Christianity fueled the movement. No one benefited more than Tim LaHaye. Sometime in the mid-eighties, LaHaye was on an airplane when he noticed that the pilot, who happened to be wearing a wedding ring, was flirting with an attractive flight attendant, who was not. LaHaye asked himself what would happen to the poor unsaved man if the long-awaited Rapture were to transpire at that precise moment. Soon, LaHaye's literary agent dug up Jerry Jenkins, a writer-at-large for the Moody Bible Institute and the author of more than 150 books, many on sports and religion. In exchange for shared billing, Jenkins signed on for the Left Behind series.

The first volume, *Left Behind*, begins with a variation of what LaHaye observed in real life. While piloting his 747 to London's Heathrow Airport, Captain Rayford Steele decides he's had just about enough of his wife's religiosity. He puts the plane on autopilot and leaves the cockpit to flirt with a "drop-dead gorgeous" flight attendant named Hattie Durham, only to be told that dozens of passengers have suddenly and mysteriously vanished, leaving behind their clothes, eyeglasses, jewelry—even their pacemakers, dental fillings, and surgical pins.

What had taken place, of course, was the Rapture. Millions of Christians who accepted Christ as their savior, including Rayford Steele's wife and young son, had been "caught up" in the air to meet Him. At the same time, people like Steele finally begin to see Christ as his Savior. Now that they have been left behind, doubters no more, they form the Tribulation Force to take on the armies of the Antichrist and win redemption by taking on the vast forces of ungodly, evolution-believing, pro-abortion secular humanists.

Kicking off the series in 1995, as the millennium clock ran down, gave it a convenient built-in marketing device. Before long, thanks to the rapid growth of evangelicalism in America, Left Behind books were selling at the astounding rate of 1.5 million copies a month. More than

10 million items of related products such as postcards and wallpaper were sold. The series, originally planned to be seven books, was extended to twelve.[37]

It wasn't just the millennial clock that fueled sales. The ascendancy of computer culture and the Internet gave way to widespread fears among End Times aficionados about the implantation of biochips as the Mark of the Beast. The birth of the European Union sparked horror about One World government forecast in the book of Revelation. Millions wondered who was the Antichrist.

For many fundamentalists, the answer could be found right in the White House. In November 1996, to the dismay of the Christian Right, Bill Clinton won reelection in a landslide. By this time, independent counsel Kenneth Starr had already been investigating the Whitewater controversy for two years. Over time, he had received authorization to investigate the White House travel office controversy, the suicide of deputy White House counsel Vince Foster, a sexual harassment lawsuit by former Arkansas state employee Paula Jones against President Clinton, and, later, perjury and obstruction of justice charges to cover up President Clinton's sexual relationship with White House intern Monica Lewinsky. The Jerry Falwell– financed pseudodocumentary *The Clinton Chronicles* had made the rounds among millions of fundamentalists, characterizing Clinton as a money-laundering drug smuggler involved in a host of shadowy criminal conspiracies and murders.[38] To millions of fundamentalists, Clinton was a veritable Antichrist, the embodiment of satanic secularism.

Seven months after Clinton's reelection, at a June 1997 meeting in Montreal of the Council for National Policy, the right-wing umbrella group founded by Tim LaHaye, the evangelicals decided to take action.[39] Televangelists Pat Robertson and James Dobson, Paul Weyrich of the Free Congress Foundation, Ralph Reed of the Christian Coalition, Phyllis Schlafly of the Eagle Forum, and Oliver North—all of the CNP—joined in, all putting the vast resources of their organizations into the movement to impeach Bill Clinton.*[40] Unbeknownst to the fundamentalists, a new savior was trying his wings, taking the first steps that would enable him to come to their rescue.

Thanks to his success with the Texas Rangers and his election as gov-

*In November 1997, Representative Bob Barr (R-Ga.), a member of the CNP, introduced HR 304, a resolution of inquiry that represented the first stage of the impeachment process in the House of Representatives.

ernor, George W. Bush had achieved partial redemption in his father's eyes. But exactly how far he could go in politics was another question. Bob Strauss, the former chairman of the Democratic National Party and a friend of George H.W. Bush, was just one of many knowledgeable pols who watched George W. grow up and concluded that he did not have the right stuff. "He was a real nice guy," said Strauss, a legendary Democratic powerbroker who served as ambassador to Moscow under Bush 41. "But if anybody told me he was going to be president, I would have thought they'd lost their mind. I don't think he's curious enough to ask about important things. That's his weakness—he's incurious."[41]

Yet now George W. set his sights on the White House—and a complicated, deeply conflicted father-son relationship began to unfold. As Bush 41 knew, his son had become a popular governor of Texas, but since Texas was a state in which gubernatorial power was severely limited, that wasn't much of a test for a potential president. More important, the elder Bush knew that his son had virtually no knowledge of foreign affairs and had barely even traveled abroad. With that in mind, in the fall of 1997, Bush senior called Prince Bandar bin Sultan, the longtime Saudi Arabian ambassador to the United States and one of his closest personal friends, to ask a favor. Bush said his son was on the verge of making an important decision and wanted some advice. "W. would like to talk to you if you have time," Bush said, according to Bob Woodward's *State of Denial: Bush at War, Part III.*[42] "Can you come by and talk to him?"

Bandar had long been the kind of friend to whom Bush 41 could turn for sensitive favors. Back in 1981, Bandar, a thirty-two-year-old grad student at Johns Hopkins University, had befriended then Vice President Bush, who helped him push through the controversial sale of a multibillion-dollar AWACS (airborne warning and control system) package to Saudi Arabia over the objections of the Israeli lobby. Over the years Bandar repaid the debt in spades, first by personally arranging back-channel financing for the Nicaraguan contras in the Iran-contra affair. Later, Bandar's Saudi friends had bailed out Harken Energy, George W. Bush's struggling oil company* in the eighties, and invested in the Carlyle Group, a large, Washington-based private equity firm with which both George Bushes and several of their colleagues were associated. During the 1991 Gulf War, Bandar had waged war against Iraq with Bush and culti-

*For a more complete account of Saudi relationships with Harken Energy and the Carlyle Group, see the author's *House of Bush, House of Saud,* pp. 117–28 and 165–69, respectively. Likewise, Prince Bandar's relationship with George H.W. Bush is discussed at length at various points throughout the same book.

vated friendships with James Baker, Brent Scowcroft, Dick Cheney, and Colin Powell. As president, Bush gave Bandar so much access to the executive branch that Colin Powell, then chairman of the Joint Chiefs of Staff, carped that he functioned as if he were a cabinet officer.[43]

Even though he and the younger George Bush were roughly the same age—forty-eight and fifty-one, respectively—Bandar did not have the same rapport with the son that he had with the father. But he was delighted to help nonetheless. An avid fan of the Dallas Cowboys and close friend of the team's owner, Jerry Jones, Bandar planned a visit to Texas with a Cowboys game as "cover."[44]

When Bandar finally landed in Austin, to his astonishment, the governor boarded the plane before the prince had a chance to disembark. Bush got right to the point: he was thinking of running for president and he already knew what his domestic agenda would be. But, he said, "I don't have the foggiest idea about what I think about international, foreign policy."

Bandar ran through his experiences with Mikhail Gorbachev, Margaret Thatcher, Tony Blair, the pope, and Ronald Reagan.[45] Finally, according to Woodward's book, Bush had a question. "There are people who are your enemies in this country who also think my dad is your friend," he said.[46]

Bush was speaking code, but to Bandar, he had made an obvious reference to supporters of Israel. In effect, he was asking Bandar what to do about the Israeli lobby and the neocons who had an aversion to both his father and the Saudis.

"Can I give you one advice?" Bandar asked.[47]

"What?"

"If you tell me that [you want to be president of the United States], I want to tell you one thing," Bandar replied.[48] "To hell with Saudi Arabia or who likes Saudi Arabia or who doesn't, who likes Bandar or doesn't. Anyone who you think hates your dad or your friend who can be important to make a difference in winning, swallow your pride and make friends of them. And I can help you. I can help you out and complain about you, make sure they understood that, and that will make sure they help you."

If George W. needed the neocons on his side to win the Republican nomination or the election, Bandar was saying, he should do whatever was necessary. "Never mind if you really want to be honest," Bandar added. "This is not a confession booth. . . . In the big boys' game, it's cutthroat, it's bloody and it's not pleasant."[49]

In other words, Bandar seemed to be saying, take them into your camp—for now.

Meanwhile, the Clinton administration continued its efforts to resolve the Israeli-Palestinian conflict. By early 1998, however, the Oslo process had basically come to a halt. Both Benjamin Netanyahu and Yasser Arafat had prepared long lists citing the other's noncompliance.[50] As a result, Netanyahu had not implemented the first scheduled withdrawal from the West Bank, and the second one was long overdue. "There is and always has been only one way to resolve the Israeli-Palestinian conflict: land for peace," wrote Anthony Lewis in the *New York Times.* "And the Netanyahu government has now made clear that it has no intention of withdrawing from enough of the land Israel occupies in the West Bank to make a deal imaginable."[51]

In January 1998, under pressure from the White House to renew the peace process with Arafat, Prime Minister Benjamin Netanyahu turned to the Christian Right once again for support—and Jerry Falwell was ready to help out. With Arafat due to arrive at the White House on Tuesday, January 22, Netanyahu wanted to have as strong a hand as possible. Arriving in Washington three days early, he was already scheduled to see Clinton, but decided to have a very public meeting with Falwell first.

"I put together one thousand people or so to meet with Bibi and he spoke to us that night," Falwell said.[52] "It was all planned by Netanyahu as an affront to Clinton."

At the time, Clinton had plenty of other things on his mind. The day before, on the ABC Sunday-morning program *This Week,* neoconservative commentator William Kristol alerted the nation to an Internet posting on the Drudge Report alleging that President Clinton had had a sexual relationship with Monica Lewinsky, a bright-eyed twenty-three-year-old White House intern.[53] Initially, many White House aides on the political side dismissed the report because Matt Drudge, its conservative proprietor, had been the source of a number of erroneous, apparently politically motivated stories. But quietly, the White House legal staff was scrambling.[54]

Meanwhile, that evening, Netanyahu met with Falwell at the Mayflower Hotel in downtown Washington. Falwell promised Netanyahu he would mobilize 200,000 pastors all over the country to resist the return of occupied West Bank territory, and would ask them to "tell President Clinton to refrain from putting pressure on Israel."[55] Televangelist John Hagee, who gave $1 million to the United Jewish Appeal the following month in hopes of hastening End Times, told the crowd that the Jewish return to the Holy Land was prophesy of the "rapidly approaching . . . final moments of history." Then he brought them to a frenzy chanting "Not one inch! Not one inch!"—a reference to how much of the West Bank should be transferred to Palestinian control.[56]

When Netanyahu spoke to the press about it, he downplayed the significance of the meeting. "I talk to liberals, I talk to conservatives," he said. "I talk to Jews, I talk to non-Jews. These meetings reflect the fact that Israel enjoys support from diverse circles in the U.S. We don't relate indifferently or condescendingly toward any of our friends."[57]

But according to Richard N. Haass, a former National Security Council official under George H.W. Bush, Netanyahu was playing hardball. "This was a way for [Netanyahu] to push back," said Haass. "If the White House was trying to make Mr. Netanyahu pay a price domestically for his lack of cooperation, essentially Mr. Netanyahu was sending a return signal: 'Two can play at this game. I can spend more time with your political opponents and this is something that can come back to bite you in 1998 and 2000.' "[58]

From the *Washington Post* to the *New York Times*, the *Chicago Tribune* to the *Los Angeles Times*, news of Netanyahu's meeting with Falwell was everywhere. When Netanyahu walked into the Oval Office, Clinton admonished him for the snub.

"Bibi [Netanyahu] told me later," Falwell recalled, "that the next morning, Bill Clinton said, 'I know where you were last night.' "[59]

Equally important, the Monica Lewinsky sex scandal had finally started to get traction. It made *Imus in the Morning*, Don Imus's nationally syndicated radio show, that morning.[60] The Manchester, New Hampshire, *Union-Leader* had written about it. *Newsweek* was working on something. "While [Netanyahu] was sitting there, he was in a very difficult spot," Falwell said. "The pressure was really on him to give away the farm in Israel. But while he was sitting there, someone came in and whispered in Mr. Clinton's ear and Mr. Clinton turned several colors. Someone was telling him that the cat was out of the bag on Monica Lewinsky. The meeting was terminated. Mr. Clinton had to save himself. The demands [to relinquish West Bank territory] that would have been forthcoming of Israel, which would have been terrible, were not made. Netanyahu flew back to Israel. He was very funny when he told me about it. He said Israel was saved by Monica Lewinsky."[61]

* * *

Four days later, on January 26, 1998, the neoconservative Project for a New American Century (PNAC) drafted a letter to President Clinton calling for Saddam's removal from power.[62] Not unlike the Cold War front groups of the seventies, PNAC's signatories included many of the usual neocon suspects and other hard-liners.* Specifically, it urged Clinton "to turn your Administration's attention to implementing a strategy for removing Saddam's regime from power. . . . If you act now to end the threat of weapons of mass destruction against the U.S. or its allies, you will be acting in the most fundamental national security interests of the country. If we accept a course of weakness and drift, we put our interests and our future at risk."

Weakened by the Monica Lewinsky sex scandal, Clinton had effectively been sandbagged by the neocons. "This was a key moment," said one State Department official. "The neocons were maneuvering to put this issue in play and box Clinton in. Now, they could draw a dichotomy. They could argue to their next candidate, 'Clinton was weak. You must be strong.'"

It was early in the presidential season, but already George W. Bush, as a popular governor in a populous state, was a leading Republican candidate. Whether or not he would be the first choice of the neocons was another question entirely. After all, the last thing they wanted was a replica of his father.

What they did not yet know was that George W. Bush also had no desire to replicate his father's presidency. Much as he loved him, a friend of the family observed, "He hated and resented being the other George Bush, having to live up to that résumé. He was constantly facing this challenge of carving out an image and identity that was his own."

Given his lack of knowledge when it came to foreign policy, his limited experience as a hands-on executive, and the extraordinary bureaucratic skills of the neocons, George W. Bush was an exceedingly easy mark. "This guy was a tabula rasa," said a State Department source who later became a critic of Bush. "He was an empty vessel. He was so ripe for the plucking."

*The complete list consisted of Elliott Abrams, Richard L. Armitage, William J. Bennett, Jeffrey Bergner, John Bolton, Paula Dobriansky, Francis Fukuyama, Robert Kagan, Zalmay Khalilzad, William Kristol, Richard Perle, Peter W. Rodman, Donald Rumsfeld, William Schneider Jr., Vin Weber, Paul Wolfowitz, R. James Woolsey, and Robert B. Zoellick.

CHAPTER ELEVEN

Dog Whistle Politics

In August 1998, former president George H.W. and Barbara Bush spent their summer vacation at Walker's Point in Kennebunkport, Maine, as usual. When Bush was president, vital matters of state such as Saddam Hussein's 1990 invasion of Kuwait had often disrupted his summer vacations. Now, however, Bush had been out of the White House for more than five years, and Bill Clinton had to deal with incessant intrusions into his summer idyll—Saddam and impeachment, Saddam and impeachment.* By contrast, the elder George Bush amused himself with his latest extravagance, a new high-speed Cigarette boat called *Fidelity.* He played golf. Kevin Costner visited him.[1] And he eagerly awaited the September publication of *A World Transformed,* the Gulf War memoir he had coauthored with his close friend, former national security adviser Brent Scowcroft.

The former president also managed to keep under wraps a delicate but vital mission close to his heart. At his invitation, two celebrated visitors were secretly ensconced at Walker's Point—his son George W. Bush and Condoleezza Rice, a protégé of Scowcroft who served as special assistant to Bush 41 on Soviet affairs from 1989 to 1991 and then had become provost of Stanford University. George W. and Condi discussed foreign policy as they ran side by side on treadmills in the gym at the Bushes' rambling compound. "We talked a lot about America's role in the

*On August 20, in hopes of killing an obscure terrorist named Osama bin Laden, Clinton ordered bombing strikes against terrorist training sites in Afghanistan and the Sudan. In response, Republican critics lashed out at him as the "Wag the Dog" president, asserting the only reason for the attacks was to distract the public from the Monica Lewinsky sex scandal that was in full bloom. The "Wag the Dog" reference is to the 1998 film starring Dustin Hoffman, Robert De Niro, and Anne Heche, about a phony war that is started to distract the public from a presidential sex scandal.

world," Rice later explained. "[George W.] was doing due diligence on whether or not to run for president."[2]

In November, George W. won reelection as governor of Texas with 69 percent of the vote. His brother Jeb won a convincing victory as governor of Florida as well, but when it came to the 2000 presidential race, George had an insurmountable four-year head start over his sibling. If he had any doubts about running for the presidency, they were soon dispelled. Shortly after the election, he and his mother went to Highland Park Methodist Church in Dallas, where Pastor Mark Craig preached a rather pointed sermon about Moses's reluctance to lead his people. As Craig put it, Moses told God, "Sorry God, I'm busy. I've got a family. I've got sheep to attend. I've got a life." Then Craig added that people are "starved for leaders who have ethical and moral courage."[3]

Barbara Bush turned to her son. "He was talking to you," she said.

Bush said it was the best sermon he'd ever heard.[4] Then he phoned James Robison, the fiery Southern Baptist evangelist who hosted the TV show *Life Today.*[*5] "I've heard the call," Bush told him.[6] "I believe God wants me to run for president."

At the same time, Iraq was back in the news. More than seven years after he had been defeated in the Gulf War, Saddam was still in power, taunting the United States. In August, he defied United Nations Special Commission (UNSCOM) weapons inspectors.

In response, at the behest of the Republican-controlled Congress, Clinton signed the 1998 Iraq Liberation Act, thereby committing support to democratic opposition to Saddam. That included Ahmed Chalabi's Iraqi National Congress, even though the administration still regarded Chalabi with suspicion. "He represents four or five guys in London who wear nice suits and have a fax machine," a Clinton official told the *Chicago Tribune.*[7]

To the dismay of Centcom (Central Command) commander General Anthony Zinni, who had operational control of U.S. combat forces in the Middle East, the Iraq Liberation Act gave Chalabi a chance to draw up new plans to overthrow Saddam. "It got me pretty angry," Zinni said. "They were saying if you put a thousand troops on the ground Saddam's regime will collapse, they won't fight. I said, 'I fly over them every day,

*Robison was expelled from the Southern Baptist Convention when he became a charismatic Christian and asserted his belief in speaking in tongues, healing, and other "gifts of the Holy Spirit."

and they shoot at us. We hit them, and they shoot at us again. No way a thousand forces would end it.' The exile group was giving them inaccurate intelligence. Their scheme was ridiculous." Zinni warned Congress that Chalabi's plan is "pie in the sky, a fairy tale."[8]

At the time, Washington was bitterly divided along partisan lines over the Clinton impeachment. But Brent Scowcroft did not join the melee. Unlike most of his neocon adversaries, he was a military man, a West Point graduate who had become a lieutenant general in the air force and who, over the years, had risen to the top of the national security infrastructure. Understated but resolute, Scowcroft had served as a military assistant to Richard Nixon when Henry Kissinger spotted him as someone who was willing to fight tenaciously for his positions.[9]

On this occasion, he supported the Clinton administration's policy of threatening to bomb Iraq—and carrying out those threats, if necessary. The president was "in a really grim position," Scowcroft said at the time, and had few options except air strikes.[10] "There's no question that Saddam is going to continue to play games," he added. "But we have to remember what our objective is. And our objective is to keep the sanctions on so he can't rebuild his conventional forces and be able to inspect to ensure that he cannot rebuild his weapons of mass destruction."*[11] In other words, he advocated containment more than regime change.

Meanwhile, George W. Bush's key advisers—Karen Hughes, Karl Rove, Don Evans, and Joseph Allbaugh—had started putting together a road map to the White House.[12] His conversations with Prince Bandar and Condoleezza Rice notwithstanding, Bush was a blank slate when it came to international relations. "Is he comfortable with foreign policy? I would say not," said Scowcroft, adding that Bush's only real experience "was being around when his father was in his many different jobs."[13]

Condoleezza Rice, however, did not judge her student particularly harshly. "I think his basic instincts about foreign policy and what needed to be done were there: rebuilding military strength, the importance of free trade, the big countries with uncertain futures," she said.[14] "Our job was to help him fill in the details."

Nevertheless, to compensate for Bush's dearth of expertise, she assembled a team of eight experienced foreign policy advisers to give the candidate a crash course about the rest of the world. They called them-

*On this occasion, Saddam once again relented and allowed inspectors unconditional access to all suspect weapons sites in Iraq.

selves the Vulcans*—Richard Armitage, who had served in a wide variety of roles for George H.W. Bush; Robert Blackwill, former presidential assistant for European and Soviet affairs for the former president; Stephen Hadley, the former U.S. assistant secretary of defense; Richard Perle, another assistant secretary of defense; Condi Rice; Paul Wolfowitz; Dov Zakheim, a former deputy under secretary of defense; and Robert Zoellick, who had been aide to James Baker when he was secretary of state. Their first meeting in Austin was also attended by Dick Cheney and former secretary of state George Shultz.[15]

Now the battle for Bush's favor began. After all, his father, James Baker, and Brent Scowcroft epitomized the policies of moderate realism that the neocons abhorred. With disciples such as Condi Rice, Robert Zoellick, and other alumni of Bush 41's administration on board, some neocons feared that if elected, Bush would merely leave them out in the cold once again. As a result, Bill Kristol, Jeane Kirkpatrick, and James Woolsey all backed Arizona senator John McCain instead.[16] But other neocons, among them Elliott Abrams, Richard Perle, Paul Wolfowitz, held out hope. After all, just because many of these advisers were hand-me-downs from his father's era didn't necessarily mean they shared all his views—or that they couldn't be outfoxed.

By this time, neoconservatism had been around for roughly thirty years and a second generation of neocons had come of age. To understand how a small cadre of ideologues had evolved into the intellectual vanguard for a radical right-wing movement, it is helpful to understand their genealogy. Take William Kristol, for example, the founding editor, with John Podhoretz, in 1994 of the Rupert Murdoch–owned, neoconservative bible *The Weekly Standard* and cofounder, with Robert Kagan, of Project for a New American Century. In addition to his work with the *Weekly Standard* and PNAC, Kristol was a commentator on Fox News, a regular talking head on political chat shows on other networks, a contributor to publications by the American Enterprise Institute, and, later, a speechwriter for George W. Bush. In addition to writing speeches for Ronald Reagan and George H.W. Bush, John Podhoretz was also a contributor to Murdoch-owned properties such as the *New York Post* and Fox News, and, later, to other conservative outlets including a blog run by the conservative *National Review.* As for Kagan, he had worked

*The term was a reference to a prominent statue of Vulcan, the god of fire and metalworking, in Rice's hometown of Birmingham, Alabama.

in the State Department in the Reagan administration, served as foreign policy adviser to former congressman Jack Kemp, been a senior associate at the Carnegie Endowment for International Peace, and wrote for the *Weekly Standard,* the *New Republic,* and the *Washington Post.*

As it happened, all three men came from prominent neoconservative families. Kristol is the son of neocon godfather Irving Kristol and Gertrude Himmelfarb, a conservative cultural critic and professor at the City University of New York. The son of neocon icons Norman Podhoretz and Midge Decter, John Podhoretz counted among his siblings and half siblings Naomi Decter, a public relations executive who wrote for the *Wall Street Journal,* the *New Republic,* and *Commentary;* Rachel Decter, a Washington lawyer who married Elliott Abrams, the neocon Iran-contra convict* who held key foreign policy positions for Ronald Reagan and George W. Bush, in what Norman Podhoretz called "the closest thing to the arranged marriage that the modern world allowed";[17] and Ruthie Podhoretz Blum, a writer for the *Jerusalem Post* and self-styled doyenne of the right-wing Bohemian set in Jerusalem. As for Robert Kagan, his father, Donald, a Yale historian, and his brother, Frederick,† a military historian at West Point and a scholar at the American Enterprise Institute, are both highly visible neocon activist-writers,[18] and, in the Bush-Cheney administration, his neocon wife, Victoria Nuland, served as ambassador to Turkey and ambassador to NATO.

In other words, even though their ideology was obscure at best to the vast majority of Americans, these three families alone promoted the neocon cause at the *Weekly Standard,* the *New York Post,* Fox News, the *Wall Street Journal* editorial pages, *Commentary,* the Coalition for the Free World, the American Enterprise Institute, Harvard, Yale, Cornell, West Point, the Project for a New American Century, the Committee on the Present Danger, the U.S. State Department, the National Security Council—not to mention the White House in no fewer than five administrations beginning in 1981. Yet their influence went much further than even that list suggests because each institution with which they were affiliated was able to nourish scores of other neoconservative protégés. Thus, at *Commentary,* the Podhoretz clan published neocon stars such as Jeane Kirkpatrick, who later became U.N. ambassador;

*Abrams was convicted on two counts of illegally withholding information from Congress.

†Frederick Kagan, along with General Jack Keane, is credited with being one of the intellectual architects of the so-called surge through which President Bush sent more U.S. troops to Iraq in 2007.

Harvard Soviet specialist Richard Pipes, of Team B fame, whose son Daniel became a hard-line Likudnik;[19] Elliott Abrams; Francis Fukuyama; Douglas Feith; Joshua Muravchik; David Wurmser; David Frum;[20] and many, many others. Likewise, the *Weekly Standard* published Fred Barnes, David Brooks, Harvey Mansfield, Charles Krauthammer, Reuel Marc Gerecht, and many more.[21]

This was the neocon infrastructure, a conglomeration of well-funded, heavily ideological institutions that were often obscure to the general public but provided secure oases for scores of right-wing policy intellectuals and one echo chamber after another to reinforce their ideology. "It is a closed loop intellectually, personally, and socially," said one Washington colleague of the neocons who watched their ascendancy. "But it operates very effectively because of all the personal relationships."

All of which made for a cozy, tightly knit affluent intellectual community in the upscale Washington suburbs of McLean, Virginia; Rockville, Maryland; Chevy Chase, Maryland; and the like in which David and Meyrav Wurmser; Michael and Barbara Ledeen; and Richard Perle and his wife, Leslie Barr, all worked together, socialized together, and shared affiliations.* Composed of Ivy League–ish intellectuals from Harvard, Yale, Columbia, University of Chicago, and USC, they were, by and large, cosmopolitan aesthetes who even named their dogs after great artists and writers from Rembrandt (Perle's)[22] to Thurber (Michael Ledeen's Airedale terrier).[23]

A hard-line hawk with a refined palate and a vacation home in Provence (where Jeane Kirkpatrick was a neighbor), Perle had tastes, for example, that ran to Beluga caviar and Monte Cristo cigars, Gauloise cigarettes and bread imported from his favorite Parisian *boulangerie.*[24] Renowned in neocon circles for whipping up sumptuous repasts—his

*Some of those affiliations include: David Wurmser (American Enterprise Institute, The Washington Institute for Near East Policy, "A Clean Break," the *Weekly Standard,* the *Washington Times,* the *Wall Street Journal;* represented by Benador Associates, the neoconservative public relations firm); Meyrav Wurmser (the Hudson Institute, cofounder of the Middle East Media Research Institute, "A Clean Break," *National Review, New York Post,* the *Weekly Standard,* the *Washington Times,* Benador Associates); Michael Ledeen (AEI, founding member of Jewish Institute for National Security Affairs, National Review Online, represented by Benador Associates); Barbara Ledeen (former executive director of neocon Independent Women's Forum, of which Lynne Cheney was a board member); and Richard Perle (AEI, the Hudson Institute, Jewish Institute for National Security Affairs, Project for a New American Century, the *Wall Street Journal,* the *Washington Times,* the *Weekly Standard,* Benador Associates).

specialty is his delicate lemon and grapefruit soufflé[25]—his kitchen, according to *The New Yorker,* is an almost exact replica of the extraordinary professional kitchen he describes in his roman à clef, *Hard Line.**[26]

"The group changes from time to time," said Meyrav Wurmser, the Israeli-American who coauthored "A Clean Break" with her husband, David Wurmser, and his mentor Richard Perle.[27] "But we are all very friendly to each other. The ones in Virginia [Elliott Abrams and his wife, Rachel, and Bill Kristol] are not our immediate circle. But [for the families in Maryland], the kids are in the same schools, we live in the same neighborhoods, and our kids know each other. . . .

"It's like a collection of really brainy people. . . . Everyone in this crowd is smart, but Doug [Feith] is one person whose sheer power of intellect . . . He's incredible. . . . And Richard [Perle] is a great cook. He can cook soufflés, he's just an amazing cook, an incredible human being, really warm, not at all like he is made out to be. . . . And I just love [Michael Ledeen]. He's got the cutest dog, named Thurber."

Meanwhile, in the heart of Texas hill country, the land of barbecue and Dr. Pepper, the next Bush presidency was taking shape. Initially, the media assumed that a new Bush administration would effectively be a continuation of the ancien régime. By and large, it portrayed Bush as a centrist who was only slightly to the right of Bill Clinton and Al Gore—"on the forty-seven yard line in one direction," as the *New Republic* quoted Lanny Davis, a White House counsel for Clinton, while Gore was "on the forty-seven yard line in the other."[28]

In late 1998 and early 1999, however, Paul Wolfowitz and Richard Perle, both of whom had been dyed-in-the-wool Scoop Jackson–Albert Wohlstetter neocons for more than two decades, semi-secretly began

*As described in *Hard Line,* the kitchen is "replete with indoor charcoal grill, restaurant stove, sixty linear feet of counter space, and a *batterie de cuisine* worthy of a good-sized restaurant. Recessed halogen lights played on a huge collection of brightly polished copper pots hanging from a large oval oak rack suspended from the ceiling. A copper marmite gurgled on the stove, the white porcelain tureen into which its contents would be placed by its side."

A review of *Hard Line* in *Library Journal* concluded, "In his first novel, a Reagan-era assistant secretary of defense (ASD) pits Michael Waterman, an ASD suspiciously like his creator, against duplicitous Russians and naive State Department functionaries. . . . Essentially a platform for voicing the author's belief that the tough policies he advocates caused the collapse of the Soviet empire, the novel features excellent prose, tiny plot, and minimal characterization. For Washington insiders and the largest public libraries."

making their pilgrimage to the governor's mansion in Austin. "They were brought in and out under very tight security," said a source who was in Bush's office at the time. "They snuck in, and snuck out. They didn't hold press conferences. Rove didn't want people to know what they were doing or what they were saying."

During these meetings, the neocons finally saw Bush as someone who would not necessarily repeat his father's policies, but as a tabula rasa who would be quite open to theirs. "The first time I met Bush 43, I knew he was different," said Richard Perle.[29] "Two things became clear. One, he didn't know very much. The other was he had the confidence to ask questions that revealed he didn't know very much. Most people are reluctant to say when they don't know something—a word or a term they haven't heard before. Not him."

At the time, Brent Scowcroft, James Baker, and other moderates allied with Bush 41 were not particularly alarmed by the presence of Wolfowitz and other neocons in Austin. "Everyone looked at it in a benign way because Wolfowitz was part of the old DOD crew," said a former State Department official who had worked with Elliott Abrams. "The idea that Wolfowitz and the neocons represented a great ideological shift away from Scowcroft's group of realists was not yet clear." Even though the neocons had coalesced into a formidable ideological force and had made plain their intentions about Iraq on a number of occasions, Scowcroft and Baker assumed the new Bush presidency would be a continuation of the realist school.

Now the education of George W. Bush got under way. The future president had only been to Europe once in his entire life, to visit his daughter in Italy,[30] and when it came to foreign policy, had no intellectual framework whatsoever. "His ignorance of the world cannot be overstated," said the State Department source.

"Then Wolfowitz and Rice started going down to Austin to tutor Bush in foreign policy," the source added. "Bush's grandiose vision emerged out of those tutorials with Rice tutoring him on global history and Wolfowitz laying out his scheme to remake the world.

"The whole view of these people was that the next president was not to be a passive actor, but was to shape the world to U.S. interests. That was the message Rice and Wolfowitz were giving to Bush. Rice was the one giving tutorials on the Westphalian nation-state concept* and the

*Many historians trace the birth of the international system that exists today to the Peace of Westphalia in 1648, which was crucial to establishing the principle of sover-

idea that we were entering some sort of 1947-like transitional period in which the United States could shape the world."

Because he had been her mentor, Scowcroft assumed that Rice would represent the realist point of view in tutoring Bush. But in fact she was far more closely aligned with the neocons than he realized. "She was certainly a fellow traveler," said the State Department official. "She came at it more with a high-level academic approach while the other guys were operational. [Her role] was a surprise to Scowcroft. She had been a protégée and the idea that she was going along with them was very frustrating to him."

Colin Powell, a key figure among Bush 41's moderates, was absent during this period. "That's a critical fact," the official added. "The very peculiar personal relationship between Rice and Bush solidified during these tutorials and Wolfowitz established himself as the intellectual face of the neocons and the whole PNAC crew."

When it came to the most explosive region in the world, Wolfowitz persuaded Bush that a broader approach to the Middle East was necessary than merely focusing on the Israeli-Palestinian conflict. "Wolfowitz had gotten to Bush, and this is where Bush thought he would be seen as a great genius. Wolfowitz convinced him that the solution to the Israeli-Palestinian problem was to leap over this constant conflict and to remake the context in which the conflict was taking place; that democracies don't fight each other. [He convinced Bush] that the fundamental problem was the absence of democracy in the Middle East and that therefore we needed to promote democracy in the Middle East, and out of that there would be a solution to the Israeli-Palestinian conflict."

Marrying American power to a new version of American exceptionalism,[31] Wolfowitz and his colleagues asserted that the United States was a powerful voice for good, and as such was duty-bound to replace brutal Middle East dictators with Western-style democracy that would sweep through the entire region. The road to peace in Jerusalem, Wolfowitz said, ran through Baghdad, Damascus, even Tehran.

It is not clear whether Bush fully grasped the implications of these proposed policies. Envisioning a domino theory that was the reverse of Cold War fantasies that led the United States into Vietnam, the neocons put forth the audacious call for a permanent "neo-war" in the Middle

---

eign states, the right of political self-determination, the principle of equality between states, and the principle of nonintervention of one state in the internal affairs of another.

East, for wars in Iraq, Syria, and even Iran, colossal wars that would sweep through the entire Middle East and affect the entire world.[32]

Meanwhile, Scowcroft and Baker were conspicuously absent from Bush's emerging brain trust.[33] It was widely known that Bush blamed Baker for the failure of his father's campaign to win reelection as president in 1992.[34] But that was the least of it. Baker and Scowcroft were not just his father's best friends and alter egos, they were the "Wise Men" of their era, commanding figures on the world stage who epitomized and implemented multilateral realist policies, who helped Bush senior bring the Cold War to a triumphant end and drive Saddam out of Kuwait.

"Then it started becoming clear to Scowcroft that they were not welcome," said the State Department source. "It was nothing specific, just that they were not being sought out, that they were not invited."

Keeping them out was anything but an oversight. "George W. did it to show his defiance," said another source, a friend of the family. "That did not reflect disrespect for his dad. It was more to have his own identity, to have his own record. He almost had to go out of his way to avoid anyone connected to his father. He constantly faced this challenge of carving out an identity of his own. . . . When he was gearing up to run and the money was flowing in and people were making these showboat trips down to Austin, he told me, 'You're not going to see any Jim Bakers around me when I'm in office.'"

As the spring of 1999 approached, Bush openly began to differentiate himself from his father. "I'll be a different candidate than the previous George Bush who ran for president. . . ." he told a Dallas reporter. "It shouldn't surprise you that there will be new names and new people involved. Second, people working for me will be conservative-minded people."[35] At the time, however, very few people realized Bush was allowing the neocons to push aside his father's realists.

In view of the ongoing problems with Iraq, Scowcroft's absence was particularly significant, especially given his commitment to containment rather than regime change. After all, both he and the elder George Bush felt strongly that history had vindicated their decision not to march to Baghdad. "Trying to eliminate Saddam . . ." they wrote in *A World Transformed*, "would have violated our guideline about not changing objectives in midstream, engaging in 'mission creep,' and would have incurred incalculable human and political costs. . . . We would have been forced to occupy Baghdad, and, in effect, rule Iraq. The coalition would have instantly collapsed, the Arabs deserting it in anger. . . .

"Furthermore, we had been self-consciously trying to set a pattern for handling aggression in the post–Cold War world. Going in and occupying Iraq, thus unilaterally exceeding the United Nations mandate, would have destroyed the precedent of international response to aggression that we hoped to establish. Had we gone the invasion route, the United States could conceivably still be an occupying power in a bitterly hostile land. It would have been a dramatically different—and perhaps barren—outcome."[36]

According to Mickey Herskowitz, who was working on George W.'s campaign autobiography at the time, this decision was one in which the son explicitly disagreed with his father. "He thought of himself as a superior, more modern politician than his father and Jim Baker. He told me, '[My father] could have done anything [during the Gulf War]. He could have invaded Switzerland. If I had that political capital, I would have taken Iraq.' "[37]

Likewise, Bush felt that military success was vital to a triumphant presidency. According to Herskowitz, Bush said, "One of the keys to being seen as a great leader is to be seen as a commander in chief. . . . I'm going to get everything passed that I want to get passed and I'm going to have a successful presidency."[38]

"We talked about it in terms of his dad's decision making during the Gulf War and how he would be different from his dad," said Herskowitz.[39] "It was clear that he felt that the presidents that are remembered and are great are the ones that got us through a war. He liked the idea of being called commander in chief."

By April 1999, key neocons had become convinced that Bush would not follow the same path as his father. "He strikes me as tougher-minded in some ways than his father," said Richard Perle. Perle added that Bush is likely to take the internationalism displayed by Clinton and Gore even further: "There will be sharp differences on almost every foreign and security issue."[40] A few months later, Norman Podhoretz expressed his delight with the neoconservative cast that was emerging among Bush's foreign policy team. "Mr. Wolfowitz has not ruled out using American troops to support Iraqi insurgents in toppling Saddam Hussein," he wrote.[41]

Bush's desire to distinguish himself from his father happened to coincide with the neoconservatives' long-held dreams of remaking the entire Middle East. But neither party fully understood the ramifications of the agenda they were asserting. "He didn't have a clue what he was letting himself in for," said a member of his father's camp. "It was like injecting a cancer into your body. He didn't have a clue that once he gave the neocons a beachhead they would spread. He had no idea what was about to happen."

Meanwhile, the Bush campaign had also begun cultivating the Christian Right as its electoral machine, which was not as easy as it may have seemed. To be sure, Bush was very much an authentic evangelical. And, thanks to his efforts in his father's 1988 presidential campaign, the names of thousands of Christian evangelical leaders were in his Rolodex—and he was in theirs. But his father's dismal relationship with the Christian Right had tarnished his credentials and Bush now faced the challenge of harnessing the evangelical vote without alienating moderate Wall Street Republicans and independents.

For the most part, he went about it secretly. On March 1, 1999, Bush spoke before the Nashville-based Southern Baptists' Ethics and Religious Liberty Commission, which was having its annual meeting in Austin at the Great Hills Baptist Church. Mary O'Grady, a radio news reporter, was the only journalist in attendance—and after she presented her press credentials, she was told that Bush's talk was closed to the media and was asked to leave the premises.[42] "There's no telling what he promised those bozos," O'Grady told a reporter.[43] "I don't think the country-club Republicans here in Texas like his association with those people."

On September 24, 1999, Bush met with a group of right-wing religious leaders called the Madison Project at the Hay-Adams Hotel on Lafayette Square across from the White House. Those in attendance included homeschooling advocate Michael Farris, Tim and Beverly LaHaye, Paul Weyrich, and Peter Marshall, author and "Christian nation" advocate, among others.[44] Well aware that he was an unknown quantity, Bush told them how Jesus came "into his heart" after his 1985 encounter with Billy Graham in Kennebunkport. "He was not a bit ashamed or reticent to [share his testimony]," Peter Marshall said.[45] "That was very encouraging to all of us."

When it came to gay rights, according to Marshall, "[Bush] said to us, 'Rest assured, . . . I would not appoint somebody to a position who was an open homosexual.' At the same time, he said that if he found out that somebody who was already doing a good job was a homosexual, 'I wouldn't necessarily can him because he's a homosexual.' "[46] After the meeting, Michael Farris, the leader of the group, told conservative columnist Cal Thomas, "I think Bush is acceptable. I'll support him if he's the nominee."[47]

About two weeks later, on October 9, 1999, Bush went to San Antonio, Texas, to address the Council for National Policy, whose membership consisted of political operatives, elected officials, military leaders, conservative broadcasters, corporate executives, financiers, and the like, and was effectively a who's who of the New Right.* "Ronald Reagan, both George Bushes, senators and cabinet members—you name it, almost anyone of consequence has been to speak before the Council," said Jerry Falwell, who was a member. "It is a group of four or five hundred of the biggest conservative guns in the country. It is the group that draws the battle lines. It is on the Right what the Council for Foreign Relations is for the Left."[48]

But when Skipp Porteous, national director of the Institute for First Amendment Studies, a watchdog group based in Massachusetts, tried to find out what Bush said, he discovered that extraordinary precautions had been taken to ensure that Bush's speech did not leak out to the press. Eager to find out what Bush had told the Council, Porteous ordered tapes from Skynet Media, the company that recorded the event.[49] But, according to Morton Blackwell, executive director of the CNP, the Bush campaign declined to release them. Ari Fleischer, spokesman for the Bush

*According to a phone book for the Council for National Policy obtained by the author, a partial list of roughly several hundred CNP members in 1996 included Richard V. Allen, former national security adviser under Reagan; Gary Bauer, former Republican presidential candidate and head of the Family Research Council; Morton Blackwell, president of The Leadership Institute; Richard Bott, of the Bott Broadcasting Company; Brent Bozell, chairman of the Media Research Institute; Larry Burkett of the Campus Crusade for Christ and Christian Financial Concepts; Congressman Dan Burton (R-Ind.); Holland Coors and Jeffrey Coors, of the Colorado beer family; Congressman William Dannemeyer (R-Calif.); James Dobson, president of Focus on the Family; Congressman Robert Dornan (R-Calif.); Jerry Falwell, Liberty University; Edwin Feulner Jr., the Heritage Foundation; George Gilder, supply-side economist; Donald Hodel, former secretary of energy and secretary of the interior; Texas billionaire Nelson Bunker Hunt; Reed Irvine, chairman of Accuracy in Media; Bob Jones III, president of Bob Jones University; David Keene, chairman of the American Conservative Union; Congressman Jack Kemp (R-N.Y.); Dr. D. James Kennedy, Coral Ridge Presbyterian Church; Congressman Alan Keyes (R-Md.); Senator Jon Kyl (R-Ariz.); Beverly LaHaye; Tim LaHaye; Marlin Maddoux, president, USA Radio Network; Ed McAteer, president, The Religious Roundtable; former attorney general Ed Meese; conservative activist Grover Norquist; Lieutenant Colonel Oliver North, North American Enterprises; Howard Phillips, chairman, The Conservative Caucus; Ralph Reed, the Christian Coalition; Pat Robertson, Christian Broadcasting Network and Regent University; Phyllis Schlafly, president, Eagle Forum; Richard Viguerie, conservative political strategist; Doug Wead; Paul Weyrich; and Donald Wildmon of the American Family Association.

campaign, declined to characterize the speech. "When we go to meetings that are private, they remain private," he told the *New York Times.*[50]

As a result, while Bush wooed the Christian Right, most of the electorate—at least the secular electorate—was in the dark as to exactly what that meant. One reason, as Stephen Bates pointed out in the *Weekly Standard,* was that over time the Christian Right had mastered "the grammar of secular politics," eschewing fiery rhetoric for less polarizing mainstream language.[51] Fundamentalist political activists working for Pat Robertson had distributed leaflets telling volunteers how to "Rule the World for God" without revealing their real agenda. "Give the impression that you are there to work for the party, not push an ideology," read one flyer. "Hide your strength. Don't flaunt your Christianity."[52]

Robertson had shown he could win millions of votes. But his affinity for supernatural Pentecostal practices such as speaking in tongues and his conspiracy theories about the Illuminati and the Freemasons scared off Republican moderates. In order for a true Christian evangelical to win the White House—not a Jimmy Carter, but a real leader of the Christian Right who shared its beliefs and politics—he would have to have the face and name of someone who was acceptable to Wall Street Republicans and moderates.

Right-wing groups had already begun to take on bland names—Concerned Women for America, Focus on the Family, the Family Research Council, the Council for National Policy, etc.—as if to mask their ideological and theological colors. In place of D. James Kennedy's inflammatory oratory ("It is dominion we are after. World conquest."), fundamentalists said they wanted "a place at the table." Instead of attacking "secular humanism," they assailed "political correctness." Indeed, Ralph Reed's 1996 book, *Politically Incorrect,* sounded more like a Bill Maher production than a book by the director of the Christian Coalition.*

Likewise, Bush and Rove practiced what Lynton Crosby, an Australian colleague of Rove, later called "dog whistle politics"—that is, the art of using coded language and campaign practices that mean one thing to the general electorate but something else entirely to the targeted base. When Bush campaigned with Dallas Cowboys quarterback Roger

*Regnery Publishing, the right-wing publishing house, took it one step further and put out an entire series of Politically Incorrect guides attacking evolution, feminism, global warming, and the like.

Staubach, secularists merely thought he was hobnobbing with a Super Bowl star. Evangelicals, on the other hand, knew he was hanging out with one of the faithful.

To most secularists, Bush's campaign slogan of "compassionate conservativism" evoked a kinder, gentler, more merciful ethos than the divisive slash-and-burn social Darwinism for which some Republicans were known. This was the theme that defined Bush in the way that Bill Clinton was "a man from Hope," in the way that Ronald Reagan promised a new "morning in America."

But when one examined the particulars, there was nothing kind or gentle about compassionate conservativism. In reality, Marvin Olasky, Bush's "compassionate conservatism" guru, based his ideas on the Christian precept of original sin. Born to a Jewish family in Massachusetts, Olasky signed up for the Communist Party in 1972,[53] long after the crimes of Stalin had been exposed. By the mid-seventies, however, Olasky fell under the influence of fundamentalist theologian Francis Schaeffer and did a sharp right turn. Later, when Olasky became a professor at the University of Texas at Austin, he joined the Redeemer Presbyterian Church[54] and became affiliated with explicitly theocratic fundamentalists and Christian Reconstructionists—fundamentalists who seek to impose biblical values on society at large.

In large measure, Olasky's compassionate conservativism was an attempt to do exactly that. "Man is sinful and likely to want something for nothing," Olasky wrote.[55] "Many persons, given the option of working, would choose to sit." As a result, Olasky argued, rather than give them a safety net, the government should "make moral demands on recipients of aid." The poor "need the internal pressure to live honored and useful lives, modeled after our perfect leader, Christ," he wrote.[56] Consequently, he argued, our tax dollars should not go to support their food, housing, or medical needs. It should go to the churches to convert them.

"The word [*compassionate*] sounds innocent enough," asserted Ira Chernus, professor of religion at the University of Colorado.[57] "Behind it, though, is a right-wing religious agenda ready and eager for war without end. Liberals make a big mistake if they dismiss 'compassionate conservatism' as just a hypocritical catchphrase. For the right, it is a serious scheme to give tax dollars to churches through so-called faith-based initiatives."

In other words, George W. Bush's "compassionate conservatism" and "faith-based initiatives" meant that federal tax dollars that might have provided a safety net for the poor would instead subsidize the Christian Right with untold millions of dollars. Bush's dog whistle effectively assured evangelicals that they would have a seat not just at the table, but that they could also pull up a chair in the Oval Office itself. At the time, however, even the most astute, tough-minded, and discerning journalists in the country bought Bush's "compassion" hook, line, and sinker. "If compassion and inclusion are his talismans, education his centerpiece and national unity his promise, we may say a final, welcome goodbye to the wedge issues that have divided Americans by race, ethnicity and religious conviction," wrote E. J. Dionne in the *Washington Post.*[58]

Or, as Dana Milbank, also of the *Post,* put it: "Bush is Clintonian in the best sense: he's got charm, he's a moderate and he's run a state with success."[59]

By seizing control of the language, by framing the debate, Bush, Karl Rove, and Olasky had managed to forge a consensus whereby the entire political spectrum—everyone from hardcore theocrats to liberal secularists—supported policies that would aide the Christian Right.

When it came to Iraq, attentive listeners could sometimes make out the real message. In late 1999, Richard Perle returned from a meeting with Bush giddy with anticipation that his long-held dreams were about to be realized. At an on-the-record breakfast in Washington sponsored by the *Christian Science Monitor* in late November, Perle told a small group of reporters that Bush wanted to give a speech that would say "it's time to finish the job. It's time for Saddam Hussein to go." Perle added that Bush felt that his father's administration underestimated the Iraqi leader's ability to stay powerful, but that that was understandable.[60]

At the time, however, few people seemed to care. After all, this was the height of the Clinton boom. The Dow was approaching the 10,000 level for the first time in history. Thanks to the dotcom bubble, NASDAQ was racing past 4,000—having soared an astounding 48 percent in just one quarter! These were the heady days of dazzlingly rich venture capitalists and IPOs, teenage zillionaires, coked-up day traders, and Humvees. "Synergy" and "burn rate" were buzz words of the dotcom culture, not to mention phrases such as "bricks and mortar" and "clicks and bricks." Foosball tables and Aeron chairs were standard office equipment. America was fat and happy. The bubble had yet to burst.

Not surprisingly, only a handful of reporters in the White House press corps were paying attention to what the next president might do about Iraq. One of them, Ann McFeatters, the Washington bureau chief for the *Pittsburgh Post-Gazette* and the *Toledo Blade,* wrote about it in the November 24, 1999, *Pittsburgh Post-Gazette*: "Texas Gov. George W. Bush is preparing a speech that will say it is time to get Saddam Hussein out of power in Iraq, an implicit admission that his father, former President George Bush, was wrong, according to a key adviser." The article quoted Perle as saying that Bush would soon call for regime change. [61]

Afterward, McFeatters spoke with former president George H.W. Bush about it. "When I talked to [Bush 41] he defended his thinking about not going after Saddam Hussein," she said.[62] "I picked up that the son was more hawkish than his father on this. . . . It wasn't that big a deal at the time."

Finally, about a week later, in early December 1999, at a little-noticed debate among Republican contenders just as the primary season was starting in New Hampshire, Bush went public and vowed to take down Saddam. "No one envisioned him still standing," Bush said.[63] "It's time to finish the task. And if I found that in any way, shape or form that he was developing weapons of mass destruction. I'd take them out. I'm surprised he's still there."

Bush's remarks made it into a few newspapers and several of the Sunday chat shows. But if this was a trial balloon, it was soon punctured. "It was a gaffe-free evening for the rookie front-runner, till he was asked about Saddam's weapons stash," wrote David Nyhan in the *Boston Globe* on December 3.[64] "It remains to be seen if that offhand declaration of war was just Texas talk, a sort of locker room braggadocio, or whether it was Bush's first big clinker."

The next day, Bush backtracked. Thanks to his Texas drawl, the phrase "take 'em out" had been misunderstood, he told the *New York Times.* It referred to the weapons—not Saddam himself. "I'm sorry," he told the paper.[65] "Them. Them. . . . My intent was the weapons—them, not him."

But Republican insiders knew otherwise. According to Colonel Patrick Lang, sometime in the spring of 2000, Stephen Hadley, who later became Bush's national security adviser, told prominent GOP policy makers that Bush's "number-one foreign-policy agenda" would be removing Saddam Hussein from power.[66] Hadley also said that a Bush administration would invest little or no political capital in trying to resolve the Israel-Palestine conflict, which had dominated the Middle East agenda of the Clinton administration.

His designs on Saddam notwithstanding, publicly Bush attacked Al

Gore as if he were the profligate foreign adventurist, laying into his Democratic rival in the first presidential debate on October 3, 2000. "The vice president and I have a disagreement about the use of troops," Bush said. "He believes in nation-building. I would be very careful about using our troops as nation builders."[67]

Eight days later, at the second debate, Bush went at it again. "Yes, we do have an obligation in the world," he said, "but we can't be all things to all people. . . . [Somalia] started off as a humanitarian mission then changed into a nation-building mission, and that's where the mission went wrong. . . . And so I don't think our troops ought to be used for what's called nation-building."[68]

To the tens of millions of voters who had their eyes trained on their televisions, Bush had put forth a moderate foreign policy with regard to the Middle East that was not substantively different from the policy proposed by Al Gore, or, for that matter, from Bill Clinton's. Only a few people might have guessed a far more radical policy was in the works. In the waning weeks of the campaign, Bush's cautionary words against nation-building went out to tens of millions of viewers again and again and again. As the November 7, 2000, election approached, Bush's message was clear: Al Gore would be likely to involve the United States in costly foreign adventures. To the media, the consensus was "Candidates Differ Little on Foreign Policy."[69] As voters went to the polls, Bush's threat to overthrow Saddam wasn't even on the radar screen.

Millions of words have been written about the closest election in the modern history of America's Electoral College—about the bitter, tortuous legal maneuvering; the vicious propaganda war; the raucous mass protests; and the thirty-six days of divisive, convoluted legal battles at the county level, at the state level, and the level of the U.S. Supreme Court. But America's greatest electoral dispute was not merely the legal and political war that the entire country saw. There was also another agenda at play—less visible but perhaps equally important in deciding who would occupy the White House.

When voters awoke on November 8, the outcome was still unresolved. The tally in the Electoral College was so close that Florida's electoral votes would tip the final balance to one candidate or the other. At 6:00 a.m., Bush led Gore by 1,784 out of nearly 6 million votes in Florida—a margin of three-one-hundredths of one percent. Even that infinitesimal lead was in question because of confusingly designed "butterfly" ballots, hanging chads, dimpled chads, and swinging chads—the tiny rectangles punched out from data cards—and scores of other irregularities. Having blithely dismissed James Baker as someone who would never find a place in his administration, Bush, much to his chagrin, ended up bringing back his father's longtime friend, one of the most astute political operatives of his era, as field general to oversee the ongoing war to win the White House.

But Katherine Harris was Bush's secret weapon. A wealthy Floridian whose maternal grandfather, Ben Hill Griffin Jr., was one of the state's great robber barons, Harris had grown up midway between Tampa and Orlando, in "a godly family," as she put it, that did evangelical missionary work in India, Africa, and various Arab countries. Her brother-in-law, Wes King, was a well-known contemporary Christian singer. But the most profound event in her life was studying under Francis Schaeffer, the influential fundamentalist theologian. "I studied under him at L'Abri," she said. "So it's a faith that is active and real. . . . It's the most important thing in my life."[70]

The petite, wasp-waisted Harris was given to short skirts and tight sweaters, perfectly coiffed, and was heavily made up enough to become the butt of jokes by late-night comedians.[71] A self-described "wannabe Jew" who compared herself to the biblical Queen Esther,*[72] she was elected secretary of state of Florida in 1998 as part of the same slate that made Jeb Bush the new governor.

As secretary of state of Florida, Harris's job normally carried astoundingly little power. But since its duties included oversight of state elections,†[73] suddenly she was in a position to determine who would be the next president of the United States. Harris was not terribly close to Jeb, but by the time of the 2000 presidential race, her political stars were clearly aligned with the ascendant Bush brothers. She and Jeb campaigned for George in the New Hampshire primary, she served as a Bush delegate at the Republican National Convention, and she was one of eight cochairs of his campaign in Florida.[74] As author Jeffrey Toobin puts it in *Too Close to Call: The Thirty-Six-Day Battle to Decide the 2000 Election,* the definitive book on the election controversy, over the next five weeks, Katherine Harris and her office "began acting as a wholly owned subsidiary of the George W. Bush campaign."[75]

*In the Bible, Queen Esther, the wife of the Persian king, also known as Xerxes, is a woman of great courage who saves the Jews by acknowledging her Jewish identity and exposing a plot by Haman to massacre the Jews.

†The office held so little power that in 1998 Florida voters voted to abolish it at the end of Harris's term.

First, there was the matter of the automatic recount. Florida law stipulated that a margin of less than .5 percent of one percent requires all counties in the state to order an automatic recount. In accordance with that law, on Wednesday morning, November 8, the day after the election, Clay Roberts of the Florida Division of Elections issued a memo saying, "Florida law requires an automatic recount of all votes cast."[76]

But Harris's office failed to follow its own instructions. Bush campaign officials repeatedly said the votes had been "counted and recounted," but that wasn't true. The votes in eighteen counties were *never* recounted. "Those counties accounted for 1.58 million votes, more than a quarter of all the votes cast in the election," wrote Toobin. "The significance of this omission can scarcely be overstated."[77]

Meanwhile, the bitter PR battle over the results ensued, with Bush forces successfully characterizing the Democrats as demanding third and fourth tallies of ballots that had already been counted—even though the complete recount never took place. As the battle continued, increasingly time became a foe for the Democrats. That was because of Katherine Harris.

As it happened, Florida election law was not without contradictions, and when it came to what was now a vital issue—the deadline for which recounts could be submitted—the law was explicitly contradictory. According to one section of Florida law, all recounts had to be submitted by 5:00 p.m. on the seventh day after the election—in this case, November 14—after which any additional results "*shall* be ignored." But another section said "such returns *may* be ignored."[78] In other words, according to this section, the deadline was a matter of discretion. Which was it—may or shall? Harris decided to go with "shall." November 14 would be the firm deadline, which effectively ruled out a complete recount. Even though Florida's elections laws called for an automatic recount, Harris's delaying tactics and discretionary decisions made that impossible.

Al Gore's forces appealed, of course. And the Florida Supreme Court overruled Harris's decision. But the Florida court was overruled by the U.S. Supreme Court. Harris had made vital decisions that helped ensure Bush would become president of the United States.

Not surprisingly, the Gore campaign expressed concern about Harris's loyalties to the Bush camp. But Harris also had an agenda that transcended her concerns about her own career or even partisan ties: she believed Christian fundamentalists had to rise up and seize the government back from secularists. "If people aren't involved in helping godly men in getting elected than we're going to have a nation of secular laws," she later told the *Florida Baptist Witness*.[79] "That's not what our founding fathers intended and that's certainly isn't what God intended . . . we need to take back this country. . . . And if we don't get involved as Christians then how could we possibly take this back? . . . If you are not electing Christians, tried and true, under public scrutiny and pressure, if you're not electing Christians then in essence you are going to legislate sin."*

She was not the only one who believed God played a central role in the election. "I think the prayers of a lot of people tipped that election," said Jerry Falwell.[80] "I really believe that with all my heart. And I believe that five to four vote in the Supreme Court, which could have gone the other way, I give God the credit, that he wasn't finished with us yet."

"I think God gave us the man we needed right now," said James Robison, the televangelist in whom Bush sometimes confided. [81]

"That's how both the evangelicals would see it and how [Bush] would see it," said Doug Wead. [82] "But he's smart enough not to say he would see it that way. . . . I'm sure that's how he would see it. That this is an opportunity to do what he thinks is right."

Once George W. became a candidate, his father had put aside his doubts and pulled out all the stops for him. But his father knew his son's weaknesses. Deep in his bones, the elder Bush must have known that Condi Rice's seminars notwithstanding, when it came to foreign policy, his son still knew next to nothing.

Now came the ultimate test as to whether George W. Bush could live up to his father's expectations. By the time Al Gore conceded the presidency on December 13, 2000, Bush—now Bush 43—had already started putting together his new administration. Because some of his appointees had served his father, to the uninitiated it appeared that Americans would be watching reruns of his father's era. But in fact nothing could be further from the truth.

*Katherine Harris was elected to Congress in 2002 but lost the 2006 senatorial race after a corporate campaign donor, a defense contractor named MZM, was tied to the bribery scandal that led to the conviction of its founder, Mitchell Wade, and California congressman Randy "Duke" Cunningham. *Radar* magazine voted her the dumbest member of Congress, noting that more than twenty-five staffers and consultants had voluntarily quit working for her, in part because of "her Stalin-esque management style, which includes attacking staffers for . . . screwing up her Starbucks order (extra-hot low-foam nonfat venti triple lattes with one packet of Sweet 'N Low)."

Bush had won his father's approval by giving up alcohol, bringing in the evangelicals to help win the 1988 election, and working with Karl Rove to put together a powerful, disciplined political machine. Now, thanks to his unfinished psychic agenda, he was intent upon going his father one better. "He had to have his own record," said one friend of the family. "So he almost had to go out of his way to avoid people who had been close to his father and instead almost sought out people who had been adversaries."

Unprepared to wield power, yet imbued with an overwhelming need to feel he embodied transcendental greatness, he was precisely the right kind of cipher to embrace a vision of American exceptionalism shared by both the evangelicals and the neocons. Consequently, the younger George Bush put together a team that was his father's worst nightmare, a team whose utopian dreams challenged the very policies at the heart of his father's legacy. For decades, both the neocons and the evangelicals had waged war against the policies of George H.W. Bush. At last they had both found a new leader—his son.

CHAPTER TWELVE

Grandmaster Cheney

If there was one single inscrutable figure who was vital to shaping the destiny of the new administration, it was Dick Cheney. In previous incarnations, as chief of staff to President Ford, as a congressman, and as secretary of defense, Cheney had projected, as one of his colleagues put it, an image "of quiet, focused competence and comfortable (if not effusive and outgoing) professional collegiality."[1] But there was a lot that was not known about the new vice president—either because he concealed it, or because it had grown within him later in life.

Neither an evangelical nor a neocon, Cheney was a nationalist whose worldview grew out of what had been called "Midwestern American isolationism" before World War II,[2] but which, over time, had evolved into a militaristic unilateralism combined with a belief in an extraordinarily powerful executive branch. In fact, he had always been much farther to the right than most people realized. His restrained demeanor belied his extreme partisanship. This was the young man who saw the Watergate scandal as a brutal assault by Democrats against the presidency and the powers of the executive. This was the White House chief of staff who had joined Donald Rumsfeld in taking on Henry Kissinger, Nelson Rockefeller, and the rest in the Halloween Massacre of 1975. This was the minority whip in the House of Representatives who had mentored Newt Gingrich, the hard-line neoconservative who succeeded him and played a key role in his ascent. Cheney's wife, Lynne, a fellow at the American Enterprise Institute, sat at the elbow of Richard Perle, Michael Ledeen, David Wurmser, and other ideologues who had mastered the darker side of the bureaucratic arts, who constituted the cadre that would help Cheney implement his idea of the imperial presidency.

But no one outside a select few seemed to know it. Even though Cheney had worked closely with Paul Wolfowitz, Scooter Libby, and Zalmay Khalilzad in his Defense Department in the days of Bush 41, the elder George Bush and his fellow realists did not particularly identify Cheney with the neocons. Always a team player who held his cards close to his vest, in the wake of the Gulf War, Cheney staunchly stood by the elder Bush's decision not to take Baghdad and overthrow Saddam. "[Going on to Baghdad] was never part of our objective," Cheney said in an August 1992 speech.[3] "It wasn't what the country signed up for. . . . Once we had rounded him up and gotten rid of [Saddam's] government, then the question is what do you put in its place? You know, you then have accepted the responsibility for governing Iraq."

But that was not what he really thought. According to Victor Gold, a former speechwriter of Bush 41, friend of the Bush family, and coauthor of a novel with Lynne Cheney, privately Cheney had urged the senior Bush to finish off Saddam, even if it meant they had to take Baghdad.[4] Later that year, Cheney also expressed delight with the new neocon strategy as put forth in the "Defense Policy Guidance" that challenged the elder Bush. "You've discovered a new rationale for our role in the world," he told Zalmay Khalilzad, one of the DPG's primary authors.[5]

But while his neocon allies dreamed of "democratizing" the entire Arab world, the vice president–elect had a different motivation. According to a memo later written by a former official who worked in the Pentagon under Cheney and who admired him, the vice president–elect now wanted to "do Iraq" because he thought it could be done quickly and easily, and because "the U.S. could do it essentially alone . . . and that an uncomplicated, total victory would set the stage for a landslide reelection in 2004 and decades of Republican Party domination." The memo, which was written in the summer of 2007 to advise interested parties about Bush administration strategy in the region, added that from Cheney's point of view, taking down Saddam "would 'finish' the undone work of the first Gulf War and settle scores once and for all with a cast of characters deeply resented by the vice president: George H.W. Bush, Colin Powell, Brent Scowcroft, and Jim Baker."[6]

Of course, as Bush 43 and Cheney prepared to take office in late 2000, Bush 41's Old Guard was mostly offstage—with one conspicuous exception, Colin Powell. During the buildup to the war in 1990, Cheney, who had avoided military service with five Vietnam War deferments, had been a gung-ho pro-war advocate, while Powell, the widely admired chairman of the Joint Chiefs of Staff, a four-star general, and a lifelong soldier, had been against it. Nothing rankled Cheney more than seeing Powell emerge as the hero of a war he had initially opposed. The press had swooned over him, as did Maureen Dowd in her *New York Times* column when Powell bowed out of presidential politics in 1995: "The graceful, hard male animal who did nothing overtly to dominate us yet dominated us completely, in the exact way we wanted that to happen at this moment, like a fine leopard on the veld, was gone."[7]

Meanwhile, Cheney was lost in the shadows. At the very end of Bush 41's administration, he pointedly left the Pentagon without even saying thank you or good-bye to Powell.[8] "I was disappointed, even hurt, but not surprised [at Cheney's silent departure]," Powell wrote in his memoir, *My American Journey.* "The lone cowboy had gone off into the sunset without even a last 'So long.' "

Now, eight years later, Cheney, who had secretly nursed presidential ambitions of his own, entered the new administration, fueled, according to the former Pentagon official, by an "anger-and-resentment-driven desire to settle scores."[9]

One can only speculate, but it is worth noting nonetheless, that Cheney's saturnine disposition may have been affected by his long-standing medical problems. Indeed, on November 22, two weeks after the election but well before the ensuing electorial controversy was resolved, Cheney suffered his fourth heart attack. According to the *New York Times,* doctors said the new heart problems were not serious enough to prevent Cheney from serving as vice president or president.[10] Cheney, who had lived with heart disease for much of his adult life, said he didn't think about it much.[11]

Nevertheless, as Dick Cheney reentered the executive branch, two things were clear: the stakes were high, and there was not much time to lose. "What seems not to have been noticed at critical junctures in Dick Cheney's political life," the former official wrote, "was just how deeply he was personally affected by events swirling around him: the seeming diminution of presidential powers in the wake of the Watergate, Vietnam and CIA scandals; and the humiliation felt by an outwardly modest but inwardly ambitious man when overshadowed by a charismatic, outgoing, articulate and widely admired chairman of the joint chiefs of staff (Colin Powell) during the first Gulf War; a war which (to Mr. Cheney's disgust) was initially opposed by the man who reaped the lion's share of credit for its ultimate success. The quiet, seemingly self-effacing Mr. Cheney would forget none of this."[12]

Much as he loathed Powell, Cheney realized that the immensely popular general—the most trusted man in America—was essential to the political perception of the administration's foreign policy decisions. As former speaker of the house Newt Gingrich put it, "If you're George Bush, and the biggest weakness you have is foreign policy, and you can have Cheney on one flank and Powell on the other, it virtually eliminated the competence issue."[13]

As a result, on December 16, 2000, three days after Al Gore conceded defeat, Colin Powell was flown to Bush's ranch in Crawford, Texas, where the president-elect announced his first cabinet appointment: Colin Powell as secretary of state. "He is a tower of strength and common sense," said Bush. "You find somebody like that, you have to hang on to them. I have found such a man."[14]

Tears filled Bush's eyes. "I so admire Colin Powell," he later explained. "I love his story."[15]

Unlike other designated cabinet appointees, Powell had not been vetted by Cheney or other campaign officials. Nor, according to *Soldier: The Life of Colin Powell*, Karen DeYoung's comprehensive biography of him, was Powell even asked any serious foreign policy questions.[16] Such discussions were not necessary. As the former Pentagon official put it, "Cheney's distrust and dislike for Mr. Powell were unbounded."[17] In other words, Powell was only there for show. Cheney immediately took measures to undermine him. The chess game began.

At the Crawford press conference on December 16, Powell was dazzling—too dazzling for his own good. As he proceeded with his lengthy discourse about the state of the world, Bush's admiring expression gradually turned to one of sour irritation. Afterward, Richard Armitage, Powell's close friend and longtime colleague, told the secretary of state–designate that he had been so comfortable in front of the cameras compared to the president-elect, that it was somewhat disturbing. "It's about domination," Armitage advised Powell. "Be careful in appearances with the president."[18]

Armitage wasn't the only one to notice. "Powell seemed to dominate the President-elect . . . both physically and in the confidence he projected," reported the *Washington Post*.[19] *New York Times* foreign affairs columnist Thomas Friedman concluded that Powell "so towered over the president-elect, who let him answer every question on foreign policy, that it was impossible to imagine Mr. Bush ever challenging or overruling Mr. Powell on any issue."[20]

None of this was lost on Cheney. Initially, Bush and he had decided that the new secretary of defense would be former Indiana senator Dan Coats, a Christian fundamentalist on the Senate Armed Services Committee who had won over the Christian Right thanks to his undiluted antipathy toward gays in the military. But now it was abundantly clear to Cheney that Coats would be no match for Powell.[21] When Coats added that he did not consider missile defense an urgent priority, Bush and Cheney dumped him immediately.

Meanwhile, Bush proceeded to pick other key cabinet officials. On December 22, he announced that his attorney general would be John Ashcroft, who had just been defeated in a bid for reelection as senator from Missouri. Ashcroft, who had preached at Jerry Falwell's Thomas Road Baptist Church, was a member of the Assemblies of God church, the denomination of Jimmy Swaggart, Jim and Tammy Faye Bakker, and Elvis Presley, which was known for charismatic practices such as faith healing and speaking in tongues.*[22]

As secretary of commerce, Bush picked Don Evans, an evangelical oil man friend from Texas who had introduced Bush to the Community Bible Studies program in Midland.[23] As chief White House speechwriter, Bush picked Michael Gerson, a graduate of Wheaton College, the so-called Harvard of evangelical colleges.†[24] These were the very people whom Neil Bush had scorned as "cockroaches" issuing "from the baseboards of the Bible-belt,"[25] and whom Bush 41 had derided as the "extra-chromosome set."[26]

As the cabinet began to take shape in late December, Colin Powell still presented the biggest potential obstacle to the ambitions of Cheney and the neocons. There was less than a month before the inauguration. Time was running out. They had to find a way to neutralize him.

According to the former Pentagon official, Cheney was convinced that even though Powell's presence was essential to the Bush administration, he "would have to be cornered bureaucratically and repeatedly reminded (even in ways involving public humiliation) that foreign policy was not something over which he presided." To accomplish that task, the official continued, Cheney "recruited Donald Rumsfeld and the neoconservatives to hammer Secretary of State Powell bureaucratically

*Ashcroft reportedly had his father anoint him with Crisco cooking oil before he took office, as a result of which critics sometimes referred to him as "the Crisco Kid." Ashcroft, who doesn't smoke, drink, dance, swear, or gamble, reportedly balked at buying a raffle ticket for a Rush Limbaugh book at one right-wing fund-raiser, reasoning that to do so constituted gambling.

†In February 2003, Wheaton bowed to pressures to liberalize and lifted its 143-year ban on student dancing. Well-known alumni include Billy Graham, former speaker of the house Dennis Hastert, and former senator Dan Coats.

while Mr. Cheney took upon himself the task of managing the President of the United States."[27]

On December 28, Donald Rumsfeld met Bush in his temporary headquarters in the Madison Hotel in Washington. To Washington cognoscenti, to Bush insiders, the idea that Rumsfeld might be invited to join a Bush administration was stunning. Rumsfeld's enmity with Bush 41 included attempts to keep Bush off the Republican ticket in 1976 and 1980 and the Team B battle with Bush's CIA. Rumsfeld openly made fun of Bush at Chicago dinner parties.[28] And when Bob Dole challenged Bush 41 for the presidential nomination in 1988, Rumsfeld had been on Dole's team. At the time, George W. Bush was the enforcer on his father's campaign. "Without question, [George W.] would have known about his father's problems with Rumsfeld," said Pete Teeley, former press secretary to Bush 41.[29] "Everybody knew."

"Real bitterness there," said another friend of Bush 41.[30] "Makes you wonder what was going through Bush 43's mind when he made him secretary of defense."

James Baker even interceded. According to Robert Draper's *Dead Certain*, he told the president-elect, "All I'm going to say is, you know what he did to your daddy." But Bush didn't listen. After all, Rumsfeld's success came from being a great courtier. Fourteen years older than his patron, vastly more experienced, Rumsfeld reportedly played to Bush's insecurity about his lack of experience, and reassured him that he was fit for command.[31] That reassurance became crucial to their relationship over the next six years.

Rumsfeld's relationship with Cheney had cooled somewhat since he and his protégé had been in the Ford White House. In 1986, Rumsfeld had made a futile stab at getting the 1988 Republican presidential nomination, and had pleaded with Cheney, unsuccessfully, for his support.[32] When George H.W. Bush won the presidency, Cheney ultimately became secretary of defense but Rumsfeld was left out in the cold.

Now that they were reunited, Cheney had a more powerful role in their partnership than before. In contrast to President-elect Bush, who had little knowledge of Washington, the two men had an unsurpassed mastery of the intricacies of the federal bureaucracy, thanks to three decades of shared experience at the highest levels of the executive branch. They knew the White House, the Pentagon, and Congress—inside and out. They knew how to make these institutions turn on a dime, when to accelerate and when to put on the brakes. Less neocon ideologues than authoritarian nationalists, they believed in an executive branch so powerful—"the imperial presidency," "the unitary executive"—that the constitutionally mandated system of checks and balances was all but negated. It was a philosophy that many neocons shared.*[33]

But in order to realize his ambitions, Cheney knew his team needed control of the entire national security apparatus. By this time, Paul Wolfowitz, a Cheney hand whose name had been widely bandied about as a potential secretary of defense, was now being touted as a possible pick to replace George Tenet as the next CIA director. If that happened, Cheney would have an ideal team in place.

Then dean of the School of Advanced International Studies (SAIS) at Johns Hopkins University—a position he had held for seven years—Wolfowitz, always intent upon proving he was the smartest guy in the room, had a cerebral style that didn't mix particularly well with Bush's frat-boy disposition. In Dick Cheney, however, he had a patron who was the most powerful voice in the new administration next to the president himself. And, during his trips to Austin, Wolfowitz had played a key role in formulating an intellectual framework through which the president-elect could craft foreign policy.

There was another problem, however, that threatened Wolfowitz's position in the new administration. His marriage was on the rocks. Worse, according to an article in the *Daily Mail* (London) by Sharon Churcher and Annette Witheridge, Wolfowitz was allegedly having an affair with a staffer at the School of Advanced International Studies.[34] Clare Wolfowitz, his wife of more than thirty years and mother of his three children, was said to be so angry that she was taking actions that might jeopardize his career.

The episode at SAIS was not the only alleged indiscretion reported about Wolfowitz. The fifty-seven-year-old Pentagon veteran had also become smitten with Shaha Ali Riza, a secular Muslim then in her forties, who had made her way through Washington's neocon network while working at the Free Iraq Foundation, a group that supported the overthrow of Saddam Hussein in the early 1990s, and the National Endowment for Democracy, a congressionally funded foundation that makes grants to promote democracy throughout the world. Born in Libya and raised in Saudi Arabia, Riza had been educated at the London School of Economics and Oxford, and had obtained British citizen-

*When Nixon resigned in the wake of the scandal, Norman Podhoretz, taking a position quite similar to Cheney's, argued that Nixon's 1972 victory had been "nullified" by a "coup d'état" by "the new leftist liberalism."

ship. According to the London *Sunday Times,* Riza shared "Wolfowitz's passion for spreading democracy in the Arab world" and "is said to have reinforced his determination to remove Saddam Hussein's oppressive regime."[35]

According to a former State Department official, Wolfowitz was quite taken with the notion that he, a secular Jew, was dating a Muslim. Their relationship put a heady, modern, and romantic face on the entire neocon project of democratizing the Middle East. As the Bush-Cheney team prepared to take office, Wolfowitz and Riza, not his wife Clare, took in the neocon social circuit together. Riza was known to Cheney. She moved in the same circles with and was admired by Ahmed Chalabi, the Iraqi exile Wolfowitz backed as a successor to Saddam.[36] "Shaha was the embodiment of the outcome of the modern Arab political system as the neocons saw it," said the State Department source. "She was the personification of the outcome they hoped for in Iraq. She was not theoretical. She was not in a burka. She was a modern Arab feminist."

Wolfowitz's critics who knew about the affair delighted in referring to Shaha Riza as "his neoconcubine." But more significant than the prurient aspects of his alleged dalliances were the questions of national security they might raise. After all, federal officials have been denied national security clearances not because of extramarital activities but because of the possibility of blackmail stemming from their nondisclosure. And if one of the women in question was a foreign national—as was Shaha Ali Riza—that raised additional serious issues about security clearances.

What hung in the balance was not merely the marriage of Paul and Clare Wolfowitz—or the sales of British tabloid newspapers. Nor was it just whether or not Paul Wolfowitz would reach the apex of his career by becoming director of the CIA. Unwittingly, Clare Wolfowitz may have put at risk Dick Cheney's dreams of the entire neocon project to remake the Middle East. After all, if Cheney, Rumsfeld, and the neocons were to outflank centrists such as Colin Powell, it was essential that they control America's intelligence apparatus. As Cheney saw it, Wolfowitz was just the man for the job. Cheney was getting all his ducks in a row—or at least trying to.

Meanwhile, just as Wolfowitz's name was being bandied about for the top job at Langley, George Tenet, the Clinton appointee who still served as CIA director, got called to a private meeting with President-elect Bush. Tenet had hoped to make it at least partway through the next administration, but the papers had been full of speculation about who might succeed him. "I guess this is the end," Tenet told a colleague as he went to meet the next president.[37]

When Tenet returned, however, he was pleasantly surprised. "[Bush] wants me to stay until he can find someone better," he said. It was not until six years later that *The Nelson Report,* a highly regarded newsletter for Washington foreign policy insiders, finally reported why Tenet had not been replaced by Wolfowitz. "A certain Ms. Riza was even then Wolfowitz's true love," the newsletter said. "The problem for the CIA wasn't just that she was a foreign national, although that was and is today an issue for anyone interested in CIA employment. The problem was that Wolfowitz was married to someone else, and that someone was really angry about it, and she found a way to bring her complaint directly to the President.

"So when we, with our characteristic innocence, put Wolfowitz on our short-list for CIA, we were instantly told, by a very, very, very senior Republican foreign policy operative, 'I don't think so.' It was then gently explained why, purely on background, of course."[38]

More specifically, the *Daily Mail,** citing a Bush administration source, reported that Clare Wolfowitz was so incensed by her husband's sexual behavior that she wrote Bush a letter suggesting that because of his infidelity her husband posed a potential national security risk.[†39] According to a memo by the former State Department official on the Washington Note website, Clare's letter "detailed her husband's extramarital affairs at SAIS and with Shaha Ali Riza. . . . Clare pointed out that her husband had a sexual relationship with a non-American citizen and that he was seeking to keep these relationships 'non-disclosed.'"[40]

Wolfowitz was now damaged goods. If Cheney and the neocons were to have control over the national security apparatus, it would not come from the CIA.

At about the same time Wolfowitz was jockeying for the CIA job, more than four thousand miles away, in Italy, a seemingly trivial and unrelated event took place. An obscure office in Rome was burglarized. Whatever

*In addition to the *Daily Mail,* reports about Clare Wolfowitz's alleged letter to Bush have appeared on several websites including *Salon,* the *Village Voice* blog, *The Washington Note,* and on worldbankpresident.com.

†Asked by the *Mail* if she had written such a letter to Bush, Clare Wolfowitz responded, "That's very interesting but not something I can tell you about."

its mysterious origins, the crime ultimately played a key role in allowing Cheney, Rumsfeld, and the neocons to implement the policies about which they had dreamed.

The break-in occurred near the Piazza Mazzini, the hub of a commercial and residential neighborhood between the Vatican and the Mussolini-era Olympic Stadium. About two blocks away stood a graffiti-scarred ten-story apartment building, Number 10 via Antonio Baiamonte. Home to scores of middle-class families and a handful of office workers, the building also housed the embassy for the Republic of Niger, the impoverished West African nation that was once a French colony and that had very little going for it economically—except for the fact that it has the world's largest deposits of uranium.

On January 2, 2001, eighteen days before Bush's inauguration, an embassy official returned there after New Year's Day and discovered that his offices had been burglarized. Evidence suggested that the thieves may have rummaged through desks and files. But little of value appeared to be missing—a wristwatch, perfume,[41] worthless bureaucratic documents, embassy stationery, and some official stamps bearing the seal of the Republic of Niger.

As unprepossessing as it is, the Niger embassy in Rome had just become the site of one of the great mysteries of our times. The consequences of the robbery would be so great that the 1972 Watergate break-in pales by comparison. After the break-in took place, information from documents that had been stolen and, later, the documents themselves made their way to people with ties to Italian military intelligence, SISMI (Servizio per le Informazioni e la Sicurezza Militare), and from there to, at various times, the CIA, other Western intelligence agencies, the U.S. embassy in Rome, the State Department, and the White House, as well as several media outlets. Soon, Western intelligence analysts would begin to hear that Saddam Hussein had sought yellowcake—a concentrated form of uranium that, if enriched, could be used in the manufacture of nuclear weapons—from Niger.

In Washington, however, no one was paying attention to a petty robbery in Italy. Bush had gone back to Crawford, Texas, with his family for Christmas, and the Cheneys to Jackson Hole, Wyoming,[42] while the rest of the Republicans began frantically preparing for the inauguration. The thirty-six-day election controversy had radically shortened the time available to organize four days and $35 million worth of festivities—not to mention getting drug companies, defense contractors, Major League Baseball, big oil, and big tobacco to pony up six-figure sums to finance it.[43]

Meanwhile, the new administration continued to staff up. With Wolfowitz still languishing in political purgatory, his protégé I. Lewis "Scooter" Libby signed on as assistant to the president, chief of staff to the vice president, and national security affairs adviser to the vice president. Heading the Pentagon transition team was Zalmay Khalilzad, who opened the doors to the disciples of Albert Wohlstetter and Scoop Jackson, members of and advisers to Team B, and various proponents of the radical policy papers calling for regime change in Iraq. Four of them—Elliott Abrams, Douglas Feith, Richard Perle, and Abram Shulsky, all Wolfowitz's colleagues from the Scoop Jackson days—waited in the wings for their positions in the administration.

Neocon firebrand Michael Ledeen met with Karl Rove. "He said, 'Anytime you have a good idea, tell me,'" said Ledeen, recalling how he initiated a relationship in which every month or so he faxed in ideas that on occasion became official policy or rhetoric.[44]

But to Cheney, Wolfowitz was still the key to making certain that the new administration would implement his policies, that the bureaucracy of the executive branch would be perfectly staffed so that he could manipulate it at will. Colin Powell had been lobbying for Richard Armitage to get the powerful post of deputy secretary of defense under Rumsfeld. If Armitage got the job, that would represent a huge victory for the centrists. The *Weekly Standard* alerted its troops with an article entitled "The Long Arm of Colin Powell: Will the Next Secretary of State Also Run the Pentagon?"

Fortunately for Cheney, Clare Wolfowitz's letter had not succeeded in ruining her husband's career. In fact, according to the State Department source, the letter that terminated Wolfowitz's appointment never even got to Bush because it was intercepted by Scooter Libby. Once Wolfowitz found out about the letter, the former official said, he "unleashed his lawyers on his wife and forced her to sign a non-disclosure agreement or forgo financial support. . . . Clare Wolfowitz signed."[45]

Just a few days later, on January 11, Donald Rumsfeld sailed through confirmation hearings on the way to becoming secretary of defense. At the same time, Rumsfeld announced that the number two job in the department, deputy secretary of defense, would go to Paul Wolfowitz. The *Washington Post* characterized the pick as "another victory for Vice

President–elect Cheney over Secretary of State–designate Colin L. Powell."[46] Only a handful of people knew about the letter.

But there was a still a problem. If Wolfowitz's personal issues disqualified him for the CIA, wouldn't they also be a problem at the Pentagon? "At this point, the White House was fully cognizant of Wolfowitz's personal habits and chose to cover up his activities," said the State Department source, adding that Rumsfeld was fully informed. "Rumsfeld told Wolfowitz to keep it zipped. He didn't want any problems. He was basically going to run the show and Wolfowitz could come on those terms."

Wolfowitz accepted. Ultimately, part of his job would entail overseeing the creation of a new and powerful operation called the Office of Special Plans, a unit designed to supply senior administration officials with raw intelligence about Iraq—in the process, making an untouched end run around the CIA.

Within a matter of days, word reached Richard Clarke, the White House counterterrorism chief, that Rumsfeld and Wolfowitz were planning for war with Iraq.[47] In certain respects, their actions were a replay of the 1976 Team B experiment, with one very important difference. This time it wasn't just a bunch of feverish ideologues presenting a theoretical challenge to the CIA. This time Team B controlled the entire executive branch of the United States.

Inauguration Day, January 20, 2001, arrived on a dreary, bone-chilling Saturday in Washington. The nation's capital did not feel overly warm toward Bush and Cheney: Its citizens had devoted only 9 percent of their vote to the Republican ticket.[48] There was considerable affection for Bill Clinton, who was frantically packing his bags, leaving the White House with an astoundingly high 68 percent approval[49] in the wake of a bitterly partisan impeachment. And, of course, millions of people felt that the November election had been stolen, plain and simple. Protesters arrived by the thousands. Their signs read "Hail to the thief."

On the other hand, among the 750,000 people who had come to Washington to celebrate the inaugural festivities, many felt they had finally found deliverance in the appearance of their "savior," as Miriam Cajiga, sixty-five, of Miami, an attendee at the Florida ball, referred to Bush. "After this, I can die happy," Cajiga said. "We have saved the world, not just the country, but the world with this president."[50]

The inaugural parade featured a special float from Wyoming for Dick Cheney, an eighty-five-foot wheeled extravaganza that included a mural of the Grand Tetons, a 162-year-old Conestoga wagon pulled by two fake horses, a stuffed buffalo, a real live, flesh-and-blood fly fisherman wearing hip boots and standing in a pool of water with a fake fish at the end of his pole, two bighorn sheep with rotating heads, and a replica of the state's official symbol, the bucking bronco.[51]

As with most inaugurals, society columnists delighted in predicting which celebs would be "in" with the new administration—Charlton Heston and Bo Derek—and who would be "out"—Barbra Streisand and Bono.[52] Whereas Bill Clinton's 1993 inaugural had featured Kim Basinger, Alec Baldwin, Maya Angelou, Aretha Franklin, Chuck Berry, Ray Charles, and Jack Nicholson, Bush and Cheney brought in Wayne Newton aka "Mr. Las Vegas," Kelsey Grammer, and Andrew Lloyd Webber.[53]

There were eight or nine inaugural balls, depending on how you counted them, and they were notable for their profusion of cowboy boots, fur coats, perfume, makeup—and oil. At the Florida ball, Katherine Harris, the Florida secretary of state who had played such a key role in determining the outcome of the election, stole the show, in her low-cut black, pleated taffeta gown, when country singer Larry Gatlin introduced her as combining the best of Joan of Arc, Rosa Parks, and Mother Teresa.[54]

The latter allusion was by no means the only religious reference of the day. To give the benediction at the Fifty-fourth Inaugural Prayer Service at the National Cathedral, Bush chose Jack Hayford, a California Charismatic who was involved in the Promise Keepers, the men's revival group, and was a supporter of Christian Reconstruction, or Dominionism. Another Dominionist believer, Anthony T. Evans of Dallas, served as the speaker at the preinaugural Washington Prayer Luncheon. According to Kevin Phillips's *American Theocracy: The Peril and Politics of Radical Religion, Oil, and Borrowed Money in the 21st Century,* Dominionist theology calls for "seizure of earthly power by 'the people of God' as the only way by which the world could be rescued. . . . A president convinced that God was speaking to him . . . might through Dominionism start to view himself as an agent called by the Almighty to restore the earth to Godly control."

The swearing-in ceremony began at the Capitol at 8:30 a.m., with the United States Marine Corps Band playing "Hail to the Chief." Reverend Franklin Graham, substituting for his father, Billy Graham, who was ill, gave the invocation. Dick Cheney was sworn in first, then Bush.

Then came the Inaugural Address. Crafted in part by chief White House speechwriter Michael Gerson, Bush's speech cast Americans in biblical terms as sinners seeking redemption—a "flawed and fallible people, united across the generations by grand and enduring ideals"—and was full of biblical references. "I can pledge our nation to a goal: When we see that wounded traveler on the road to Jericho, we will not pass to the other side," Bush said, referring to Jesus's parable of the Good Samaritan.[55]

Then Bush turned to the global horrors the country might face. "We will confront weapons of mass destruction," he said, "so that a new century is spared new horrors. The enemies of liberty and our country should make no mistake."

On occasion, Bush had privately said that he felt he was an instrument of God; now, citing a letter from the Virginia statesman John Page to Thomas Jefferson, Bush told the nation: "We know the race is not to the swift nor the battle to the strong. Do you not think an angel rides in the whirlwind and directs this storm?"[56]

Cheney, the man who had already filled many key executive branch positions, stood by as Bush spoke. No doubt he understood that Bush believed their new policies would have the force of God behind them. "We are not this story's author, who fills time and eternity with His purpose," Bush continued. "Yet His purpose is achieved in our duty, and our duty is fulfilled in service to one another."[57]

Bush's reference was to a verse from the Bible's book of Nahum,[58] the great ethical lesson of which is that God is not only a just God, but a "furious" and wrathful God who "hath his way in the storms," who "will not at all acquit the wicked."[59] The passage asks, "Who can stand before his indignation? And who can abide in the fierceness of his anger?[60]

Such would be the guiding force behind the government of the United States of America, its new president was saying to the assembled listeners and millions on television. "This work continues," George W. Bush finished. "This story goes on. And an angel still rides in the whirlwind and directs this storm."[61]

CHAPTER THIRTEEN

Cheney's Gambit

Less than eight months into George W. Bush's presidency came the signal event of the early twenty-first century, the September 11 attacks on America. For the most part, what has been written about that period has understandably been framed in the context of 9/11, specifically in terms of what the Bush administration was told about the threat of Al-Qaeda and what it did or did not do in response. Such questions, of course, will continue to bear scrutiny in the years to come. But at the same time, from January 2001 to September 2001, other unseen agendas were under way. Ultimately, they would have an impact on history even greater than 9/11.

Although much of Bush's electoral support had come from fundamentalist Christians, the media didn't pay much attention to the religious imagery in his inaugural address. As usual, secular America was deaf to the language of the religious right. But when the new president spoke directly to his religious brethren, the message came through loud and clear. Rabbi Daniel Lapin, a politically conservative Orthodox rabbi met with Bush not long after he became president,* and saluted him for see-

*Lapin was close friends with House Majority Leader Tom DeLay and lobbyist Jack Abramoff, who was convicted on three felony counts of defrauding American Indian tribes and corruption of public officials. Lapin became embroiled in the Abramoff scandal in 2005 when Senate hearings revealed e-mails between him and Abramoff in which Lapin was asked to provide phony awards to Abramoff to help him gain membership to Washington's exclusive Cosmos Club. According to the *Washington Post,* Abramoff wrote Lapin: "I hate to ask your help with something so silly, but I have been nominated for membership in the Cosmos Club." Abramoff observed that the club has "Nobel Prize winners, etc. Problem for me is that most prospective members have received awards and I have received none. I was wondering if you thought it possible that I could put that I have received an award from Toward Tradition [an institution run by Lapin] with a sufficiently academic title, perhaps something like Scholar of Talmudic Studies? . . . Indeed, it would be even better

ing "that the lens that has played a dominant role in nearly 300 years of American history is the lens of Old Testament Christianity that the Founders brought to this country." Bush, he added, was "a fitting heir to that tradition."[1]

When Rabbi Lapin referred to the Founding Fathers, of course, he meant Puritans such as Cotton Mather and not Thomas Jefferson, Thomas Paine, and other men of the Enlightenment. In other words, insofar as America had always been a nation caught between modernism and fundamentalism, the election of George W. Bush was a historic turning point. "President Bush clearly believes that God is involved in American history and that America has a divinely ordained mission," Lapin said. "In that belief he is no different from the Founders who actually saw themselves replaying the [role of the] Israelites crossing the Red Sea to find religious freedom in the Promised Land."[2]

The secularism of Bill Clinton and Al Gore had given way to the Bible-based presidency of George W. Bush. The new president of the United States saw himself as an instrument of God, his path as divinely ordained. The absolute nature of his faith meant that the values of secular democracy would take a backseat to his Christian beliefs.[3] Ironically, in terms of the dark metaphors of the millennium—the End of Days, Armageddon—Bush may have been exactly the right man for the times.

As a new president, he needed to staff not just the White House but many key positions in other departments, and the new administration soon had a religious cast. From the Kennedy era to the Clinton administration, the White House had been the redoubt of graduates from Ivy League universities and other top-tier schools. But now, Bush's Harvard and Yale degrees notwithstanding, such credentials were black marks

---

if it were possible that I received these in years past, if you know what I mean. Anyway, I think you see what I am trying to finagle here!"

Lapin reportedly talked to Abramoff by phone and then responded by e-mail: "I just need to know what needs to be produced . . . letters? plaques? Neither?" Abramoff replied: "Probably just a few clever titles of awards, dates and that's it. As long as you are the person to verify them [or we can have someone else verify one and you the other], we should be set. Do you have any creative titles, or should I dip into my bag of tricks?"

Abramoff's official biography subsequently listed two 1999 awards from Toward Tradition and another institution run by Lapin, the Cascadia Business Institute. Lapin subsequently denied having given Abramoff the awards and said he assumed the e-mails were a joke. His official statement said, "Anyone familiar with Abramoff's jocular and often fatally irreverent e-mail style won't be surprised that I assumed the question to be a joke. . . . I regret the exchange."

against prospective applicants. Instead, White House interns came from the likes of Patrick Henry College, a college for evangelicals founded in 2000 in Purcellville, Virginia, the vast majority of whose students had been homeschooled by their fundamentalist parents to avoid exposure to the evils of secularism.[4] John Ashcroft's Justice Department was flooded with young evangelical lawyers who had graduated from Pat Robertson's Regent University (formerly CBN University) Law School.*[5]

Such overt religiosity was not confined to the lower levels of the administration. "Cabinet sessions began with prayers," said Bush speechwriter David Frum. "There were regular Bible study programs for those who wanted to attend. . . . Sometimes in the lunch room, you would see people before they eat bow their heads and say a few words silently to themselves."[6]

The Council for National Policy, which had been cofounded by Tim LaHaye, now had regular access to the Oval Office. "Within the council is a smaller group called the Arlington Group," said Jerry Falwell.[7] "We often call the White House and talk to Karl Rove while we are meeting. Everyone takes our calls."

Bush regularly asked people to pray for him. Early on in the administration, Charles Colson, the convicted Watergate felon who became a born-again Christian, met Bush in a receiving line and said he was praying for him. "He stopped me and put both hands on me," said Colson, "and he said, 'Nothing means more to me. I am sustained in this place by the prayers of people.' It's the real deal with George Bush."[8]

The Christian Right was not alone in having a special relationship with the president. Not long after the inauguration, according to the *Washington Post,* Dan Quayle, who had been vice president under George H.W. Bush, paid Dick Cheney a visit.

"Dick, you know, you're going to be doing a lot of this international

*Monica Goodling, a Justice Department official who was a liaison to the White House and became a key figure in the 2007 Attorneygate scandal, was one of 150 graduates from Regent hired by the Justice Department.

According to Frederick Clarkson, the author of *Eternal Hostility: The Struggle Between Theocracy and Democracy,* the school changed its name from Christian Broadcasting Network University to Regent because it better reflected the purpose of the institution: "Robertson explained that a 'regent' is one who governs in the absence of a sovereign and that Regent U. trains students to rule, until Jesus, the absent sovereign, returns. Robertson says Regent U. is 'a kingdom institution' for grooming 'God's representatives on the face of the earth.'"

traveling, you're going to be doing all this political fund-raising . . . you'll be going to the funerals," Quayle said. "We've all done it."

As Quayle recalled, Cheney got "that little smile," then he replied dryly with a typically cryptic understatement.

"I have a different understanding with the president," Cheney said.[9] That understanding encompassed not just Cheney's vastly superior foreign policy experience but, in effect, the sum and substance of the two men's personalities. Bush approached policy matters in a manner that all but demanded that someone else take charge. He preferred broad strokes. As he put it himself, "I don't do nuance."[10] Once, when Prince Bandar succinctly explained that North Korea was important because the United States had stationed 38,000 American troops across the border, Bush responded, "I wish those assholes would put things just point-blank to me. I get half a book telling me about the history of North Korea."[11]

Another example of Bush's short attention span could be found in how he handled his paperwork. Presidents often began their day by reading the PDB, the Presidential Daily Briefing, prepared every morning by the CIA, a report about twelve pages long. Bush, however, preferred a shorter PDB, seven to ten pages long, that, according to a former senior intelligence official, was written with the understanding that Bush was a "multi-modality learner" who processed information better through questions and answers as he reads.[12]

Given Bush's unfamiliarity with the functions of the White House, Cheney was perfectly positioned to redefine the job of vice president, a role most famously described as "not worth a pitcher of warm spit." He not only enjoyed unparalleled access to Bush but was his gatekeeper. A master of the federal bureaucracy, he was working with a president who had no experience whatsoever with foreign policy or the executive branch of the federal government, a president who worked most naturally, as the *Washington Post* put it, "at the level of broad objectives, broadly declared."[13]

Bush, according to the memo by the former Pentagon official who had once worked with Cheney, "would allow others to define the key 'policy' choices to make without any preliminary framing guidance from him. Then he would 'decide,' in the sense of choosing from a multiple choice menu whose entrees were drafted by others. He was (and is) 'the decider.' " The memo, which was written for a select readership in the private sector, continued, "What Vice President Cheney has managed to do . . . is to become the sole framer of key issues—issues deemed important by Dick Cheney—for President Bush."

Exactly what went on behind the scenes in framing these issues was difficult to say because secrecy was paramount to Cheney. To store the everyday business of the office of the vice president he had installed a man-size Mosler safe, used elsewhere in government to guard highly classified secrets, a large vault made by the same company that manufactures the vaults holding the government's gold in Fort Knox. The names of Cheney's staffers, in many cases, were kept secret. His office location in the West Wing of the White House notwithstanding, it was even unclear which branch of government Cheney worked in.*[14] Even such mundane matters as talking points for reporters were sometimes stamped "Treated As: Top Secret/SCI." According to the *Washington Post,* it was a designation Cheney's office appears to have invented. SCI means "sensitive compartmented information," the most closely guarded category of government secrets of all. In effect, Cheney was protecting unclassified work as though its disclosure would cause "exceptionally grave damage to national security."[15]

Cheney's core concerns, "the iron issues," as his aide Mary Matalin referred to them, were the economy, energy, the White House legislative agenda—and, of course, national security.[16] To address those concerns, Cheney put together a staff that was unparalleled in the history of the vice presidency, in the process expanding its powers accordingly. When Al Gore was vice president, he usually had four or five staffers who were experts in national security. By contrast, Cheney hired at least fourteen on his own staff alone, and had scores of loyalists scattered throughout other departments.[17] Cheney's chief of staff, Scooter Libby, who had experience both at the Pentagon and the State Department, was given the additional title of "national security adviser to the vice president" to give

*At times, when asked to comply with various congressional requests, Cheney claimed executive privilege, thereby suggesting the vice presidency was in the executive branch, which it is. In 2007, however, hoping to get around a presidential order giving the National Archives' Information Security Oversight Office the right to make sure that Cheney and his office had demonstrated proper security safeguards, Cheney's legal staff argued that because the vice president serves as president of the Senate, he is in the legislative branch. Curiously, during the Bush administration, *The Plum Book,* a government directory, published a new definition of the vice presidency, which had been ascribed speculatively to Cheney attorney David Addington: "The Vice Presidency is a unique office that is neither a part of the executive branch nor a part of the legislative branch, but is attached by the Constitution to the latter. The Vice Presidency performs functions in both the legislative branch (see article I, section 3 of the Constitution) and in the executive branch (see article II, and the amendments XII and XXV, of the Constitution, and section 106 of title 3 of the United States Code)."

him more clout at interagency meetings.[18] Another special adviser to the vice president for national affairs was William Luti, who, according to the *Washington Post,* in the early 1990s, while serving as deputy director of the chief of naval operations' executive panel, became a follower of Albert Wohlstetter, who in turn provided him with entrée to Wolfowitz and Perle.[19]

As counsel to the vice president, Cheney brought in David Addington, a loyal acolyte who had been counsel for the House Committee on Intelligence and the Committee on Foreign Affairs in the mid-eighties, and followed Cheney to the Pentagon during the presidency of Bush 41. Addington, according to a *New Yorker* profile by Jane Mayer, was extremely private. He kept his office door locked at all times for national security reasons. Like others on Cheney's staff, he left almost no public paper trail, and refused to speak to the press or allow himself to be photographed.[20]

Methodical, analytical, immersed in intelligence and national security law for more than two decades, Addington, like Cheney, believed that the Watergate scandal involving Richard Nixon had weakened the American presidency. Like Cheney, he was determined to restore those powers to the executive branch in the interests of national security. When it came to pursuing those ends, Addington, according to Colonel Lawrence Wilkerson, chief of staff to Colin Powell, was "utterly ruthless." Added a former top national security lawyer, "He takes a political litmus test of everyone. If you're not sufficiently ideological, he would cut the ground out from under you."[21]

There were many more Cheney men—virtually all long-term neocon loyalists who were committed to overthrowing Saddam—including underlings in other departments such as Sean O'Keefe, the deputy head of the Office of Management and Budget, and Stephen Hadley, the deputy head of the National Security Council, both alumni from Cheney's days running the Pentagon.[22] Unknown to all but a few Washington insiders, they would prove to be extremely important to the vice president.

Even before they had finished moving into their offices, the neocons were confronted with their first national security issue, one that revealed the slant of their new foreign policy. The issue stemmed from events in the last months of the Clinton years. In October 2000, the USS *Cole,* a U.S. Navy destroyer, had been the target of a successful suicide bombing mission orchestrated by Osama bin Laden and Al-Qaeda in the

Yemeni port of Aden. Seventeen sailors had been killed. In the aftermath of the bombing, counterterrorism czar Richard Clarke—officially, head of the Counterterrorism Security Group of the National Security Council (NSC)—felt acutely that the threat of Islamic terrorism was greater than ever.

Thanks to Clarke's book, *Against All Enemies: Inside America's War on Terror,* and other reports about the Bush administration's national security failures, there have been many accounts of how the Bush administration declined to act on Clarke's plan to destroy Al-Qaeda.* Clarke was renowned as an abrasive, sharp-elbowed bureaucratic infighter, but on January 30, 2001, at the administration's first National Security Council meeting, it became clear why he would get nowhere. When the president opened his mouth, it was as if he were channeling Wolfowitz. As far as the Middle East was concerned, the new administration was listening to the neocons, not Richard Clarke. "We're going to correct the imbalances of the previous administration on the Mideast conflict," Bush announced. "We're going to tilt it back toward Israel."

Bush added that America should abandon its efforts to resolve the Israeli-Palestinian conflict. When Colin Powell said that might have dire consequences for the Palestinians, Bush brushed him off. "Sometimes a show of force by one side can really clarify things," he said.[23]

Next, with National Security Adviser Condoleezza Rice parroting the neocon line, the meeting's discussion turned to Iraq. Without mentioning any threat, Rice asserted that "Iraq might be the key to reshaping the entire region."[24] According to Secretary of the Treasury Paul O'Neill, who attended the meetings, no one even questioned why Iraq should be invaded. "From the very beginning, there was a conviction that Saddam Hussein was a bad person and that he needed to go," said O'Neill. "It was all about finding a way to do it. That was the tone of it. The president saying 'Go find me a way to do this.'"[25]

These were the policies that even Benjamin Netanyahu and the Israeli right had not dared to implement. As one senior administration official told the *Washington Post,* "The Likudniks are *really* in charge now."[26]

As part of the neoconservatives' grand vision of democratizing the Middle East, there was another powerful rationale fueling the drive toward war—oil. After Saudi Arabia, Iraq had the next largest oil

*For a more detailed account of Clarke's presentation of his plan and its rejection by the Bush administration, see the author's *House of Bush, House of Saud,* pp. 219–22.

reserves on the planet, with 115 billion barrels of proven reserves.[27] On February 1, 2001, two days after the NSC meeting, Bush officials circulated a memo titled "Plan for post-Saddam Iraq" and began discussing what to do with Iraq's oil wealth.[28]

Cheney was particularly sensitive to the geopolitical ramifications of oil. When he had served as chief of staff to Gerald Ford, he watched Ford's electoral hopes vanish in part because of the soaring inflation that followed the 1973 Arab oil embargo orchestrated by the Organization of Petroleum Exporting Countries (OPEC). Cheney knew, given America's close relations with Saudi Arabia, the world's biggest oil producer, that a new, friendly pro-West regime in Iraq, with someone like Chalabi in charge, would put the United States in a much stronger position with respect to OPEC. In addition, as long as Saddam was at the helm, the United Nations sanctions committee had the authority to approve or veto all oil deals[29] with Iraq—a situation that was less than ideal for American oil companies. As Chevron CEO Kenneth Derr remarked in a 1998 speech, "Iraq possesses huge reserves of oil and gas—reserves I'd love Chevron to have access to."[30]

Cheney himself had spoken publicly about the issue. "By 2010 we will need on the order of an additional fifty million barrels a day," he said in a speech before the London Institute of Petroleum in 1999, when he was CEO of Halliburton, the giant energy services company. "So where is the oil going to come from? . . . While many regions of the world offer great oil opportunities, the Middle East with two thirds of the world's oil and the lowest cost, is still where the prize ultimately lies."[31]

Cheney convened a highly secretive Energy Task Force and went to work. Altogether that spring, Cheney and his aides had at least forty meetings with a total of three hundred groups and individuals, including ExxonMobil vice president James J. Rouse, who was a major donor to the Bush inauguration; Enron chief Kenneth Lay, a longtime Bush friend; Duke Energy; Constellation Energy Group; British Petroleum; and major energy industry lobbying groups, including the National Mining Association, the Interstate Natural Gas Association of America, and the American Petroleum Institute.

At one of the meetings, Daniel Yergin, the head of Cambridge Energy Research Associates and author of *The Prize: The Epic Quest for Oil, Money, and Power*, the Pulitzer Prize–winning history of the oil industry, presented data to Cheney showing that for the first time increased drilling had failed to raise U.S. production.[32] In light of recent intimations that the entire planet was approaching Peak Oil, the point at which the world's crude oil production peaks, and then enters irrevocable decline, and in the context of China's and India's extraordinary ascendancy and soaring consumption of oil, secure access to Middle East oil for the United States was more vital than ever before. In that context, Iraq had the potential to become a pivot point not just for the neoconservatives' dreams of transforming the Middle East, but also for America's geostrategic position in the world.

According to documents obtained by Judicial Watch, a conservative watchdog group, a map of Iraq was presented at the meetings with a list of "Iraq oil foreign suitors." The map reportedly had eliminated all features of the country except for its oil deposits and an accompanying list of dozens of foreign companies that were negotiating for them.[33]

In addition, according to *The New Yorker*, a top-secret document directed National Security Council staffers to cooperate fully with the Energy Task Force as it considered "melding" two areas of policy that appeared to be unrelated: "the review of operational policies toward rogue states," such as Iraq, and "actions regarding the capture of new and existing oil and gas fields."[34]

"People think Cheney's Energy Task Force has been secretive about domestic issues," said Mark Medish, senior director for Russian, Ukrainian, and Eurasian affairs at the National Security Council during the Clinton administration. "But if this little group was discussing geostrategic plans for oil, it puts the issue of war in the context of the captains of the oil industry sitting down with Cheney and laying grand, global plans."[35]

At the same time, all this took place within the context of the neocons' grandiose dreams. "The motivation to invade Iraq was breathtaking in its presumptuousness," said a former Bush administration official who worked with the neocons and who became highly critical of them. "Oil had to be a part of that calculation. Having a Chalabi in charge of Iraq meant that the new Iraq would recognize Israel, that it would not be 'OPECing' the United States. You would have a bastard in charge, but he would be our bastard, in all senses. His policies would be compatible with the interests of the United States."

Late one night in February 2001, according to *The New Yorker*, Francis Brooke, the unofficial lobbyist for Chalabi's Iraqi National Congress, got a call from Deputy Secretary Paul Wolfowitz promising him that Saddam would be overthrown. Wolfowitz has denied making the call. But, according to Brooke, the deputy secretary said he was so deeply committed to deposing Saddam that he would resign if he couldn't accomplish it.[36]

* * *

To moderates in the State Department, however, at this point, it was not yet clear that a historic and radical shift in foreign policy was under way. After all, their champion, Colin Powell, held the top job in the department and was the most admired man in the entire country. "I felt safe with Powell," said one State Department official who worked with him. "With Powell in charge, there did not seem to be an ideological threat from the neocons. There were inroads, but we seemed to be somewhat insulated." Immediately under Powell was his close friend Richard Armitage. Other moderates held highly visible positions: Richard N. Haass as director of policy planning; Marc Grossman as under secretary of state for political affairs; and Grant S. Green as under secretary of state for management.[37]

Then things began to change. In late March, after Wolfowitz was confirmed as deputy secretary of defense, *The New Republic* reported that his former aide Zalmay Khalilzad and Douglas Feith would be joining the administration. John Hannah, a neoconservative lawyer from the Washington Institute for Near East Policy (WINEP), a pro-Israel think tank, was also joining Cheney's staff "to handle the vice-president's Iraq portfolio"[38] and to act as a liaison with Ahmed Chalabi and the Iraqi National Congress.

Moderates worried by these appointments might have taken solace in the fact that word was out that Bush was going to appoint Brent Scowcroft to be head of the president's Foreign Intelligence Advisory Board.[39] "Scowcroft thought he was in perfectly good odor," said an ally of his. "It was not at all clear that he was about to be first marginalized and then ostracized." At about the same time, John Bolton, a hawk with close ties to Cheney, was also sworn in as under secretary of state for arms control and international security. Powell wasn't happy about it, but most moderates assumed that Bolton was boxed in and isolated.

They were wrong. In June 2001, Elliott Abrams, yet another graduate from Scoop Jackson's staff and a staunch, take-no-prisoners hard-liner, was appointed special assistant to the president and senior director on the National Security Council for Near East and North African Affairs. To State Department insiders, the appointment of Abrams, who had been convicted in 1991 on two counts of unlawfully withholding information from Congress during the Iran-contra investigation, was a tipping point. In 1982, during the Reagan administration, Congress had passed the Boland Amendment 411–0 outlawing funding contra rebels in Nicaragua for the purpose of overthrowing the Nicaraguan government. Abrams had been a key figure in the cabal that circumvented that law. In his guilty plea, Abrams admitted that he had withheld knowledge of Oliver North's illegal aid to the contras and that he had also withheld from Congress information that he had solicited $10 million in aid for the contras from the sultan of Brunei.[40]

In other words, Abrams had been a key operative in creating parallel foreign policy channels that operated outside the official bureaucracy—and in direct opposition to a unanimous congressional mandate. "Elliott embodied the hubris of the neocon perspective," said a State Department official who worked with him. "His attitude was 'All the rest of you are pygmies. You don't have the scope and the vision we have. We are going to remake the world.' His appointment meant that good sense had been overcome by ideology."*

With Abrams in the administration, signatories to the Project for the New American Century's (PNAC) 1998 call for regime change in Iraq began to join the administration en masse.† "I don't think that most people in State understood what was going on," said the official. "I understood what this was about, that PNAC was moving from outside the government to inside. In my view, it was an unfriendly takeover."

"Clean Break" coauthor Richard Perle, another PNAC signatory, was appointed chairman of the Defense Policy Board Advisory Committee, a powerful advisory board that focused on strategic planning for the Pentagon. Perle, of course, had a role in shaping Team B as it took on Bush 41's CIA. A staunch foe of the realist policies championed by Scowcroft and James Baker, he had begun to develop a second generation of neocon foreign policy hands who were ready to join the administration.

By midsummer 2001, Wolfowitz, Perle, Feith, Abrams, and Khalilzad had taken key positions in the national security apparatus. They had worked together on and off for more than twenty years, crusading ideologues who were finally in a position to realize their dreams. As their patrons, they had none other than Donald Rumsfeld and Dick Cheney,

*Elliott Abrams declined a request for an interview.

†In all, thirteen out of the eighteen signatories to the PNAC letter won appointments in the Bush-Cheney administration: Elliott Abrams, Richard Armitage, Assistant Secretary of State Jeffrey Bergner, John Bolton, Under Secretary of State Paula Dobriansky, Francis Fukuyama (President's Council on Bioethics), Zalmay Khalilzad, Richard Perle, Assistant Secretary of Defense Peter Rodman, Donald Rumsfeld, Defense Science Board appointee William Schneider, Paul Wolfowitz, and U.S. Trade Representative Robert Zoellick.

two of the savviest people in Washington when it came to manipulating the bureaucracy.

One of Cheney's chief problems was that he wasn't happy about having George Tenet, a Clinton holdover, still in place as head of the CIA. To circumvent that obstacle, Cheney assembled his own mini–National Security Council. He had moles in the State Department such as John Bolton to keep Colin Powell in check. Wolfowitz, though under Rumsfeld's aegis in the Pentagon, was a Cheney man.

In order to implement their policies, however, one thing was missing. The neocons had to make a persuasive case that Saddam posed a powerful threat to the West. In fact, they had started a relentless drumbeat for war as soon as Bush was in office. On January 21, William Kristol and Robert Kagan asserted in the *San Diego Union-Tribune* that U.S. military action "could well be necessary to bring Saddam down."[41] In February, on CNN, Richard Perle advocated "removing Saddam."[42] During his confirmation hearings, Wolfowitz testified that he was in favor of overthrowing the Iraqi dictator.[43] In April, *Washington Post* columnist Jim Hoagland championed Ahmed Chalabi as "a dedicated advocate of democracy" in Iraq.[44] On July 30, in the *Weekly Standard*, Reuel Marc Gerecht characterized the United States as a "cowering superpower" under Bush 41's presidency and demanded that Bush explain "how we will live with Saddam and his nuclear weapons."[45]

For the most part, however, such columnists spoke to foreign policy insiders. The vast majority of Americans had other concerns on their minds. The baseball season had begun. The real estate market was frothy. Bush had pushed through a tax cut that many voters liked, but otherwise his presidency appeared to be going nowhere. Most Americans were still oblivious to the neocons—much less their designs on Iraq.

Meanwhile, though few people knew about it, the prospects for overthrowing Saddam were being furthered thanks to the New Year's break-in at the Niger embassy in Rome.

Because the Niger break-in happened before Bush took office, it is often assumed that the robbery was initiated as a small-time job. But according to Colonel Patrick Lang, a former Middle East analyst for the Defense Intelligence Agency, it is also possible that from the start the Niger operation was aimed at provoking an invasion of Iraq. The scenario, he cautions, is merely speculation on his part. But he says that the neocons wouldn't have hesitated to reach out to Italian intelligence even before Bush took office. "There's no doubt in my mind that the neocons had their eye on Iraq," he said.[46] "This is something they intended to do, and they would have communicated that to SISMI (Servicio per le Informazione e la Sicurezza Militare, i.e., Italian military intelligence) or anybody else to get the help they wanted."

In Lang's view, SISMI would also have wanted to ingratiate itself with the incoming administration. "These foreign intelligence agencies are so dependent on us that the urge to acquire IOUs is a powerful incentive by itself," he said.[47] "It would have been very easy to have someone go to Rome and talk to them, or have one of the SISMI guys here [in Washington], perhaps the SISMI officer in the Italian embassy, talk to them."

A key figure in the dissemination of the Niger documents was Rocco Martino, an elegantly attired man in his sixties with white hair and a neatly trimmed mustache. Martino had served with SISMI until 1999 and had a long history of peddling information to other intelligence services in Europe, including France's Direction Générale de la Sécurité Extérieure (DGSE). By 2000, Martino had fallen on hard times financially. Then a longtime colleague from SISMI, Antonio Nucera, came to him with a lucrative proposition.[48]

A colonel specializing in counterproliferation and WMDs, Nucera told Martino that Italian intelligence had long had an "asset" in the Niger embassy in Rome: a woman who was about sixty years old, had a low-level job, and occasionally sold embassy documents to SISMI. But now SISMI had no more use for the woman, who later became known in the Italian press as "La Signora" and was subsequently identified as the ambassador's assistant, Laura Montini. Perhaps, Nucera suggested, Martino could use La Signora as Italian intelligence had, paying her to pass on documents copied or stolen from the embassy.

There have been many conflicting accounts of the story behind the New Year's break-in at the Niger embassy. But the first comprehensive report, based on interviews with Rocco Martino, SISMI director Nicolò Pollari, and other sources came from an investigative series by Carlo Bonini and Giuseppe D'Avanzo that appeared in 2005 in *La Repubblica,* a left-of-center daily in Italy. According to the series in *La Repubblica,* Martino denied participating in the robbery itself, and said he only became involved when SISMI had La Signora give him documents from the burglary. "I was told that a woman in the Niger embassy in Rome had a gift for me. I met her and she gave me documents," he said.[49]

Later, Rocco Martino said, SISMI dug into its archives and added new papers. In addition to a codebook, there was a dossier with a mixture of fake and genuine documents. Among them was an authentic telex dated

February 1, 1999, in which Adamou Chékou, the ambassador from Niger, wrote another official about a forthcoming visit from Wissam al-Zahawie, Iraq's ambassador to the Vatican.

The last document Martino says he received, and the most important one, was, however, not genuine. Dated July 27, 2000, it was a two-page memo purportedly sent to the president of Niger concerning the sale of five hundred tons of pure uranium, or "yellowcake," per year by Niger to Iraq.

Early in the summer of 2001, about six months after the break-in, information from the forged documents was given to U.S. intelligence for the first time. Details about the transfer are sketchy, but it is highly probable that the reports were summaries of the documents rather than the documents themselves. It is standard practice for intelligence services, in the interests of protecting sources, to share reports but not original documents with allies. Without seeing the documents, of course, it was impossible to examine them to determine if they were authentic.

Nevertheless, to many WMD analysts in the CIA, the State Department, and the Pentagon, the initial reports sounded absurd. "The reports made no sense on the face of it," said Ray McGovern, a CIA analyst for twenty-seven years.[50] "Most of us knew the Iraqis already *had* yellowcake. It is a sophisticated process to change it into a very refined state and they didn't have the technology."

"The idea that you could get that much yellowcake out of Niger without the French knowing, that you could have a train big enough to carry it, much less a ship, is absurd," said Colonel Larry Wilkerson, former chief of staff to Secretary of State Colin Powell.[51]

"Yellowcake is unprocessed bulk ore," explained Lieutenant Colonel Karen Kwiatkowski, who served in the Pentagon's Near East and South Asia division in 2002 and 2003. "If Saddam wanted to make nuclear bombs, why would he want unprocessed ore when the best thing to do would be to get processed stuff in the Congo?"[52]

"When it comes to raw reports, all manner of crap comes out of the field," McGovern added. "The CIA traditionally has had experienced officers. They are qualified to see if these reports make sense. For some reason, perhaps out of cowardice, these reports were judged to be of such potential significance that no one wanted to sit on it."

Since Niger was a former French colony, French intelligence was the logical choice to vet the allegations. "The French were managing partners of the international consortium in Niger," said former ambassador

Joseph C. Wilson, who later traveled to Niger to investigate the uranium claim. "The French did the actual mining and shipping of it."[53]

So Alain Chouet, then head of security intelligence for France's DGSE, was tasked with checking out the first Niger report for the CIA. He recalled that much of the information he received from Langley was vague, with the exception of one striking detail. The agency had heard that in 1999 the Iraqi ambassador to the Vatican, Wissam al-Zahawie, had made an unusual visit to four African countries, including Niger. Analysts feared that the trip may have been a prelude to a uranium deal.

However, Chouet soon learned that the al-Zahawie visit was no secret. It had been covered by the local press in Niger at the time, and reports had surfaced in French, British, and American intelligence. Chouet had a seven-hundred-man unit at his command, and he ordered an extensive on-the-ground investigation in Niger.

"In France, we've always been very careful about both problems of uranium production in Niger and Iraqi attempts to get uranium," Chouet told the *Los Angeles Times* in 2005.[54] Having concluded that nothing had come of al-Zahawie's visit and that there was no evidence of a uranium deal, French intelligence forwarded its assessment to the CIA. But the Niger affair had just begun.

Meanwhile, in June 2001, George W. Bush faced a new foreign policy quandary. Just as Bush had promised at his initial NSC meeting, the United States had withdrawn from the Israeli-Palestinian peace process. But the change in policy had deeply angered Saudi Arabia. Historically, the Saudis had refrained from intervening as forcefully as they might have on the side of the Palestinians. But Israeli forces had killed more than three hundred Palestinians and wounded more than eleven thousand the previous year. The previous February, Likud's Ariel Sharon had been elected to replace Ehud Barak as prime minister. Throughout the spring, Israeli soldiers had attacked Palestinians day after day. On March 3 alone, Israelis killed Palestinians in three separate incidents. In April, they shot dead a fourteen-year-old Palestinian. Yet Bush blamed the violence on Arafat.

Given the Bush family's intimate ties with the Saudis, the irony was inescapable.* In just five months as president, however, Bush had man-

*For an extended account of the Bush family's financial relationships with the Saudis, see the author's *House of Bush, House of Saud.*

aged to jeopardize a relationship with an oil-rich ally of the United States, at a time when America was more profoundly dependent on foreign oil than ever.

The Saudis were so irate at Bush's uncritical acceptance of Sharon that in May, Crown Prince Abdullah, the de facto ruler of Saudi Arabia,* even turned down an invitation to the White House. In an interview with the *Financial Times,* Abdullah said, "We want them [the United States] to consider their own conscience. Don't they see what is happening to the Palestinian children, women, the elderly, the humiliation, the hunger?"[55]

Among those who echoed Abdullah's assessment was Brent Scowcroft, who reportedly said that moderate Arab countries were "deeply disappointed with this administration and its failure to do something to moderate the attitude of Israel."[56] He added that unless there was the promise of a viable Palestinian state, the Palestinians would not stop their violence.

In effect, in their own constrained fashion, the father and son had drawn swords. Scowcroft's comments were perceived as having been made with the assent of George H.W. Bush, who continued to receive regular briefings from the CIA. It was a privilege granted to all former presidents, but one that Bush used far more than the others—presumably because his son had so little foreign policy experience. At the CIA, the briefings were jokingly referred to as the "president's daddy's daily briefings."[57]

On the weekend of July 7 and 8, Scowcroft flew to the Bush compound in Kennebunkport, Maine, for a frank discussion with both father and son. Advisers close to Bush senior told the *New York Times* that these overtures were initiated by the father, who had made foreign policy the centerpiece of his administration, rather than by his son, still a novice in the field.[58]

Then, in mid-July, Bush 41 intervened yet again, this time calling Crown Prince Abdullah to assure him that his son's "heart" was in the right place. The *New York Times* reported that he made the call "to vouch for his son and to assuage concerns of the crown prince . . . that some strains in the United States' relations with Saudi Arabia over the Israeli-Palestinian conflict do not represent a permanent breach."[59]

*Abdullah's half brother, Fahd, was king, but due to a debilitating stroke he had suffered in 1995, he was unable to perform his duties.

According to the *Times* article, President Bush was in the room when his father made the call.

Bush senior succeeded in easing the Saudis' concerns, at least temporarily. But he had opened another can of worms: his relationship with his son. When the two articles about 41's interventions appeared in the *New York Times,* George W. Bush, so sensitive about his father, was particularly unhappy. "Both were upset about it," said a former adviser to Bush 41. "It looked like Poppy* was helping out the son. They made a solemn oath from that day forth that no one would know from then onwards."

The president had other problems, too. By mid-summer, his postinaugural honeymoon was ending. His approval ratings had sunk to 50 percent, close to a historic low for a president during his first six months in office.[60] Ridiculed for his malapropisms,† dependent on his father, Bush had lost the popular vote in 2000. There were already intimations that he would be a one-term president.

On August 4, he went out to the new Western White House, his sixteen-hundred-acre ranch in Crawford, Texas. In the past, the media had enjoyed covering George H.W. Bush and Bill Clinton at Kennebunkport and Martha's Vineyard, respectively, taking in clambakes, lobster rolls, beaches, and picturesque scenery as diversions at the two elite resorts. But George W. Bush had chosen the parched environs of Crawford instead. The media hated it. With a population of 705 people, Crawford had only one sidewalk and it was still under construction. There was nothing to do. Nearby Waco was famous for its cricket infestations and the Dr Pepper Museum.‡[61] Other than George and Laura Bush, Crawford's lone celebrity was rock guitarist Ted Nugent, of

*George H.W. Bush was sometimes referred to as Poppy by his wife, Barbara, members of the family, and close friends.

†Books such as *Bushism* and *Dumbass,* and scores of websites itemized hundreds of malapropisms by Bush. Among them: "I know how hard it is for you to put food on your family."—Greater Nashua, N.H., January 27, 2000; "I know the human being and fish can coexist peacefully."—Saginaw, Mich., September 29, 2000; "Rarely is the question asked: Is our children learning?"—Florence, S.C., January 11, 2000; "Our enemies are innovative and resourceful, and so are we. They never stop thinking about new ways to harm our country and our people, and neither do we." —Washington, D.C., August 5, 2004.

‡The museum's slogan: "Paris has the Louvre, New York has the Guggenheim, Waco has the Dr Pepper. Belching is not encouraged. We are a Museum after all."

"Wang Dang Sweet Poontang" fame.*[62] The temperature was a hundred degrees day after day. The disgruntled press corps waited interminably in a stifling gymnasium, from which they called in to presidential aides in hopes of garnering some news.

There were a number of items on Bush's agenda to satisfy his evangelical constituency. The White House had pushed through the House of Representatives a bill to give government grants to faith-based groups for drug treatment and domestic abuse counseling, but it still required Senate approval.[63] In addition, Bush had decided to ban federal funding for research on human embryonic stem cells because right-to-life fundamentalists viewed such research as tantamount to abortion. Bush had his staff prepare a statement for release later that week. He was coming out against stem cell research even though his own sister, Robin, had died at the age of three of leukemia, a disease that promised to benefit from such research.[64]

In the daily papers, however, reporters spilled more ink grousing about Bush's interminable vacation and the searing heat. It was so hot at the sun-scorched ranch that Bush's favorite fishing hole was evaporating and first lady Laura was begging to go back to Washington early.[65] Not only was Bush taking the longest presidential vacation in thirty-two years, but, by the time he was scheduled to return to the White House, he would have spent forty-two percent of his early presidency at or traveling to vacation spots.

We now know, of course, that there were other vital issues on the agenda. On the hot Texas morning of August 6, after a four-mile run, Bush received a Presidential Daily Briefing[66] with a title that is now difficult to forget: "Bin Laden Determined to Strike in US." Specifically, the memo concluded that FBI information indicated "suspicious activ-

*A die-hard Republican and member of the National Rifle Association, Nugent famously addressed an NRA Convention in Houston, telling delegates, "Remember the Alamo! Shoot 'em! To show you how radical I am, I want carjackers dead. I want rapists dead. I want burglars dead. I want child molesters dead. I want the bad guys dead. No court case. No parole. No early release. I want 'em dead. Get a gun and when they attack you, shoot 'em."

Nugent met Bush at an inauguration party in 2000. "When he noticed me," Nugent said, "he was surrounded by these huge bankrollers from his campaign. He literally swept past all of them and said, 'Laura! Look who's here! It's Ted!' Then he hugged me and took me by the shoulders. He said, 'Just keep doing what you're doing. Don't think that we don't know what you're up to out here. Stay on course.' "

Later, however, Nugent became critical of Bush for not taking a more aggressive course in Iraq. "Our failure has been not to Nagasaki them," Nugent said.

ity in this country consistent with preparations for hijackings or other types of attacks, including recent surveillance of federal buildings in New York."[67] But Bush spent the rest of the day fishing for bass, and there were no further meetings about terrorism during the next thirty days. Even though the threats were more serious than ever before, Richard Clarke's frantic pleas to take action against Al-Qaeda went unheard.

At the time, of course, few Americans had ever heard of Richard Clarke or even Osama bin Laden. Nor did they realize that the neocons were making inroads on taking over America's foreign policy. However, a few astute observers noticed that something was wrong. If there was one man who had promised to be a figure of great stature in the Bush-Cheney administration—a star of global magnitude, a giant on the world stage—it was Colin Powell. Yet where was he? Deflated, marginalized, invisible, neutralized, he was nowhere to be seen.

One top diplomat, when asked to provide an adjective for the phrase "Colin Powell is a 'blank' Secretary of State," replied, "Yes, he is."

"I've been struck by how not struck I am by him," added a senior administration official.[68]

A *Time* magazine story on the incredible shrinking secretary of state observed that his centrist politics "made Powell chum in the water for the sharks in Dubya's sea," namely Cheney and Rumsfeld. More to the point, when Wolfowitz was asked why he took the number two position [in the Pentagon], he reportedly gave a one-word answer: Powell.[69]

Wolfowitz later denied making the terse response, perhaps because it was too revealing. It gave the game away. Wolfowitz, as *Time* noted, was "a zealous advocate of 'regime change' in Baghdad." He was there to neutralize Powell, to implement the hard-line neocon vision. "Enthusiasm," the magazine reported, "is building inside the Administration to take down Saddam once and for all."

The publication date of that issue of *Time* was September 10, 2001.

"What if" is the great parlor game of historians. What if nothing special had happened on September 11? Would there have been an Iraq War? Seen through the prism of the most important political and cultural event of the twenty-first century, the question is fascinating.

Some Washington insiders have concluded there would have been an Iraq War no matter what. "Absolutely," said a former State Department official. "They would have done it under any circumstance. Absolutely no doubt about it. They had the road map. They had the plan." On the other hand, a paper published in 2000 by the Project for the New American Century asserted that the process of changing American policy to regime change in Iraq "is likely to be a long one, absent some catastrophic and catalyzing event like a new Pearl Harbor."[70]

We will never really know the answer. But one thing was certain. Cheney had made his opening gambit. He had accumulated extraordinary powers for a vice president. His men were in position throughout the executive branch. They were ready to pounce.

Donald Rumsfeld (left) and Dick Cheney (right) executed dazzling bureaucratic sleights of hand in the Ford White House—and again under George W. Bush. (David Hume Kennerly/Gerald R. Ford Library/Corbis)

James Baker and Dick Cheney with President George H.W. Bush. Brent Scowcroft, who later became Cheney's nemesis, is behind them. As Bush 41's close friend, Scowcroft later discreetly tried to curb Bush 43's neoconservative ways. (Jean-Louis Atlan/Sygma/Corbis)

Colin Powell, General Norman Schwarzkopf, and Paul Wolfowitz during Desert Storm in 1991. Wolfowitz later became a key architect of the Iraq War. (PH2 Susan Carl/Corbis)

Bush was initially converted to evangelical Christianity not by Billy Graham as he claimed, but by Arthur Blessitt (seen here with Bush), who won fame for carrying a huge wooden cross around the globe. (Courtesy of Arthur Blessitt)

George W. Bush was widely seen as a religious leader by his Christian evangelical supporters. (Brooks Kraft/Corbis)

Albert Wohlstetter was one of several role models for Stanley Kubrick's *Dr. Strangelove* and an inspirational father figure for many of the neoconservatives who helped start the war in Iraq, including Paul Wolfowitz, Richard Perle, Ahmed Chalabi, and Zalmay Khalilzad. (Courtesy of RAND Corporation)

Neoconservative superhawks Richard Perle (left) and Michael Ledeen confer at the American Enterprise Institute, the think tank most closely tied to the war. Perle served as chairman of Bush's Defense Policy Advisory Board, while Ledeen allegedly launched unauthorized foreign policy initiatives in secret meetings in Rome. (James A. Parcell/*Washington Post*)

A key figure in the 1987 Iran-contra scandal who was convicted of withholding information from Congress, neocon Elliott Abrams wielded immense power behind the scenes in the Bush White House. (Richard A. Bloom/Corbis)

Once America's ally against Iran, Saddam Hussein was depicted as posing a grave and imminent threat to America as war fever heated up. Captured by American troops in December 2003, he was executed in Iraq three years later. (Lazim Ali/Reuters/Corbis)

The Oslo Accords brokered by President Clinton in September 1993 were marked by Yitzhak Rabin's handshake with PLO leader Yasir Arafat. The agreement and the handshake led to Rabin's assassination by Yigal Amir in November 1995. (Reuters/Corbis)

An acolyte of theologian Francis Schaeffer, Florida secretary of state Katherine Harris felt that "godly men" should hold high office and helped put George W. Bush in the White House. Florida governor Jeb Bush looks on. (Scott Audette/AP Photo)

Condoleezza Rice and Paul Wolfowitz with Bush at his ranch in Crawford, Texas. Although she was a protégée of realist Brent Scowcroft, Rice ended up helping implement Wolfowitz's neoconservative war against Iraq. (John Healey/Reuters/Corbis)

In the new Bush administration, Cabinet meetings were preceded by a moment of prayer. (Brooks Kraft/Corbis)

The inner circle of Dick Cheney, Scooter Libby, and Donald Rumsfeld knew how to get things done its way in Washington, D.C. Libby was convicted on five felony counts of obstructing justice, perjury, and making false statements to investigators; Rumsfeld resigned after the 2006 midterm elections showed how unhappy Americans were with the war in Iraq. (David Hume Kennerly/Getty Images)

Though perceived as a voice of moderation, Secretary of State Colin Powell was ultimately used by Cheney to give credibility to the case for war. (Brooks Kraft/Corbis)

Paul Wolfowitz enjoyed the close confidence of Dick Cheney. (Chuck Kennedy/KRT/Newscom)

Paul Wolfowitz's romantic relationship with Shaha Ali Riza may have been a factor that prevented him from becoming director of the CIA. (Courtesy of World Bank)

Rumsfeld with Ahmed Chalabi, whom neoconservatives hoped to install as the leader of Iraq. In June 2004, intelligence officials told reporters that Chalabi had leaked U.S. state secrets to Iran. He denied the allegations. (Rabih Moghrabi/AFP/Getty Images)

When Colin Powell presented the case for war before the U.N. on February 5, 2003, he insisted that CIA director George Tenet put his credibility on the line by appearing right behind him. The presentation was full of lies. (Elise Amendola/AP Photo)

After former ambassador Joseph C. Wilson IV found evidence that Iraq had *not* sought to buy uranium from Niger, Bush administration officials leaked the identity of his wife, Valerie Plame Wilson, as a CIA operative. (Lawrence Jackson/AP Photo)

Six weeks after the U.S. invasion of Iraq, President Bush, in Tom Cruise–like *Top Gun* garb, landed on the USS *Kitty Hawk,* where he announced the end of combat operations beneath a banner reading MISSION ACCOMPLISHED. In fact, the war had barely begun. (Tyler J. Clements/ U.S. Navy photo/Reuters/Corbis)

At Abu Ghraib prison in Iraq, torture and abuse of detainees became part of administration policies that included secret rendition of captives and disregard of the Geneva Convention. (AP Photo)

Lee Hamilton (left) and James Baker (right) led the Iraq Study Group and made hard-nosed recommendations to Bush 43 about extricating the United States from Iraq. (Pablo Martinez Monsivais/AP Photo)

In December 2006, with the war in Iraq a disaster and George W. Bush's presidency in political free fall, his father, George H.W. Bush, broke down in tears at the Florida State Capitol in Tallahassee, where his son Jeb was governor. (Courtesy of the *Today Show*/NBC)

CHAPTER FOURTEEN

In the Shadows

From the moment American Airlines Flight 11 hit the North Tower of the World Trade Center at 8:46 a.m. on September 11, officials at the highest levels of the American intelligence apparatus—CIA director George Tenet, counterterrorism czar Richard Clarke, and countless analysts—knew that what was taking place was a terrorist attack orchestrated by Osama bin Laden and executed by Al-Qaeda. Nevertheless, the neocons, their patrons, and the Christian Right just as quickly came up with different villains and began to act accordingly.

The neocon propaganda machine went into action nearly as quickly as the Pentagon did. By 1:15 p.m., just four hours after the attacks, Michael Ledeen, the neocon operative who had won notoriety in the Iran-contra scandal, filed a dispatch on the *National Review*'s website attacking the remaining realists in the administration and urging someone to remind Bush that "we are still living with the consequences of Desert Shame, when his father and his father's advisers—most notably Colin Powell and Brent Scowcroft—advised against finishing the job and liberating Iraq."[1]

At 2:40 p.m., Secretary of Defense Donald Rumsfeld ordered the military to put together retaliatory plans to go after not just bin Laden, but Saddam Hussein as well. According to notes taken by a Rumsfeld aide, the secretary of defense wanted "best info fast, judge whether good enough to hit SH [Saddam Hussein] at the same time, not only UBL," the initials used to identify Osama bin Laden. "Go massive, sweep it all up, things related and not," the notes said.[2]

Meanwhile, Laurie Mylroie, a darling of the neocons who had falsely reported that Saddam was behind the 1993 World Trade Center bombing, turned in an opinion piece that was published in the *Wall Street*

Journal the next day, blaming the Iraqi dictator for 9/11 as well.*[3] Intelligence analysts had already begun to tell journalists that Iraq had nothing to do with it.[4] Yet Mylroie's theory was cited by and given credence by the *Washington Post*,[5] CBS News,[6] ABC,[7] Fox News,[8] *U.S. News & World Report*,[9] CNN,[10] *The Times* (London),[11] the *Dallas Morning News*,[12] and more than a hundred other media outlets—some of them repeatedly. "In my view, yesterday's events were the latest step in Saddam's war against the United States," Mylroie told CBS News.[13]

Mylroie's baseless theories would have been harmless but for two things. They dovetailed perfectly with the policy aims of her associates who had won powerful positions in the Bush administration. In addition, at the time, few Americans had even heard of Osama bin Laden or Al-Qaeda. It scarcely seemed credible that a handful of jihadists in Afghanistan could possibly have put together such a spectacular operation. Saddam Hussein, on the other hand, thanks to the Gulf War, was already clearly defined in the public imagination as a villain straight out of central casting. As a result, Saddam would soon be indelibly linked to 9/11 in the minds of millions of Americans.

*Mylroie was widely discredited by terrorism analysts such as CNN's Peter Bergen, who called her a "crackpot" for arguing that Saddam was behind not just the 1993 World Trade Center bombing, "but also every anti-American terrorist incident of the past decade, from the bombings of U.S. embassies in Kenya and Tanzania to the leveling of the federal building in Oklahoma City to September 11 itself." Likewise, Daniel Benjamin, a fellow at the Center for Strategic and International Studies, notes that "Mylroie's work has been carefully investigated by the CIA and the FBI. . . . The most knowledgeable analysts and investigators at the CIA and at the FBI believe that their work conclusively disproves Mylroie's claims."

In a piece on Mylroie in the *Washington Monthly*, Bergen noted that her book, *Study of Revenge: Saddam Hussein's Unfinished War Against America*, which was published by AEI in 2000, "makes it clear that Mylroie and the neocon hawks worked hand in glove to push her theory that Iraq was behind the '93 Trade Center bombing. Its acknowledgements fulsomely thanked John Bolton and the staff of AEI for their assistance, while Richard Perle glowingly blurbed the book as 'splendid and wholly convincing.' Lewis 'Scooter' Libby, now Vice President Cheney's chief of staff, is thanked for his 'generous and timely assistance.' And it appears that Paul Wolfowitz himself was instrumental in the genesis of *Study of Revenge.*"

Bergen pointed out that "the most comprehensive criminal investigation in history—involving chasing down 500,000 leads and interviewing 175,000 people—has turned up no evidence of Iraq's involvement" in 9/11, nor had the occupation of the country by a large American army turned up any such link. Mylroie was also the coauthor of *Saddam Hussein and the Crisis in the Gulf* with Judith Miller, the *New York Times* reporter who touted false information about Saddam's WMD program during the run-up to war.

In the administration, the neocons began acting publicly as if Saddam was one of the perpetrators. At a Pentagon press briefing on September 13, Paul Wolfowitz told reporters that "ending states who sponsor terrorism" would be a priority for the administration.[14] An irate Colin Powell asked General Henry Shelton, chairman of the Joint Chiefs of Staff, "What are these guys thinking about? Can't you get these guys back in the box?"[15]

Regime change in Iraq now became the order of the day. On September 16, Richard Perle told CNN, "Even if we cannot prove to the standards we enjoy in our own civil society that [Iraq] was involved [in 9/11], we do know that Saddam Hussein has ties to Osama bin Laden."[16] The next day, also on CNN, conservative Republican senator George Allen[17] broached the topic of regime change in Iraq, Iran, and Syria. On September 18, Wesley Clark, the former supreme allied commander of NATO, warned CNN that the Bush administration might "think it's time for regime change" in Iraq.[18]

Ten days after the attacks, on September 21, 2001, President Bush was told in a highly classified briefing that there was no credible evidence linking Saddam Hussein to the attacks, nor was there strong reason to believe Iraq had collaborated with Al-Qaeda.[19] But the neocon pressure mounted. On September 25, as the United States prepared for military action in Afghanistan, William Kristol advocated in the *Washington Post* for "regime change where possible," and went after Colin Powell for being against it.[20] The next day, the *Washington Times* chimed in against Powell and for regime change.[21] On September 26, on PBS, Richard Perle was at it again, calling for destroying regimes in "countries like Iraq, like Syria, Sudan—parts of Lebanon and others."[22]

At the same time, the Christian Right had identified its own villain as being responsible for 9/11. "The ACLU had to take a lot of blame for this . . ." said Jerry Falwell in a televised interview on Pat Robertson's *700 Club*.*[23] "Throwing God out of the public square, out of the schools, the abortionists have got to bear some burden for this because God will not be mocked and when we destroy 40 million little innocent

*In a phone call to CNN, Falwell later apologized for his statements and said only the terrorists should be blamed for the attacks. But he added that the ACLU and other groups "have attempted to secularize America, have removed our nation from its relationship with Christ on which it was founded. . . . I therefore believe that that created an environment which possibly has caused God to lift the veil of protection which has allowed no one to attack America on our soil since 1812."

babies, we make God mad. . . . I really believe that the pagans and the abortionists and the feminists and the gays and the lesbians who are actively trying to make that an alternative lifestyle, the ACLU, People for the American Way, all of them who try to secularize America . . . I point the finger in their face and say you helped this happen."

To fundamentalists, the atrocity of 9/11 was a harbinger of End Times. On raptureready.com, a website for followers of biblical prophecy, the Rapture Index, "the Dow Jones Industrial Average of end time activity" preceding the Rapture,[24] hit a historic high of 182, suggesting the end was near.* "Are these the End Times?" asked Don Poage, a member of Bush's Bible study group in Midland. "You read in the Bible about wars, rumors of war, earthquakes, fires and floods. . . . We have those all the time. This was unique."[25]

Evangelicals who believe in "a real and personal" God, believe in "a real and personal" Devil as well—and now he had come to life in the persons of Osama bin Laden and Saddam Hussein. "We believe," said Jerry Falwell, "it is Satan who works through the Stalins and the Adolf Hitlers and the Saddam Husseins and Osama bin Ladens," designating them to be, in the biblical phrase, "instruments of evil,"[26] as Bush himself described the terrorists a few weeks later.

To evangelicals, Bush included, the attacks epitomized the struggle of good and evil, of God versus Satan, with the hijackers standing in for the latter. "Behind them is a cult of evil which seeks to harm the innocent and thrives on human suffering," Bush said. "Theirs is the worst kind of cruelty, the cruelty that is fed, not weakened, by tears. Theirs is the worst kind of violence, pure malice, while daring to claim the authority of God. We cannot fully understand the designs and power of evil. It is enough to know that evil, like goodness, exists. And in the terrorists, evil has found a willing servant."[27]

Now Satan had shown his hand. "We have come to know truths that we will never question," Bush told the nation.[28] "Evil is real. And it must be opposed."

"This will be a monumental struggle of good versus evil," Bush said. "But good will prevail."[29] Never had his demeanor been more solemn.

If Saddam was Satan's man for the hour, to American evangelicals, it followed naturally that George W. Bush was God's. A few days after the attacks, Bush had a small group of evangelical leaders to the White House to discuss the theological implications of the terrorist attacks.

*According to the website, a reading of 145 means "Fasten your seat belts."

According to an article by journalist Max Blumenthal, James Merritt, president of the Southern Baptist Convention, offered the president a few words of hope.

"Mr. President, you and I are fellow believers in Jesus Christ," Merritt said. Bush nodded.

"We both believe there is a sovereign God in control of this universe."

Again, Bush agreed.

"Since God knew that those planes would hit those towers before you and I were ever born," Merritt said, "since God knew that you would be sitting in that chair before this world was ever created, I can only draw the conclusion that you are God's man for this hour."[30]

With that, Bush reportedly began to cry.

In the horror of a moment that struck the deepest chords of America's sense of nationalism and patriotism, Bush's stark moral clarity resonated powerfully across the country. When American troops arrived in Afghanistan in early October, and Operation Enduring Freedom began, his approval ratings shot through the roof—hitting 92 percent, according to an ABC poll. He now had the highest approval rating of any president in American history—topping his father's stratospheric ratings after the first Gulf War.[31]

Meanwhile, Dick Cheney quietly went to work behind the scenes. A groundbreaking four-part series on the vice president in the *Washington Post* by Barton Gellman and Jo Becker in June 2007, the first journalism to penetrate the clandestine and mystifying Office of the Vice President, provides the most revealing glimpse at how adroitly Cheney navigated the bureaucratic shoals of the executive branch to get exactly what he wanted.* According to the series, the vice president's reaction to the shocking collapse of the South Tower of the World Trade Center showed a man who, while watching one of the great tragedies in American history, silently and instantly turned his complete attention to operating the levers of power—to the exclusion of any apparent emotional response whatsoever.

*Perhaps as much as any single piece of journalism, the Gellman-Becker series illustrated the problems of informing the American people about what is really going on in their government today. Even though the series was hailed on the Internet in the liberal blogosphere and featured prominently in the *Washington Post,* the essential but arcane details behind Cheney's bureaucratic machinations were largely ignored by the mainstream press.

With thousands dead or dying and millions of people across the country in a state of panic, Cheney made no sound and his expression remained unchanged. "What [his colleagues] saw," the *Post* reported, "was extraordinary self-containment and a rapid shift of focus to the machinery of power. . . . Cheney began planning for a conflict that would call upon lawyers as often as soldiers and spies."[32]

A master bureaucrat who had long believed in an exceptionally powerful executive branch, Cheney immediately moved to expand the authority of the White House to fight a new kind of war, one that required new tactics, strategies, and tools. This was the beginning of what was sometimes called the Unitary Executive or the New Paradigm, doctrines that called for vastly enhanced presidential authority and extreme flexibility in its use.

Cheney knew well that the principle of checks and balances, of the separation of powers, involved far more than Congress having the power to go toe to toe with the White House. Over the years, Congress and scores of government agencies had devised hundreds of bureaucratic mechanisms backed by statutory power to ensure that officials could be held responsible by the people who elected them. Cheney, however, often viewed such mechanisms as unnecessary encumbrances. His first step was to use his loyal cadre to secretly implement a series of profound and far-reaching legal changes.

Specifically, on the morning of 9/11, that meant summoning back to work his trusted counsel David Addington, then making the long trek home by foot, along with thousands of other panicked evacuees.[33] Within a few hours, Addington had joined forces with deputy White House counsel Timothy Flanigan*[34] and John Yoo, the deputy chief of the Office of Legal Counsel in the Justice Department who later became known for his work on the Patriot Act and for being the author of memos advocating the possible legality of torture and denying enemy combatants the protection afforded by the Geneva Convention. White House counsel Alberto Gonzales, who had been the president's personal lawyer in Texas, joined them later.

As the *Post* series reported it, with Yoo playing the role of chief theorist, the trio got right to work drafting the document that was later approved by the U.S. Senate to authorize the use of military force in

*When Flanigan left his job in late 2002 to become general counsel at Tyco, one of his first hires was lobbyist Jack Abramoff. His mission: to lobby Congress and the White House so that Tyco's Bermuda tax haven would be preserved.

Afghanistan. Specifically, the document allowed the president "to use all necessary and appropriate force against those nations, organizations, or persons he determines planned, authorized, committed, or aided the terrorist attacks that occurred on September 11, 2001, or harbored such organizations or persons, in order to prevent any future acts of international terrorism against the United States. . . ."[35] The extraordinarily broad language was used, Yoo said, because "this war was so different, you can't predict what might come up."[36]

Cheney made sure that his legal team could work in complete secrecy. As he himself explained on NBC's *Meet the Press* just after the attacks, "We also have to work, through, sort of the dark side, if you will. We've got to spend time in the shadows in the intelligence world. A lot of what needs to be done here will have to be done quietly, without any discussion. . . ."[37] Information flowed into Cheney's office, the *Post* reported, but nothing flowed out. On Cheney's orders, the Secret Service destroyed visitor logs into the Office of the Vice President.[38] Meanwhile, Yoo went to work to give the president even more powers. Since 1978, the warrantless interception of communications to and from the United States had been prohibited by federal law. But Yoo's September 25 memo authorized such surveillance, in secret, "incident to" the authority Congress had just granted.[39] The memo dramatically enhanced the power of the executive branch by leaving the courts and Congress out of the loop.

As the ranking national security lawyer in the White House, John Bellinger should have had direct line responsibility for authorizing the new surveillance program. According to the *Post,* however, David Addington had contempt for Bellinger, and thought he was likely to object to the surveillance. As a result, to avoid a potential conflict with Bellinger, Addington, armed with Cheney's proxy, kept the program secret not just from Congress and the courts, but from Bellinger as well.[40] In other words, an operative of the vice president was keeping the White House lawyer in the dark.

According to Bruce Fein, a constitutional lawyer who served as associate deputy attorney general under Ronald Reagan, the new domestic spying program that eventually emerged from this authorization flouted the Foreign Intelligence Surveillance Act of 1978 (FISA) by directing the National Security Agency to subject American citizens to electronic surveillance "on [the president's] say-so alone." Writing in the *Washington Monthly,* Fein characterized the program as being based on "an imperial theory of inherent constitutional power that would empower [the president] to open mail, break in and enter homes, or torture detainees, even in violation of federal criminal statutes."[41]

About a month later, toward the end of October, Cheney called on his legal team to resolve another significant issue. The vice president had become frustrated by a key question that confronted U.S. forces fighting the war on terror: Should captured fighters from Al-Qaeda and the Taliban face federal trials, military courts-martial—or other procedures such as military commissions?[42] Federal trials, of course, would grant captured combatants all the legal privileges afforded by the U.S. criminal justice system. Military courts-martial would give them the rights granted by the Geneva Conventions. Military commissions, on the other hand, essentially served as tribunals designed to operate outside either civilian or military law and offered far fewer protections. Secret evidence could be used in a trial but withheld from the defendant, if it was deemed appropriate. Hearsay evidence was admissible. So was coerced testimony. Prisoners could be held indefinitely.

An interagency group led by Pierre Prosper, U.S. ambassador at large for war crimes, had already been assembled to deal with the thorny legal issues involved in resolving which forum to use, but it had yet to come up with a definitive answer. Prosper's group reportedly invited Cheney's legal team to join their meetings, but the vice president's lawyers did not respond. They were sick of waiting and knew what their boss wanted anyway. "The interagency was just constipated," a Cheney colleague told the *Post.*[43]

Instead of working with Prosper, the vice president's legal team undertook a dazzling bureaucratic tour de force that revealed Cheney at his vintage best. According to the *Post* series, Timothy Flanigan and David Addington decided it would be constructive to show the bureaucrats that the president could act "without their blessing—and without the interminable process that goes along with getting that blessing."[44] On November 6, John Yoo wrote another opinion, this one declaring that President Bush had the constitutional authority to institute military commissions without consulting either Congress or the federal courts. Yoo told the *Post* that he saw no need to consult Congress or the Justice Department.

Over the next week, Cheney's team adroitly guided its new policy to the Oval Office, avoiding along the way the oversight provided by normal bureaucratic channels in the State Department, the Justice Department, Congress, and potentially hostile White House lawyers and presidential advisers. On Veterans Day, the *Post* reported, Cheney faced down an irate Attorney General John Ashcroft, who was enraged that as the senior law enforcement officer in the land, he had been entirely cut out of the loop. After the meeting, Cheney not only barred Ashcroft from having an audience with Bush about the subject, he also delivered the coup de grâce.

As Gellman and Becker reported it:

> Three days after the Ashcroft meeting, Cheney brought the order for military commissions to Bush. No one told [White House national security lawyer John] Bellinger, [National Security Adviser Condoleezza] Rice, or Powell, who continued to think that Prosper's working group was at the helm.
>
> After leaving Bush's private dining room, the vice president took no chances on a last-minute objection. He sent the [unsigned] order on a swift path to execution that left no sign of his role. After Addington and Flanigan, the text passed to [Bradford] Berenson, the associate White House counsel. Cheney's link to the document broke there: Berenson was not told of its provenance.[45]
>
> Berenson rushed the order to deputy staff secretary Stuart W. Bowen Jr., bearing instructions to prepare it for signature immediately—without advance distribution to the president's top advisers. Bowen objected, he told colleagues later, saying he had handled thousands of presidential documents without ever bypassing strict procedures of coordination and review. He relented, one White House official said, only after "rapid, urgent persuasion" that Bush was standing by to sign and that the order was too sensitive to delay.

Three days later, on November 14, Cheney spoke before the U.S. Chamber of Commerce. Asserting that terrorists were not the same as lawful combatants, Cheney declared, "They don't deserve to be treated as a prisoner of war. They don't deserve the same guarantees and safeguards that would be used for an American citizen going through the normal judicial process. . . . They will have a fair trial, but it'll be under the procedures of a military tribunal."[46]

At the time, Bush's poll numbers were still in the stratosphere. In Afghanistan, the United States had just carpet-bombed enemy forces in the city of Mazari Sharif. Immediately afterward, the pro-U.S. Northern Alliance swept in and routed the Taliban there, leading to a collapse of

Taliban positions throughout the country. The pro-U.S. Northern Alliance swept through one province after another at will. On November 12, the Taliban fled the capital, Kabul. They were in full retreat throughout Afghanistan. American casualties were minuscule. Al-Qaeda and Taliban forces, possibly including Osama bin Laden himself, appeared to have been cornered, forced into the Tora Bora cave complex on the Pakistani border. Determined to fight back, and doing so victoriously, the American people were united as one.

But only a handful of Americans knew what was really going on in Washington. By concocting "theories for evading the law and Constitution that would have embarrassed King George III," as Bruce Fein put it,[47] Cheney had authorized domestic surveillance programs to spy on American citizens, thereby contravening the Foreign Intelligence Surveillance Act. By authorizing military commissions, he had given the president "the functions of judge, jury, and prosecutor in the trial of war crimes," according to Fein, claiming "the authority to detain American citizens as enemy combatants indefinitely . . . a frightening power indistinguishable from King Louis XVI's execrated *lettres de cachet* that occasioned the storming of the Bastille."[48]

He had shredded cherished constitutional guarantees, including the right of habeas corpus. On PBS's *Bill Moyers Journal*, Fein was more specific. "We're talking about assertions of power that affect the individual liberties of every American citizen," he said. "Opening your mail, your e-mails, your phone calls. Breaking and entering your homes. Creating a pall of fear and intimidation [so that] if you say anything against the president you may find retaliation very quickly."[49]

Perhaps most astonishing of all, somehow or other, Cheney had managed to accomplish all of this by dancing through the executive bureaucracy, leaving almost no footprints.

For the few remaining allies of the elder George Bush in the administration, the ongoing machinations of Cheney and the neocons, and George W. Bush's seeming acquiescence with them, created a difficult situation. Brent Scowcroft was a case in point.

In an administration in which the president was bereft of foreign policy expertise, no one had more experience than Scowcroft. As senior military assistant to Richard Nixon at the end of the Vietnam War, Scowcroft had impressed Henry Kissinger, then serving as Nixon's national security adviser. "I saw Scowcroft disagreeing with [Nixon chief of staff H. R.]

Haldeman, and Haldeman very imperiously tried to insist on his point of view," Kissinger told *The New Yorker's* Jeffrey Goldberg. "[Scowcroft] was a terrier who had got hold of someone's leg and wouldn't let it go. In his polite and mild manner, he insisted on his view, which was correct. It was some procedural matter, but he was challenging Haldeman at the height of Haldeman's power."[50]

Scowcroft's modest demeanor was something of an anomaly in a world dominated by swaggering foreign policy hands intoxicated by the allure of juggling the fate of nations.[51] With Kissinger as his patron, however, the soft-spoken, understated Scowcroft rose to become national security adviser to President Gerald Ford and held the same position later under George H.W. Bush. To the latter especially, he became a trusted counselor. The two men became so close that Scowcroft got a condominium in Kennebunkport, Maine, near the Bush family compound. "[Scowcroft is] not a blowhard. . . ." Bush 41 wrote in an e-mail to *The New Yorker.* "I mean someone I can depend on to tell me what I need to know and not just what I want to hear, and at the same time he is someone on whom I know I always can rely and trust implicitly."[52]

During the Gulf War, Scowcroft, perhaps more than any adviser in the administration, had convinced Bush 41 that Saddam's occupation of Kuwait demanded that the United States go to war. Likewise, he had been equally forceful in declaring that America's military response had to be tempered. A true realist, Scowcroft believed power must be used when necessary, but it should be used judiciously, appropriately, and pragmatically.

Now seventy-six and still at the top of his game, Scowcroft found himself in an especially awkward position. In his few months serving Bush 43 as head of the Foreign Intelligence Advisory Board, a bipartisan council established to advise the president on how effectively the intelligence community was serving the country, Scowcroft had already figured out that George W. Bush was not paying much attention to the board.[53] Given his decades of experience, he had expected to play a bigger role in the administration.

After 9/11, Scowcroft had revised a report he started before the attacks, and concluded that the U.S. intelligence apparatus had been designed to meet the needs of the Cold War era and therefore needed to be overhauled. The terrorism of 9/11 merely made his point more relevant. After all, the attacks had come from rogue Islamist terrorists based in underdeveloped Afghanistan, a country about which the United

States knew little—not a giant superpower like China or the former Soviet Union.

Rumsfeld, however, "was strongly opposed" to Scowcroft's ideas,[54] presumably because they would give more power to the CIA rather than the National Security Agency, which is under the authority of the Pentagon. Not long after discussing the issue with Rumsfeld, according to Ron Suskind's *The One Percent Doctrine: Deep Inside America's Pursuit of Its Enemies Since 9/11*, Scowcroft went to the Office of the Vice President to see what Cheney's take would be. Well aware that Cheney and Rumsfeld had a long friendship, shared everything, and often operated as a duo, Scowcroft sought to give Cheney an easy out. "Dick, look," he said. "My proposals would be disruptive. There's no question about it. If you think this is a bad time for it, I'll just fold my tent and go away. I don't want to. . . . But I'll be guided by you."[55]

Cheney now had the choice of calling Scowcroft off, thereby defusing a potential conflict within the administration, or sending Scowcroft out on a fool's errand, pitting Bush 41's close friend, as Suskind noted, against Bush 43's cabinet secretary, who just happened to be 41's lifelong nemesis. Cheney chose the latter. "Go ahead, submit the report to the President," Cheney said. Scowcroft had once been Cheney's mentor, his patron. Now the vice president was just humoring him.

By October, the Bush-Cheney administration was nine months old and increasingly it appeared to Scowcroft that he was being set up for failure or irrelevancy. It was hard not to notice the rise of the neocons and the increasing antipathy toward realists like him and Colin Powell.

In the State Department, tension was mounting daily. "There had been a sense of euphoria when Powell arrived," recalled one veteran State Department official, "because this was the guy who was going to lead us and bring prestige to the department. But suddenly there was realization that Powell wasn't driving foreign policy, that the State Department's role was being usurped by the Pentagon."

At the same time, in the immediate aftermath of the attacks, Under Secretary of Defense Douglas Feith put together what became known as the Counterterrorism Evaluation Group—a duo really—consisting of "Clean Break" coauthor David Wurmser and Michael Maloof, who was then with the Pentagon's Technology Security Operations, to work in a small windowless room in the Pentagon where they dug through raw, uncorroborated classified data that had not made it into the CIA's final reports. "We discovered tons of raw intelligence," Maloof told the *New York Times*'s James Risen. "We were stunned that we couldn't find any mention of it in the CIA's finished reports."[56] Each week, they reported to Steve Cambone, a fellow neocon and PNAC signatory who was then Feith's chief aide.*[57]

Maloof said that Cheney questioned the CIA's abilities because in 1990 the Agency had failed to uncover Saddam's ongoing nuclear program, which was exposed after the Gulf War. "After the war, when [former head of the Iraq Survey Group] David Kay and the inspectors went in," said Maloof, "they reported that Iraq was only six months away."[58]

Richard Perle, who was close to Feith, Wurmser, and Maloof, minced no words. "I think the people working on the Persian Gulf at the CIA are pathetic," he said. "They have just made too many mistakes. They have a record over 30 years of being wrong." He explained that the CIA's analyses did not allow for the possibility that Iraq was working with Al-Qaeda.[59]

Instead of relying on traditional intelligence techniques and sources, Maloof and Wurmser went to already discredited sources such as INC founder Ahmed Chalabi. "I knew Chalabi from years earlier," said Maloof, "so I basically asked for help in giving us direction as to where to look for information in our own system in order to be able to get a clear picture of what we were doing. They were quite helpful."[60]

But veteran intelligence analysts looked askance at their sourcing. "I don't have any problem with them bringing in a couple of people to take another look at the intelligence and challenge the assessments," said Patrick Lang, the former analyst for the DIA. "But the problem is that they brought in people who were not intelligence professionals, people were brought in because they thought like them. They knew what answers they were going to get."[61]

By October, the cry for toppling Saddam went out again and again from all the usual suspects—Robert Kagan,[62] William Kristol, Charles Krauthammer,[63] Richard Perle, William Safire, and Paul Wolfowitz in the *Weekly Standard*, the *New York Times*, the *Washington Post*, and the

*In November 2001, Paul Glastris wrote in the *Washington Monthly*: "It would be hard to exaggerate how much Secretary of Defense Donald Rumsfeld and his top aide Stephen Cambone were hated within the Pentagon prior to September 11. Among other mistakes, Rumsfeld and Cambone foolishly excluded top civilian and military leaders when planning an overhaul of the military to meet new threats, thereby ensuring even greater bureaucratic resistance. According to the *Washington Post*, an Army general joked to a Hill staffer that 'if he had one round left in his revolver, he would take out Steve Cambone.'"

National Review.[64] "It is absurd to claim . . ." wrote Safire, "that Iraq is not an active collaborator with, harborer of, and source of sophisticated training and unconventional weaponry for bin Laden's world terror network."[65]

The neocons begged for Bush to widen the war on terror. At times, these pleas for war took the form of taunts assailing Scowcroft's lack of machismo. "[R]egime change . . . makes [Scowcroft's] palms sweaty with anxiety," wrote Rich Lowry in the *National Review.* "Please, Condi," Lowry counseled the national security adviser, "stop taking his calls."[66]

As the close friend of Bush 41, and his public voice, Scowcroft was particularly sensitive to criticizing Bush 43 in public. Judicious to a fault, he generally preferred to work behind the scenes, discreetly giving his advice and letting people take or leave it. But by mid-October it was clear that they were choosing the latter. His recent meeting with Cheney would turn out to be his last.

So, on October 16, 2001, for the first time, Scowcroft went public in the *Washington Post*. Harking back to the Gulf War of 1991, he asserted that if the United States had gone on to Baghdad, "Our Arab allies . . . would have deserted us, creating an atmosphere of hostility to the United States . . . [that] might well have spawned scores of Osama bin Ladens. . . . [W]e already hear voices declaring that the United States is too focused on a multilateral approach. The United States knows what needs to be done, these voices say, and we should just go ahead and do it. Coalition partners just tie our hands. . . ." For the war on terror to be successful, he concluded, it would have to be "even more dependent on coalition-building than was the Gulf War."[67]

Scowcroft's words were measured, but they were exactly what the neocons did not want to hear. By now, it was clear to him that they were trying to use 9/11 as a springboard to invade Iraq. "He knew they were going to try to manipulate the president into thinking there was some unfinished business," said an administration official. "For him to say something publicly was a watershed. This was where the roads diverged."

At about the same time, in Italy, the Niger operation went into overdrive. The details of how this happened and who did what remain murky. Accounts from usually reputable newspapers in Great Britain, Italy, and the United States, from the U.S. Senate Intelligence Committee, and other outlets are at great variance with one another on a number of points. But in October 2001, the Italian intelligence service, SISMI, which had already sent reports about the alleged Niger deal to French intelligence, finally had the information forwarded to British and U.S. intelligence.[68] The exact dates of the distribution are unclear, but according to the British daily newspaper *The Independent*, Rocco Martino, the freelance SISMI operative, personally delivered the dossier to the Vauxhall Cross headquarters of the British Secret Intelligence Service (MI6) in south London.[69] Early that month, according to *La Repubblica*, SISMI also gave a report about the Niger deal to Jeff Castelli, the CIA station chief in Rome.[70]

Then, on October 15, 2001, Nicolò Pollari, the newly appointed chief of SISMI, made his first visit to his counterparts at the CIA in the United States. Under pressure from conservative Italian prime minister Silvio Berlusconi to turn over information that would be useful for America's Iraq war policy, Pollari met "with top CIA officials to provide a SISMI dossier indicating that Iraq had sought to buy uranium in Niger,"[71] according to an article by Philip Giraldi in the *American Conservative*.

Pollari's dossier, however, apparently did not include the actual forged documents themselves. According to the Senate Intelligence Committee, the analysts saw the report as "very limited and lacking needed detail."[72] Nevertheless, the State Department directed the U.S. embassy in Niger to check out the alleged uranium deal. On November 20, 2001, the U.S. embassy in Niamey, the capital of Niger, disseminated a cable reporting on a meeting between the American ambassador, Barbro Owens-Kirkpatrick, and the director general of Niger's French-led consortium. According to the Senate report, the head of the consortium told Ambassador Owens-Kirkpatrick that "there was no possibility" that the government of Niger had diverted any yellowcake to Iraq.[73]

Shortly afterward, State Department analysts also concluded that the Niger deal was a fraud. In December 2001, Greg Thielmann, director for strategic proliferation and military affairs at the State Department's Bureau of Intelligence and Research (INR), reviewed Iraq's WMD program for Secretary of State Colin Powell. "A whole lot of things told us that the report was bogus," said Thielmann. "This wasn't highly contested. There weren't strong advocates on the other side. It was done, shot down."[74]

Who was directing reports of an Iraqi-Niger yellowcake deal to American intelligence officials? And why did the reports keep coming back to U.S. officials even after they had been discredited? The answer to those questions may never be known. But some knowledgeable observers wondered if one of the people involved might be Michael Ledeen, the neocon analyst in Washington who had connections both to Italian intelligence and the top neocons in the Bush administration.

Waving an unlit cigar in his eleventh-floor office at the American Enterprise Institute in Washington on a December afternoon in 2005, Ledeen insisted otherwise. After nineteen years at the AEI, Ledeen holds the institute's Freedom Chair and rates a corner office decorated with prints of the Colosseum in Rome, the Duomo in Florence, and other mementos of his days in Italy when he studied Italian fascism in the seventies and was Rome correspondent for the *New Republic* in the eighties. Having served at various posts in the Pentagon, the State Department, and the National Security Council, Ledeen relished playing the role of the intriguer.*[75] During the Iran-contra scandal of the eighties, Ledeen won notoriety for introducing Oliver North to his friend the Iranian arms dealer Manucher Ghorbanifar, who was labeled "an intelligence fabricator" by the CIA. Ledeen has made his share of enemies along the way. According to Larry C. Johnson, a former CIA officer who was deputy director of the State Department Office of Counterterrorism from 1989 to 1993, "The C.I.A. viewed Ledeen as a meddlesome troublemaker who usually got it wrong and was allied with people who were dangerous to the U.S., such as Ghorbanifar."[76]

Apprised of such views, Ledeen, no fan of the CIA, responded, "Oh, that's a shock. Ghorbanifar over the years has been one of the most accurate sources of understanding what is going on in Iran. . . . I have always thought the CIA made a big mistake."[77]

Bearded and balding, the sixty-seven-year-old Ledeen made for an unlikely 007. On the one hand, he could be self-deprecating, describing

*Ledeen's fascination with fantasy and intrigue, albeit of a very different variety, dates back to his childhood in California where his father designed the air-conditioning system for Disney Studios. As Ledeen wrote in "Remembering My Family Friend, Walt Disney," he grew up having forged "a special bond" with the legendary Walt Disney himself, surrounded by Disney creatures, "from Mickey and Donald to Pooh and Eeyore." Family legend had it, he wrote, that "my mother was the model for Snow White." And for intrigue, Ledeen says that he "got to see Walt's 'secret room,' which you got to by pushing a button under his desk and then a wall panel opened and revealed a playroom full of all kinds of toys and gadgets. And his house was really a playhouse; there was a model train that ran from the kitchen out to the backyard, and on a good day the train would come puffing out with hamburgers and Cokes. Such fun."

himself as "powerless . . . and, well, schlumpy."[78] On the other, one of his bios grandiosely proclaimed that he has executed "the most sensitive and dangerous missions in recent American history."[79]

Propping his feet up on his desk next to an icon of villainy—a mask of Darth Vader—Ledeen explained, "I'm tired of being described as someone who likes fascism and is a warmonger. I've said it over and over again. I'm not the person you think you are looking for. . . . I think it's obvious I have no clout in the administration. I haven't had a role. I don't have a role." He barely knew Karl Rove, he said, and "had no professional relationship with any agency of the federal government during the Bush administration. That includes the Pentagon."

However, considerable evidence suggests that Ledeen has had far more access to the highest levels of the Bush administration than he lets on. Even before Bush took office, Rove asked Ledeen to funnel ideas to the White House.[80] According to U.S. Airforce Lieutenant Colonel Karen Kwiatkowski, who worked in the Pentagon during the run-up to the Iraq War, Ledeen "was in and out of [the Pentagon] . . . all the time." "Clean Break" signatory Meyrav Wurmser described Ledeen as being a close friend of her and her husband, David, who held key posts in the Pentagon and the State Department and went on to become Dick Cheney's chief Middle East adviser.[81] Through his ties to Rove and Deputy National Security Adviser Stephen Hadley,[82] Ledeen was also wired into key figures in the White House.

To Ledeen, Iraq was just one part of a larger war. When Scowcroft warned that a war with Iraq might turn the region into a cauldron, Ledeen wrote, "It's always reassuring to hear Brent Scowcroft attack one's cherished convictions; it makes one cherish them all the more. . . . One can only hope that we turn the region into a cauldron, and faster, please."[83]

"Faster, please" became his mantra, repeated incessantly in his *National Review* columns. Rhapsodizing about war week after week, in the aftermath of 9/11, seemingly intoxicated by the grandiosity of his fury, Ledeen became chief rhetorician for neoconservative visionaries who wanted to remake the Middle East. "Creative destruction* is our middle name, both within our own society and abroad," he wrote just

*The phrase *creative destruction* is most often associated with the economist Joseph Schumpeter, who used it to describe the transformation whereby entrepreneurs introduce radical innovations that create long-term growth but destroy the value of older companies in the process.

nine days after the attacks. "We must destroy [our enemies] to advance our historic mission."[84]

The U.S. must be "imperious, ruthless, and relentless," he argued, until there has been "total surrender" by the Muslim world. "We must keep our fangs bared," he wrote, "we must remind them daily that we Americans are in a rage, and we will not rest until we have avenged our dead, we will not be sated until we have had the blood of every miserable little tyrant in the Middle East, until every leader of every cell of the terror network is dead or locked securely away, and every last drooling anti-Semitic and anti-American mullah, imam, sheikh, and ayatollah is either singing the praises of the United States of America, or pumping gasoline, for a dime a gallon, on an American military base near the Arctic Circle."[85]

If Ledeen seems like too marginal a figure to have played a role in the run-up to war, and Rome an unlikely site for unseemly events, it is worth recalling that Ledeen has a long history of ties to Italian intelligence. In addition, it is important to understand that in Italy ultraconservative Cold Warriors battled the Communists not just electorally but through undercover operations in the intelligence world. "In addition to the secret service, SISMI, there was another, informal, parallel secret service," said Guido Moltedo, an editor at *Europa*, a center-left daily newspaper in Italy. "It was known as Propaganda Due."[86]

Led by a neo-fascist named Licio Gelli, Propaganda Due, or P-2, with its penchant for secret rituals and exotic covert operations, was the stuff of conspiracy fantasies—except that it was real. According to the *Sunday Times* of London, until 1986 members agreed to have their throats slit and tongues cut out if they broke their oaths. Subversive, authoritarian, and right wing, the group was sometimes referred to as the P-2 Masonic Lodge because of its ties to the secret society of Masons, and it served as the covert intelligence agency for militant anticommunists. It was also linked to Operation Gladio, a secret paramilitary wing of NATO that supported far-right military coups in Greece and Turkey during the Cold War.

In 1981 the Italian Parliament banned Propaganda Due, and all secret organizations in Italy, after an investigation concluded that it had infiltrated the highest levels of Italy's judiciary, parliament, military, and press, and was tied to assassinations, kidnappings, and arms deals around the world. But before it was banned, P-2 members and their allies participated in two ideologically driven international black-propaganda schemes that foreshadowed the Niger embassy job twenty years later. The first took place in 1980, when Francesco Pazienza, a charming and sophisticated Propaganda Due operative in SISMI, allegedly teamed up with Michael Ledeen, who was then the Rome correspondent for the *New Republic.*

According to the *Wall Street Journal,* Pazienza said he first met Ledeen that summer, through a SISMI agent in New York who was working under the cover of a U.N. job.[87] Ultimately, the investigation in *The Wall Street Journal,* articles in *The Nation,* and a book by Frank Brodhead, *The Rise and Fall of the Bulgarian Connection,* linked Ledeen and SISMI to two major international disinformation scams. One of the operations targeted Billy Carter, President Jimmy Carter's hard-drinking younger brother, for having financial ties to Libyan dictator Muammar Qaddafi in what became known as the Billygate scandal. The second one, which was known as the Bulgarian Connection, involved a series of articles by Ledeen and others that falsely tied the 1981 attempted assassination of Pope John Paul II to the KGB.

According to the 1985 investigation by Jonathan Kwitny in the *Wall Street Journal,* Ledeen's Billygate articles, which appeared in *The New Republic* in America and *Now* magazine in Great Britain, were part of a SISMI disinformation operation[88] to tilt the 1980 American presidential election. Billy Carter admitted accepting a $220,000 loan from Qaddafi's regime, but the *Journal* reported that he "wasn't the only one allegedly getting money from a foreign government." According to Pazienza, Kwtiney reported, Michael Ledeen had received at least $120,000 from SISMI[89] for his work on Billygate and other projects. Ledeen even had a coded identity, Z-3, and allegedy was paid via a Bermuda bank account.

Ledeen told the *Journal* that a consulting firm he owned, ISI, worked for SISMI and may have received the money.[90] He said he did not recall whether or not he had a coded identity.

Pazienza was subsequently convicted in absentia on multiple charges, including having used extortion and fraud to obtain embarrassing facts about Billy Carter. Ledeen was never charged with any crime, but he was cited in Pazienza's indictment, which read, "With the illicit support of the SISMI and in collaboration with the well-known American 'Italianist' Michael Ledeen, Pazienza succeeded in extorting, also using fraudulent means, information . . . on the Libyan business of Billy Carter, the brother of the then President of the United States."[91]

Ledeen denied having worked with Pazienza or Propaganda Due as part of a disinformation scheme. "I knew Pazienza," he explained. "I didn't think P-2 existed. I thought it was all nonsense—typical Italian fantasy."[92] He added, "I'm not aware that anything in [the Billygate] story turned out to be false."[93] Asked if he had worked with SISMI, Ledeen said, "No." But then he added, "I had a project with SISMI—one project." He described it as a simple "desktop" exercise in 1979 or 1980, in which he taught Italian intelligence how to deal with U.S. officials on extradition matters. His fee, he said, was about $10,000. (For more on Ledeen's role in these disinformation operations, see Note 94 in Chapter 14.)[94]

When Reagan took office, Ledeen was made special assistant to Alexander Haig, Reagan's secretary of state. Ledeen later took a staff position on Reagan's National Security Council and played a key role in initiating the illegal arms-for-hostages deal with Iran that became known as the Iran-contra scandal.

In 1981, P-2 was outlawed and police raided the home of its leader, Licio Gelli. Authorities found a list of nearly a thousand prominent public figures in Italy who were believed to be members. Among them was a right-wing billionaire media mogul who had not yet entered politics—Silvio Berlusconi.[95] Officially, Propaganda Due was dead, but in some ways, it was becoming more powerful than ever.

In 1994, Berlusconi was elected prime minister. Rather than distance himself from the organization, he told a reporter that "P-2 had brought together the best men in the country," and he began to execute policies very much aligned with it.

Among those Berlusconi appointed to powerful national security positions were two men known to Ledeen. A founding member of Forza Italia, Berlusconi's right-wing political party, Minister of Defense Antonio Martino was a well-known figure in Washington neocon circles and had been close friends with Ledeen since the 1970s. Ledeen also occasionally played bridge with the head of SISMI under Berlusconi, Nicolò Pollari. "Michael Ledeen is connected to all the players," said Philip Giraldi, who was stationed in Italy with the CIA in the 1980s and has been a keen observer of Ledeen over the years.[96]

Exactly how much clout Ledeen carried within the Bush administration has been a matter of debate, but one measure of his influence may be a series of secret meetings he set up—with the approval of then deputy national security adviser Stephen Hadley, he claims—in Rome in the second week of December 2001. Among those in attendance were Harold Rhode, a protégé of Ledeen, and Larry Franklin, who later worked in the Pentagon's Office of Special Plans supplying Bush officials with raw intelligence regarding Iraq that had not been vetted by veteran intelligence analysts. (In a separate matter, Franklin has since pleaded guilty to passing secrets to Israel and been sentenced to twelve years in prison.) Manucher Ghorbanifar was also present and helped Ledeen arrange the meetings.

Ledeen claims that Ghorbanifar and his sources produced valuable information at the 2001 meetings about Iranian plans for attacking U.S. forces in Afghanistan. "That information saved American lives in Afghanistan," Ledeen said.[97]

But there were also reports that there was some discussion of destabilizing Iran. According to an article in the *Washington Monthly* by Joshua Micah Marshall, Laura Rozen, and Paul Gastris, the meetings raised the possibility "that a rogue faction at the Pentagon was trying to work outside normal U.S. foreign policy channels to advance a 'regime-change' agenda."*[98] One indication that such plans might be under consideration was the reported presence of Members of the Mujahideen e-Khalq, or MEK, an urban-guerrilla group of Iranian dissidents that practiced a brand of revolutionary Marxism.[99]

To old hands in the State Department who had survived the Iran-contra scandal, the appearance of Ledeen on the scene was like a recurring nightmare. The reappointment of Elliott Abrams earlier that year was bad enough, but then came Ledeen with Ghorbanifar right behind. "One of the truly remarkable elements of the neocon story is their addiction to Ghorbanifar," said a State Department official. "It is part of their 'we are smarter, you are stupid' attitude." The key players in Iran-contra were back in business.[100]

Meanwhile, Italian intelligence was frustrated by the CIA's refusal to legitimize the report of the Niger yellowcake deal. According to *La Repubblica,* at an unspecified date, SISMI director Nicolò Pollari discussed the issue with Ledeen's longtime friend Minister of Defense Antonio Martino. Martino, the paper reported, told Pollari to expect a visit from "an old friend of Italy,"[101] namely Ledeen. Soon afterward, according to *La Repubblica,* Pollari allegedly took up the Niger matter with Ledeen when he was in Rome. Ledeen denied having had any such conversations with Pollari. Pollari declined to be interviewed and

*Asked if there was any talk at the Rome meeting about destabilizing Iran, Ledeen replied by e-mail, "Please mr unger, have the good manners to go away and stay away."

has denied playing any role in the Niger affair. Antonio Martino has declined to comment.

Freelance operative Rocco Martino, no relation to Antonio Martino, said his only motive in distributing the documents was money. "He was not looking for great amounts of money—$10,000, $20,000, maybe $40,000," said Carlo Bonini, who coauthored the Nigergate stories for *La Repubblica.*

SISMI director Nicolò Pollari acknowledged that Rocco Martino has worked for Italian intelligence in the past. But he claimed that Italian intelligence played no role in the Niger operation. "[Nucera] offered [Martino] the use of an intelligence asset [La Signora]—no big deal, you understand—one who was still on the books but inactive—to give a hand to Martino," Pollari told a reporter.[102]

Rocco Martino, however, said SISMI had another agenda: "SISMI wanted me to pass on the documents, but they didn't want anyone to know they had been involved."[103]

Whom should we believe? Characterized by *La Repubblica* as "a failed carabiniere and dishonest spy," a "double-dealer" who "plays every side of the fence,"[104] Martino has reportedly been arrested for extortion and for possession of stolen checks, and was fired by SISMI in 1999 for "conduct unbecoming." Elsewhere he has been described as "a trickster" and "a rogue." He is a man who traffics in deception.

On the other hand, operatives like Martino are highly valued precisely because they can be discredited so easily. "If there were a deep-cover unit of SISMI, it would make sense to use someone like Rocco," said Patrick Lang. "His flakiness gives SISMI plausible deniability. It's their cover story. That's standard tradecraft with the agencies."[105] In other words, Rocco Martino may well have been the cutout for SISMI, a "postman," in the parlance of the intelligence world, who, if he dared to go public, could be disavowed.

Martino declined to talk to the press after an initial spate of interviews. Before going silent, however, he gave interviews to Italian, British, and American journalists characterizing himself as a pawn who distributed the documents on behalf of SISMI and believed that they were authentic. "I sell information, I admit," Martino told the *Sunday Times* of London, using his pseudonym, Giacomo. "But I sell only good information."

There was, however, a problem with what he was selling. Documents in the Niger dossier were not just forged, they were full of errors.

A letter dated October 10, 2000, was signed by Minister of Foreign Affairs Allele Elhadj Habibou—even though he had been out of office for more than a decade. Its September 28 postmark indicated that somehow the letter had been received nearly two weeks before it was sent.[106] In another letter, President Tandja Mamadou's signature appeared to be phony. The accord signed by him referred to the Niger constitution of May 12, 1965, when a new constitution had been enacted in 1999. One of the letters was dated July 30, 1999, but referred to agreements that were not made until a year later. Finally, the agreement called for the five hundred tons of uranium to be transferred from one ship to another in international waters—a spectacularly difficult feat.

In time, the Niger documents and reports based on them made at least three journeys to the CIA. They also found their way to the U.S. embassy in Rome, to the White House, to British intelligence, to French intelligence, and to Elisabetta Burba, a journalist at *Panorama,* the Milan-based newsmagazine. Each of these recipients in turn shared the documents or their contents with others, in effect creating an echo chamber that gave the illusion that several independent sources had corroborated an Iraq-Niger uranium deal.

"It was the Italians and Americans together who were behind it. It was all a disinformation operation," Rocco Martino told a reporter at England's *Guardian* newspaper. He called himself "a tool used by someone for games much bigger than me."[107]

What exactly might those games have been? Berlusconi defined his role on the world stage largely in terms of his relationship with the United States, and he jumped at the chance to forge closer ties with the White House when Bush took office in 2001. In its three-part series on Nigergate, *La Repubblica* charged that Berlusconi was so eager to win Bush's favor that he "instructed Italian Military Intelligence to plant the evidence implicating Saddam in a bogus uranium deal with Niger." (The Berlusconi government, which lost power in April 2006, denied the charge.)

The idea that the Niger documents were part of disinformation rings especially true to Frank Brodhead, coauthor of *The Rise and Fall of the Bulgarian Connection,* the book about the Cold War disinformation plot to blame the attempted assassination of Pope John Paul II on the Soviet Union's KGB—and Michael Ledeen's role in it. "When I read that the Niger break-in took place before Bush took office, I immediately thought back to the Bulgarian Connection," said Brodhead.[108] "That job had been done during the transition [between presidential administrations] as well. Ledeen . . . saw himself as making a serious contribution to the Cold War through the Bulgarian Connection. Now, it was possible, twenty years later, that he was doing the same to start the war in Iraq."

Brodhead is not alone. Several media outlets—the *San Francisco Chronicle,* United Press International, and the *American Conservative*—and many more bloggers—Daily Kos,[109] TPM Café,[110] the Left Coaster,[111] and Raw Story[112] among them—have addressed the question of whether Ledeen was involved with the Niger documents. But none has found any hard evidence. One highly regarded intelligence analyst, Tyler Drumheller, who served more than twenty-five years in the CIA, thinks too much has been made of them. "I don't buy the conspiracy theories that have erupted . . . suggesting that someone in the United States faked the documents in order either to discredit the president or generate intelligence that would back up the war," Drumheller argued in *On the Brink: An Insider's Account of How the White House Compromised American Intelligence.* "It would have been far easier just to leak the story to the press. There would have been no need for all the intrigue. . . ."[113]

On the other hand, most serious intelligence veterans seem to believe that the Niger documents were part of a secret plot to start the war. "This wasn't an accident," said Milt Bearden, a thirty-year CIA veteran who was a station chief in Pakistan, Sudan, Nigeria, and Germany, and the head of the Soviet–East European division. "This wasn't 15 monkeys in a room with typewriters."[114]

To help trace the path of the Niger documents and unravel its mysteries, a number of former intelligence and military analysts who have served in the CIA, the State Department, the Defense Intelligence Agency (DIA), and the Pentagon were interviewed for this book. Some of them referred to the Niger documents as "a disinformation operation," others as "black propaganda," "black ops," or "a classic psy-ops [psychological-operations] campaign." But whatever term they used, at least nine of these officials believed that the Niger documents were part of a covert operation to deliberately mislead the American public and start a war with Iraq.

The nine officials were Milt Bearden; Colonel W. Patrick Lang; Colonel Larry Wilkerson, former chief of staff to Secretary of State Colin Powell; Melvin Goodman, a former division chief and senior analyst at the CIA and the State Department; Ray McGovern, a CIA analyst for twenty-seven years; Lieutenant Colonel Karen Kwiatkowski, who served in the Pentagon's Near East and South Asia division in

2002 and 2003; Larry C. Johnson, a former CIA officer who was deputy director of the State Department Office of Counterterrorism from 1989 to 1993; former CIA official Philip Giraldi; and Vincent Cannistraro, the former chief of operations of the CIA's Counterterrorism Center.

"In the world of fabrication, you don't just drop something and let someone pick it up," Bearden added. "Your first goal is to make sure it doesn't find its way back to you, so you do several things. You may start out with a document that is a forgery, but you use a photocopy of a photocopy of a photocopy, which makes it hard to track down. You go through cut-outs so that the person who puts it out doesn't know where it came from. And you build in subtle, nuanced errors so you can say we would never do that. If it's very cleverly done, it's a chess game, not checkers."[115]

By early 2002, U.S. military and intelligence professionals had seen the Niger reports repeatedly discredited, and assumed that the issue was dead. But they were still alive. "[The neocons in the Pentagon] delighted in telling people, 'You don't understand your own data,'" said Patrick Lang.[116] "'We know that Saddam is evil and deceptive, and if you see this piece of data, to say just because it is not well supported it's not true is to be politically naïve.'"

Nevertheless, the CIA had enough doubts about the Niger claims initially to leave them out of the President's Daily Briefing (PDB). On February 5, 2002, however, for reasons that remain unclear, the CIA issued a new report on the alleged Niger deal, one that provided significantly more detail, including what was said to be "verbatim text" of the accord between Niger and Iraq. In the State Department, analysts were still suspicious of the reports.[117]

But in the Pentagon, the neocons pounced on the new material. On February 12, the DIA issued "a finished intelligence product," titled "Niamey Signed an Agreement to Sell 500 Tons of Uranium a Year to Baghdad,"[118] and passed it to the office of Vice President Dick Cheney.

Even though the CIA had dismissed the Niger claims, Cheney was able to give them new life. "The [CIA] briefer came in. Cheney said, 'What about this?,' and the briefer hadn't heard one word, because no one in the agency thought it was of any significance," said Ray McGovern, whose job at the CIA included preparing and delivering the PDB in the Reagan era.[119] "But when a briefer gets a request from the vice president of the United States, he goes back and leaves no stone unturned."

The CIA's Directorate of Operations, the branch responsible for the clandestine collection of foreign intelligence, immediately tasked its Counterproliferation Division (CPD) with getting more information. According to the Senate Select Committee on Intelligence report,[120] just hours after Dick Cheney had gotten the Niger report, Valerie Plame Wilson, who worked in the CPD, wrote a memo to the division's deputy chief that read, "My husband has good relations with both the PM [prime minister] and the former Minister of Mines (not to mention lots of French contacts), both of whom could possibly shed light on this sort of activity."

Her husband, as the world soon learned, was Joseph Wilson, who had served as deputy chief of mission at the U.S. embassy in Baghdad and as ambassador to Gabon under George H.W. Bush. Once he got the assignment, Wilson approached the task with a healthy skepticism. "The office of the vice president had asked me to check this out," Wilson said.[121] "My skepticism was the same as it would have been with any unverified intelligence report, because there is a lot of stuff that comes over the transom every day."

At the time, Wilson was oblivious to any schism in the intelligence world. "I was aware that the neocons had a growing role in government and that they were interested in Iraq," he said. "But the administration had not articulated a policy at this stage."[122]

Before leaving for Niger, Wilson was not given the Niger documents, nor was he aware of their history. "To the best of my knowledge the documents were not in the possession of the agency at the time I was briefed," he said. "The discussion was whether or not this report could be accurate. During this discussion, everyone who knew something shared stuff about how the uranium business worked and I laid out what I knew about how the government works in Niger, what information they could provide."

He arrived in Niger on February 26, 2002. "Niger has a simplistic government structure," he said. "Both the minister of mines and the prime minister had gone through the mines. The French were managing partners of the international consortium. The French mining company actually had its hands on the product. Nobody else in the consortium had operators on the ground."[123]

In addition, Wilson personally knew Wissam al-Zahawie, the Iraqi ambassador to the Vatican, whose visit to Niger had raised suspicions. "Wissam al-Zahawie was a world-class opera singer, and he went to the Vatican as his last post so he could be near the great European opera houses in Rome," said Wilson. "He was not in the Ba'athist inner circle.

He was not in Saddam's tribe. The idea that he would be entrusted with this super-secret mission to buy five hundred tons of uranium from Niger is out of the question."[124]

On March 1, the State Department weighed in with another cable, this one headed "Sale of Niger Uranium to Iraq Unlikely." Citing "unequivocal" control of the mines, the cable asserted that President Tandja of Niger would not want to risk good relations with the United States by trading with Iraq, and cited the prohibitive logistical problems in a transaction requiring "25 hard to conceal 10-ton-tractor trailers" that would have to travel a thousand miles and cross one international border before reaching the sea.[125]

A few days later, Wilson returned from Niger and told CIA officials that he had found no evidence to support the story about the alleged uranium deal. By now the Niger reports had been discredited more than half a dozen times—by the French in 2001, by the CIA in Rome and in Langley, by the State Department's INR, by some analysts in the Pentagon, by the ambassador to Niger, by Wilson, and yet again by the State Department.

But the top brass at the CIA knew what Cheney wanted. They went back to French intelligence again—twice. According to the *Los Angeles Times,* the second request that year, in mid-2002, "was more urgent and more specific." The CIA sought confirmation of the alleged agreement by Niger to sell five hundred tons of yellowcake to Iraq. Alain Chouet, the French intelligence official, reportedly sent five or six men to Niger and again found the charges to be false. Then his staff noticed that the allegations matched those brought to him by Rocco Martino. "We told the Americans, 'Bullshit. It doesn't make any sense.'"[126]

The formal cables sent to the CIA did not use precisely those words, Chouet said, but the language had been candid. "We had the feeling we had been heard," he said.

By late summer 2002, Brent Scowcroft, in his efforts to stop the neocon onslaught, had resembled a lone figure standing on a railroad track in hopes of stopping a powerful oncoming locomotive. Minutes of a July 20, 2002, meeting between British prime minister Tony Blair and his intelligence chiefs succinctly explained the challenge Scowcroft faced, with Sir Richard Dearlove, head of MI6, the British intelligence service, saying, "Bush wanted to remove Saddam, through military action, justified by the conjunction of terrorism and [weapons of mass destruction]. But the intelligence and the facts were being fixed around the policy."[127]

In June, Ghorbanifar had had a second unauthorized meeting with unnamed Pentagon officials and Iranians, this time in Paris. In July and August, the Pentagon's Counterterrorism Evaluation Group (CTEG) under Doug Feith went toe to toe with the CIA to make its baseless case that Iraq had supported Al-Qaeda. Defectors such as Khidir Hamza, who played a role in Iraq's nuclear program, came out of the woodwork in droves with frightening, though questionable, stories about Saddam's WMDs. All of which were being shaped and delivered to the press. "To them the press was a bunch of marionettes whose strings are there to be pulled," said one administration official.

As for Ahmed Chalabi, when he wasn't feeding dubious intelligence to Feith's group in the Pentagon, he was off to Tehran to meet with Ayatollah Sayed Mohammed Baqir al-Hakim, the leader of the Supreme Council for the Islamic Revolution in Iraq, a radical group dedicated to overthrowing Saddam that had close ties to Iran. "Everybody was using everybody," said a State Department official. "Chalabi was using the neocons, the neocons were using Chalabi. Iran was using Chalabi, Chalabi was using Iran. Cheney and Rumsfeld were using the neocons, the neocons were using them. Everybody had his own little thing going."

White House aides openly mocked journalists who actively sought to determine what was really taking place. An unnamed senior adviser to Bush, unhappy about an article in *Esquire* by Ron Suskind, patiently told the journalist that guys like him are "in what we call the reality-based community"—people who "believe that solutions emerge from your judicious study of discernible reality. That's not the way the world really works anymore. We're an empire now, and when we act, we create our own reality. And while you're studying that reality—judiciously, as you will—we'll act again, creating other new realities, which you can study too, and that's how things will sort out. We're history's actors . . . and you, all of you, will be left to just study what we do."[128]

It was in this brazen and grandiose context that Brent Scowcroft, perhaps not conniving enough for his times, stepped out of the shadows and, in his quiet deliberate manner, began his own mini–media campaign to try to head off war with Iraq—one man against the neocons. Scowcroft discussed with the former president what he was doing, not just because he was on a collision course with his good friend's son, but because it was such a delicate matter. In short order, he appeared on Fox News, BBC, and, on August 4, CBS's *Face the Nation*—with the elder Bush's blessing.

Each time, Scowcroft's case was essentially the same one he and the senior George Bush had crafted during the Gulf War and in *A World Transformed.* On *Face the Nation,* he warned that a unilateral invasion of Iraq could destabilize the Middle East and undermine efforts to defeat international anti-American militant groups. "It's a matter of setting your priorities," he said. "There's no question that Saddam is a problem. He has already launched two wars and spent all the resources he can working on his military. But the president has announced that terrorism is our number one focus. Saddam is a problem, but he's not a problem because of terrorism."

According to a source who was familiar with Scowcroft's efforts, Bush senior "appreciated the effort that was being made and understood what was going on." For the most part, however, Scowcroft, though still chairman of the Presidential Foreign Intelligence Advisory Board, was working at an abstract level trying to create a different discourse about the Middle East. After more than two decades in the inner sanctum of the White House, he had been shut out of key policy discussions, and was now almost entirely removed from the machinations that had set the policy machine in motion.

Worse, in an administration that boasted high-profile centrists such as Colin Powell and Condoleezza Rice, Scowcroft stood alone. His qualms about the rush to war notwithstanding, Powell, Hamlet-like, did not act to slow the momentum. By May 2002, appearing on ABC's *This Week,* Powell had effectively bowed to the pressures of the neocons. "The United States reserves its option to do whatever it believes might be appropriate to see if there can be a regime change. . . ." he said. "U.S. policy is that regardless of what the inspectors do, the people of Iraq and the people of the region would be better off with a different regime in Baghdad."[129]

So, when it came to framing the essential foreign policy questions of the day, the questions that would ultimately mean war or peace, life or death to thousands of people, and the fate of America's strategic position in the world, Scowcroft was relying on his old protégée National Security Adviser Condoleezza Rice, who had become very close to the president. But Rice, either because she lacked Scowcroft's convictions or because she did not have the bureaucratic savvy to outwit the neocons, or perhaps both, failed to frame Scowcroft's policies as an option for Bush to consider.

Even with his decades of experience and close friendship with the president's father, Scowcroft could not get his ideas into the Oval

Office. Worse, the clearer it was that Scowcroft's criticism was sanctioned by Bush 41, the more resentful Bush 43 became.

On August 15, Scowcroft went at it again, with an opinion piece in the *Wall Street Journal,* "Don't Attack Saddam," that made its way into the Oval Office. When George W. Bush found out about it, he was neither receptive to its message nor generous toward its writer, his father's loyal friend. "Scowcroft has become a pain in the ass in his old age," the president said.[130]

CHAPTER FIFTEEN

Fear: The Marketing Campaign

Throughout most of the summer of 2002, the American people remained largely oblivious to the Bush administration's emerging plans for war against Iraq. But in August 2002, Douglas Feith put together the Office of Special Plans (OSP) in the Pentagon. A larger, more powerful successor to the Policy Counterterrorism Evaluation Group that Feith ran with David Wurmser and Michael Maloof, the OSP's mission was to ferret out evidence that Saddam had close ties to Al-Qaeda and a huge stockpile of WMDs, including a nuclear weapons program, and as such posed a grave threat both to the region and to the United States. The existence of the OSP effectively meant that Cheney, Rumsfeld, and the neocons had declared war on the CIA by creating a bureaucratic operation whose sole purpose was to circumvent and subvert the nation's statutorily authorized intelligence apparatus.

Astoundingly, CIA director George Tenet, ever anxious to ingratiate himself with the White House, didn't make a peep about the new rogue intelligence outfit that was challenging the Agency. "That's totally unacceptable for a CIA director," said Greg Thielmann, chief of the State Department's Bureau of Intelligence and Research (INR).[1] Deeply wary of Cheney, Tenet by now had gotten the message loud and clear from the White House that he should do everything within his power to make sure the CIA got the goods on Saddam. But in effect, by remaining silent about the OSP, Tenet was betraying his own men at the CIA—and the Agency's mission.

The pressure on Tenet was heightened by a September 5, 2002, meeting of the Senate Intelligence Committee in which he was asked what basis there was in the National Intelligence Estimate (NIE) for a preemptive war against Iraq.[2] "We've never done a National Intelligence

Estimate on Iraq, including its weapons of mass destruction," Tenet replied.[3]

Senator Bob Graham (D-Fla.), the ranking Democrat on the Intelligence Committee, was stunned. War loomed on the horizon and Congress couldn't really know what was going on without an NIE. "We do these on almost every significant activity—much less significant than getting ready to go to war," he said. "We were flying blind."[4]

Even more astounding, when the Senate asked Tenet for an NIE, the CIA director was reluctant to provide one. "We're doing a lot of other things," Tenet said.[5] "Our staff is stretched thin."

But with a Senate vote to authorize military force against Iraq just three weeks away, Graham insisted. "We said: 'We don't care. This is the most important decision that we as members of Congress and that the people of America are likely to make in the foreseeable future. We want to have the best understanding of what it is we're about to get involved with.'"[6]

One reason for Tenet's objection may have been that the intelligence the CIA was uncovering wasn't exactly the kind of material that would endear him to the White House. As it happened, the CIA had just accomplished what may have been its greatest single intelligence achievement with regard to Iraq. It had actually penetrated Saddam's inner sanctum by "turning" Iraqi foreign minister Naji Sabri, a high-level cabinet official. Tenet delivered the news personally to Bush, Cheney, and other top officials in September 2002. Initially, the White House was ecstatic about this coup.

That reaction, however, changed dramatically when they heard what Sabri had to say. "He told us that they had no active weapons-of-mass-destruction program," said Tyler Drumheller, the CIA's chief of operations in Europe until 2005, in an interview on CBS's *60 Minutes*.[7] On September 18, 2002, George Tenet personally delivered news to President Bush. "Tenet told me he briefed the president personally," a former CIA officer told *Salon*.[8] According to the article, Bush insisted that Sabri was simply telling them "the same old thing." "The president had no interest in the intelligence," the CIA officer said. Another officer said, "Bush didn't give a fuck about the intelligence. He had his mind made up."

The CIA officers, however, continued to corroborate material given them by Sabri. When French intelligence listened in on his telephone conversations, they shared them with the CIA and found that the phone taps backed up Sabri's claims, according to one of the CIA officers. Even though Bush had been told firsthand that Saddam had no WMDs, according to Drumheller, the White House was "no longer interested.... They said, 'Well, this isn't about intel anymore. This is about regime change.'"

In the first year and a half of the Bush administration, Tenet's back-slapping affability had won the favor of the president and helped him survive Cheney's attempts to replace him. But as a result, he was trapped between pleasing the White House, on the one hand, and, on the other, his loyalty to the CIA.

Now the conflict-averse Tenet had to do precisely what he had so assiduously avoided—deliver the NIE, the definitive word on Saddam's WMD program. Worse, the Agency had only three weeks to come up with it.[9] This was the kind of document that should have been prepared at the outset of the Bush administration and updated whenever there were significant developments.[10]

All of which meant the CIA frantically had to verify dubious bits of intelligence, including one questionable nugget that Tenet had indiscreetly divulged to the Senate Select Intelligence Committee and the Senate Armed Services Committee. "We know Iraq has developed a redundant capability to produce biological warfare agents using mobile production units," he had asserted.[11]

The source of that charge was an Iraqi émigré in Germany code-named Curveball, who described himself as a former chemical engineer who had worked at the Chemical Engineering and Design Center near the Rashid Hotel in central Baghdad.[12] Curveball's reports had been flowing into the Pentagon since the spring of 2000, a year before the Bush-Cheney administration took office.[13] Over time, he had delivered about a hundred reports on secret WMD programs to his contacts at BND (Bundesnachrichtendienst), the German intelligence agency.[14] His accounts were highly prized because he claimed to have actually produced biological weapons in a mobile lab. No other informant had ever been inside one, and his story was exactly what Cheney was hoping to find.

Most significantly, Curveball was a proprietary source of the BND, which passed its information from him to the Pentagon's Defense HUMINT Service.[15] In other words, even though the United States had no direct access to Curveball, Tenet was so anxious to please the White House that he had given the Senate this explosive, but unsubstantiated, revelation. But now, with the crucial Senate vote over the war imminent, Tenet had to make sure Curveball was for real. Not long after meeting with the Senate Intelligence Committee, Tenet asked Tyler Drumheller to get direct access to Curveball.

According to a report in the *Los Angeles Times* by Bob Drogin and

John Goetz, Drumheller soon met with the BND station chief in Washington at a seafood restaurant in Georgetown,[16] but the results were disappointing. Not only did his German counterpart decline to make Curveball available, but he also told Drumheller something disturbing. "I think the guy is a fabricator," he said. "We also think he has psychological problems. We could never validate his reports."[17]

This kind of news was not terribly unusual in the intelligence world. Hundreds of Iraqi refugees fled to Germany every month because of its generous benefits, and a handful of them tried to improve their conditions by offering exaggerated intelligence reports. "The Iraqis were adept at feeding us what we wanted to hear," said an official at the Pentagon's Defense Intelligence Agency (DIA) who debriefed Iraqi émigrés in Germany. "Most of it was garbage."[18]

Other allied analysts were dubious about Curveball's authenticity as well. The British intelligence service MI6 had told the CIA that his behavior was "typical of . . . fabricators."[19] According to the *Los Angeles Times,* satellite photos taken in 1997 conflicted with Curveball's description of the Iraqi warehouse where he supposedly worked.[20] Analysts at the DIA had also determined that Curveball was a liar,[21] and in May 2002 had issued a "burn notice" warning to agents to steer clear of the man.

More to the point, when Drumheller reported his doubts to his superiors, to his astonishment he found himself embroiled in the most contentious meetings he had ever seen at the CIA during his entire career. "People were cursing. These guys were absolutely violently committed to it," Drumheller said.[22]

Unbeknownst to Drumheller, a draft of the National Intelligence Estimate had already been written containing Curveball's accounts, which meant that its authors were deeply invested in Curveball's legitimacy.[23] That meant failure to validate Curveball would be an unwanted complication—very unwanted. The chief of staff to John McLaughlin, the deputy director of Central Intelligence, was especially distraught when Drumheller told him that Curveball might be a fabricator.

"Man, I hope not," he said, "because this is really the only substantive part of the NIE."[24]

Suddenly Drumheller understood the magnitude of the problem. The Iraqi émigré he was investigating—who was said to be not just unreliable, but also an alcoholic and someone to whom no American agent had even spoken firsthand—was the *only* source for the vital intelligence about mobile weapons vans, and his reports were playing a key role in starting a war.[25]

Drumheller went to his group chief and told her he had assumed the administration had other sources. "No," she told him. "This is why they're fighting so ferociously to validate this source."[26]

Drumheller sent his work on Curveball to John McLaughlin's office. But nothing happened. A decision had already been made to invade Iraq, so that meant the only acceptable intelligence was intelligence that supported that decision. "Politicization, real politicization, rarely [takes the form of] blatant, crude arm twisting . . ." explained Paul Pillar, the CIA's former national intelligence officer for the Near East and South Asia, who helped compile the intelligence leading up to the Iraq War.[27] "It's always far more subtle. . . . Intelligence assessments that conform with what is known to be the policy [have] an easier time making it through."

To the extent the CIA had any clout with the White House at all, it was because the Agency had demonstrated considerable success leading the American effort in Afghanistan. But now that the push toward war in Iraq was in high gear, Cheney moved quickly to undercut the CIA's power through a series of adroit bureaucratic sleights of hand that were almost completely unseen by administration officials—much less the public at large. Many senior intelligence officers, for example, were unintentionally kept in the dark about the OSP. "I didn't know about its existence," said Greg Thielmann.[28] "They were cherry-picking intelligence and packaging it for Cheney and Rumsfeld to take to the president. That's the kind of rogue operation that peer review is intended to prevent."

As the summer came to an end, Cheney continued to move pieces on the chessboard, making sure all his men were advantageously positioned. William Luti, the Wohlstetter acolyte who had become special adviser to Cheney, moved to Feith's Office of Special Plans. Abram Shulsky, another alumnus from Scoop Jackson's staff and a colleague of Richard Perle, joined him there. Stephen Cambone,* an associate of

*In March 2003, Cambone became the first under secretary of defense for intelligence. He first came to the attention of the general public when Major General Antonio Taguba testified before the Senate Armed Services Committee in the Abu Ghraib scandal, and Cambone disputed Taguba's statement that prison guards were under the effective control of military intelligence. In November 2006, Germany announced it would prosecute Cambone for his alleged role in the abuse of prisoners in Abu Ghraib prison in Iraq. However, on April 27, 2007, the German federal prosecutor announced that his country had decided not to pursue the prosecution for jurisdictional reasons.

both Rumsfeld and Cheney, went to the Pentagon as a special assistant to Rumsfeld. With Rumsfeld, Wolfowitz, Feith, and Cambone, Cheney effectively had four key men at the highest levels of the Pentagon. On his own staff he had Scooter Libby interacting with the Pentagon and leaking its scoops to selected reporters. David Addington oversaw Cheney's aggressive and highly secretive legal staff.[29]

Stephen Hadley kept Condoleezza Rice in check at the National Security Council, lest she fall back into Scowcroft's orbit. Wurmser had been moved out of Doug Feith's unit at the Pentagon and joined John Bolton at the State Department to keep an eye on Colin Powell.[30] Cheney also had John Yoo as an ally in the Justice Department. All told, Cheney's loyal operatives were stationed in key positions in the State Department, the Pentagon, the National Security Council, and the Justice Department. And that didn't include vital allies in Congress such as Senator Pat Roberts (R-Kans.) on the Senate Intelligence Committee.

It was a team with astounding ideological cohesiveness, certitude, discipline, and experience. For many of them, after twenty years of struggle, the policies they had long hoped to implement were finally about to become operational.

When it came to the Oval Office, Cheney himself took care of its occupant. In his memo about the president and the vice president, the former Pentagon official who had worked with Cheney compared Bush's approach to the presidency of the United States to the way he ran the Texas Rangers baseball team. "George W. Bush reportedly did not involve himself in the essence of baseball operations: player development, marketing and the business aspects of running a Major League Baseball franchise," the official wrote. "Instead he served . . . as a corporate master-of-ceremonies, attending to the morale of the management team and focusing on narrow issues . . . that interested him."

By becoming the sole framer of key issues for Bush, the memo continued, Cheney "rendered the policy planning, development and implementation functions of the interagency system essentially irrelevant. He has, in matters he has deemed important, governed. As a matter of protocol, good manners, and constitutional deference he has obtained the requisite 'check-mark' of the president, often during one-on-one meetings after a Potemkin 'interagency process' had run its often inconclusive course."[31]

At about the same time that the OSP was launched, the White House established the White House Iraq Group, with chief of staff Andrew

Card, deputy chief of staff Karl Rove, Cheney aide Mary Matalin, deputy national security adviser Stephen Hadley, and Cheney's chief of staff Scooter Libby. Its mission was simple: sell the war. The group's plan was to open a full-fledged marketing campaign in the media after Labor Day, featuring images of nuclear devastation and threats of biological and chemical weapons and a rogue Iraq with a brutal and insane dictator at the helm. A key piece of the evidence was the Niger dossier. Test marketing began in August, with Cheney and his surrogates asserting repeatedly that "many of us are convinced that Saddam will acquire nuclear weapons fairly soon." Making Cheney seem moderate by comparison, a piece by Michael Ledeen appeared in the *Wall Street Journal* on September 4, suggesting that, in addition to Iraq, the governments of Iran, Syria, and Saudi Arabia should be overthrown.

But the real push was delayed until the second week of September. As Card famously put it, "From a marketing point of view, you don't introduce new products in August."[32] The first anniversary of the 9/11 attacks presented the perfect opportunity.

The opening salvo was fired on Sunday, September 8, 2002, in the form of a star-studded extravaganza on the Sunday morning chat shows featuring appearances by Dick Cheney, Colin Powell, Donald Rumsfeld, and Condoleezza Rice on NBC, Fox News, CBS, and CNN. "There will always be some uncertainty about how quickly [Saddam] can acquire nuclear weapons," National Security Adviser Condoleezza Rice told CNN. "But we don't want the smoking gun to be a mushroom cloud."[33] The smoking-gun-mushroom-cloud catchphrase was such a hit that Bush, Cheney, and Rumsfeld all picked it up in one form or another, sending it out repeatedly to billions of people all over the world.

Far less obvious, however, was Cheney's most Machiavellian flourish. All four of the officials—Cheney, Powell, Rice, and Rumsfeld—attributed their latest information to a story that appeared that very day in the nation's most prestigious newspaper. "There's a story in the *New York Times* this morning . . ." Cheney told Tim Russert on NBC's *Meet the Press*, "and I want to attribute the *Times*. I don't want to talk about, obviously, specific intelligence sources, but it's now public that, in fact, he has been seeking to acquire . . . the kind of tubes that are necessary to build a centrifuge."[34]

Specifically, Cheney was referring to a story by Judith Miller and Michael Gordon. Miller, in particular, has been the subject of considerable critical attention, most notably in an article by Michael Massing in the *New York Review of Books*,[35] on PBS's *Bill Moyers Journal*,[36] and in

Slate, The Nation, Editor & Publisher, the *American Journalism Review,* the *Columbia Journalism Review,* and many other publications.

Coauthor with neocon conspiracy theorist Laurie Mylroie of a 1990 book, *Saddam Hussein & the Crisis in the Gulf,* "a dear friend" of Richard Perle, as Perle described their relationship,[37] and the author of at least one WMD story based on information from Ahmed Chalabi,[*38] Miller, in the past, had been an eager recipient of information from the administration and its allies, information that had provided her with six major stories in the *Times* about WMDs. In 2001, Chalabi had arranged for her to visit Thailand to interview Adnan Ihsan Saeed al-Haideri, a civil engineer who had defected from Iraq.[†39] Another dubious source was Khidhir Hamza, a scientist who had been a senior official in Iraq's nuclear program until the late 1980s, and who was being shopped around by Chalabi's Iraqi National Congress as an interview subject.[‡40]

*In an e-mail to fellow *Times* reporter John Burns, Miller said that she had been covering Chalabi for about ten years and that Chalabi "has provided most of the front-page exclusives on WMD to our paper." But, Michael Massing wrote, Miller later said that her assertion was "part of 'an angry e-mail exchange with a colleague.' In the heat of such exchanges, Miller said, 'You say things that aren't true. If you look at the record, you'll see they aren't true.'"

Massing concluded: "This seems a peculiar admission. Yet on the broader issue of her ties to Chalabi, the record bears Miller out. Before the war, Miller wrote or co-wrote several front-page articles about Iraq's WMDs based on information from defectors; only one of them came via Chalabi. An examination of those stories, though, shows that they were open to serious question. The real problem was relying uncritically on defectors of any stripe, whether supplied by Chalabi or not."

†Al-Haideri, a Kurd from northern Iraq, told Miller the Iraqis had hidden chemical and biological weapons right under Saddam's "presidential sites," and after Miller's article on him in the *Times,* the story spread all over the world. But, as Jonathan Landay of Knight Ridder told *Bill Moyers Journal,* parts of the story were difficult to believe. "The first was the idea that a Kurd—the enemy of Saddam—had been allowed into his most top secret military facilities. I don't think so. That was, for me, the biggest red flag. And there were others, like the idea that Saddam Hussein would put a biological weapons facility under his residence. I mean, would you put a biological weapons lab under your living room?" Added Landay's colleague at Knight Ridder, Warren Strobel, "The first rule of being . . . a journalist . . . is you're skeptical of defectors, because they have a reason to exaggerate. They want to increase their value to you. They probably want something from you."

‡The coauthor of *Saddam's Bombmaker: The Daring Escape of the Man Who Built Iraq's Secret Weapon,* Hamza was also introduced to reporters by Chalabi. However, David Albright, the author of several books on nuclear proliferation and the founder of the Institute for Science and International Security, told the *Financial Times* that Hamza's claims were "often inaccurate . . . He sculpts his message to get the message across . . . [He] wants regime change [in Iraq] and what interferes with that is just ignored."

On this occasion, Miller and Gordon* reported that over the last fourteen months, "Iraq has sought to buy thousands of specially designed aluminum tubes, which American officials believe were intended as components of centrifuges to enrich uranium." The article added that American intelligence experts were persuaded the tubes were for use in Iraq's nuclear program.[41]

The *Times* story was attributed to unnamed administration officials. In other words, during his *Meet the Press* appearance, Cheney, in graciously attributing the story to the *New York Times,* was actually citing information, or rather disinformation, his office had leaked to the *Times.*

Gordon and Miller told Michael Massing that the information about the aluminum tubes was not a leak, and that they had pried it out of the administration. "The administration wasn't really ready to make its case publicly at the time," Gordon said. "Somebody mentioned to me this tubes thing. It took a lot to check it out." However, if the Bush administration had wanted to keep a lid on the story, it's hard to believe it would have put Cheney, Rice, Powell, and Rumsfeld on national television simultaneously to cite the *Times* story.

Moreover, Cheney's team had not merely planted the story in the *Times,* it had orchestrated a spectacular media campaign around the leak by sending out its best spokespeople—Cheney, Rumsfeld, Rice, and Powell, terrific hard-to-get catches for the people who book guests on *Meet the Press, Face the Nation,* etc.—to promote the story on national television. As a result, it appeared that Cheney and his surrogates had independent corroboration from the *New York Times.*

In addition, given the *Times*'s special place in American journalism, that corroboration carried with it extraordinary authority. Network television news producers, wire service editors, and editors at major newspapers across the globe read the *Times* daily to pick up stories deemed important by the paper's editors, and pass them on to their readers and viewers. On that one day alone, nearly five hundred newspaper articles and broadcasts all over the country, indeed the world, discussed Iraq as a nuclear threat, largely as a result of the Miller-Gordon story in the *Times.*

But the real impact of the story was not merely that the *Times* had spread it far and wide. It was one thing for the conservative Rupert Murdoch–owned Fox News to serve as a mouthpiece for the Bush

*In 1995, Gordon coauthored, with Bernard Trainor, *The Generals' War: The Inside Story of the Conflict in the Gulf,* a sharp critique of the U.S. decision to leave Saddam in power. The book noted how stunned U.S. intelligence analysts had been after the Gulf War to discover how advanced Iraq's nuclear weapons program was.

administration. But it was quite another for the *New York Times,* the very embodiment of the dreaded liberal press, to play that role. Together, Fox News and the *Times* effectively defined the right and left parameters of the American political conversation. If Fox News and the *Times* agreed on the conventional wisdom, every other possible point of view was marginalized. There could be no more debate.

Eleven days later, in the *Washington Post,* Joby Warrick wrote a piece that took issue with Miller and Gordon's article, citing independent experts "who question whether thousands of high-strength aluminum tubes recently sought by Iraq were intended for a secret nuclear weapons program."[42] But the piece was buried on page eighteen. No one paid attention. Once the conventional wisdom had been forged, mere facts did not suffice to change things.[*43]

Somehow, in the wake of 9/11, there had been a radical shift to the right in the American consciousness. Fear was everywhere and unabashed patriotism appeared to provide the answer. The word was out at CNN not to do stories that were critical of the Bush administration. "[T]here was even almost a patriotism police . . . sort of picking anything a Christiane Amanpour, or somebody else would say as if it were disloyal," Walter Isaacson, the chairman and CEO of CNN at the time, told *Bill Moyers Journal.* "There was a real sense that you don't get that critical of a government that's leading us in wartime . . . big people in corporations were calling up and saying, 'You're being anti-American here.'" Isaacson sent his staff a memo making sure they did not focus too much on civilian casualties in Afghanistan.[44] Newspapers ordered photo editors to keep pictures of civilian casualties off page one.[45] Presidential press conferences were scripted with a list of prescreened reporters on whom Bush could call, safe in the knowledge they would not grill him on the evidentiary specifics behind the rush to war with Iraq.[†46]

*As an example of the impenetrability of this newly forged conventional wisdom, *Times* columnist William Safire called the alleged meeting in Prague between 9/11 hijacker Mohammed Atta and an Iraqi agent an "undisputed fact," and wrote about it ten different times in his op-ed column even though the news pages of the *New York Times* raised serious doubts about it. In all, Bill Moyers reported, Safire wrote "27 opinion pieces fanning the sparks of war."

†On at least one occasion, such scripting had embarrassing results, as when the president called on a woman reporter who had not even raised her hand, but who, when prompted, lobbed a decidedly softball question at the president:

PRESIDENT BUSH: Let's see here . . . Elizabeth . . . Gregory . . . April . . . Did you have a question or did I call upon you cold?

As for the *Washington Post,* according to media critic Howard Kurtz of both CNN and the *Post,* between August 2002, when the Iraq campaign was not even under way, and March 2003, when the war started, it gave its front page over to about 140 articles making the Bush administration's case for war,[47] and 27 pro-war editorials on its editorial pages. But when 100,000 people demonstrated in Washington against the war, in one of the biggest antiwar demonstrations since the Vietnam era, the *Post* buried the story in the Metro section. When Ted Kennedy delivered an impassioned speech asserting that the administration had presented "no persuasive evidence that Saddam is on the threshold of acquiring the nuclear weapons," according to Eric Boehlert, the author of *Lapdogs: How the Press Rolled Over for Bush,* the *Post* gave it a total of 36 words, out of roughly one million words it published on Iraq in 2002.[48]

Almost overnight, war fever was everywhere. President Bush himself took up the "smoking gun–mushroom cloud" catchphrase. As the November midterm elections approached, it was endlessly repeated on CNN, ABC, NBC, CBS, and Fox. Again and again Saddam was conflated with 9/11. A CBS news poll showed that 51 percent of Americans thought Saddam Hussein "was personally involved in the Sept. 11 attacks."[49]

Her ties to Scowcroft and Bush 41 notwithstanding, Condoleezza Rice joined in. Asked on Fox News whether or not there were links between Saddam Hussein and Osama bin Laden, Rice, initially circumspect, said that Iraq "clearly has links to terrorism . . . links to terrorism [that] would include al-Qaeda. . . ."[50]

The campaign became, in effect, "all fear, all the time." Color-coded threat levels dramatically issued by the Department of Homeland Security* became as ubiquitous as weather reports. There were arcane

---

APRIL: No, I have a question [laughter].

PRESIDENT BUSH: Okay. I'm sure you do have a question.

APRIL: How is your faith guiding you?

PRESIDENT BUSH: My faith sustains me because I pray daily. I pray for guidance.

*The use of the word *homeland* itself was a striking addition to the American discourse, evocative of a patriotic, romantic nationalism and somewhat suggestive of words such as "motherland," "fatherland," or Mother Russia—terms used by Germany and the Soviet Union to summon forth images and feelings of national identity. In a 2002 column, Republican speechwriter Peggy Noonan wrote a column suggesting that the Bush administration change the name of the Department of Homeland Security. "The name Homeland Security grates on a lot of people, understandably," she wrote. "*Homeland* isn't really an American word, it's not something we used to say or say now."

chemical and biological weapons, exotic viruses, anthrax, unmanned predator planes, suitcase bombs, dirty bombs, and nukes around every corner. "Hussein has vowed revenge," said ABC's George Will. "He has anthrax, he loves biological weapons, he has terrorist training camps, including 747's to practice on."[51]

By September 24, just sixteen days after the Bush-Cheney administration launched its smoking gun–mushroom cloud campaign, 72 percent of Americans believed that Saddam Hussein would use weapons of mass destruction against the United States. According to Republican pollster David Winston, "The reaction that you're getting from the American people is, for the first time, their personal safety and security is threatened in a way that it's never been before, and they want action taken."[52]

Bush fanned the fires tying Saddam to Osama. "The danger is, is that al-Qaeda becomes an extension of Saddam's madness and his hatred and his capacity to extend weapons of mass destruction around the world," Bush told reporters that day. "I can't distinguish between the two, because they're both equally as bad, and equally as evil and equally as destructive."[53]

In this atmosphere of fear, merely to question this newly forged consensus was to court treason. When Congressman Jim McDermott (D-Wash.) returned from a trip to Iraq and said that President Bush was willing to "mislead the American people," he was promptly labeled a traitor.[54] Likewise, former U.N. arms inspector Scott Ritter, who discounted the Iraqi threat and dared to go to Iraq and address the Iraqi National Assembly, was the target of a massive smear campaign. "I don't know what accounts for his Dr. Jekyll and Mr. Hyde personality," said Dick Spertzel, an ex-U.N. inspector who worked with him. "But in my book, aiding and abetting the enemy is treason."[55] MSNBC talk-show host Phil Donahue was assailed as a traitor,[56] and on CNN, Richard Perle later called *The New Yorker*'s Seymour Hersh, then sixty-five, "the closest thing American journalism has to a terrorist."[57]

On Fox News, right-wing pundit Ann Coulter called Senator Ted Kennedy part of "the treason lobby."[58] NPR reported that Secretary Rumsfeld wanted to jail leakers who worked for the government, that a Defense Department analyst proposed sending SWAT teams into reporters' homes to help catch the leakers, and the FBI was investigating congressmen over leaks and had proposed giving them lie detector tests.[59]

The Christian Right joined in, staging a Stand By Israel rally[60] on the mall in Washington in October, and prepared its followers for the imminent religious war. "The God of Islam is not the same God," said Reverend Franklin Graham, son of Billy Graham and heir to his father's huge ministry. "He's not the son of God of the Christian or Judeo-Christian faith. It's a different God, and I believe it is a very evil and wicked religion."[61]

Reverend Jerry Vines, former head of the massive Southern Baptist Convention (SBC), did not hold back. For him, the prophet Muhammad was "a demon-obsessed pedophile."[62] The day after Vines's comments, President Bush addressed the SBC, and praised the group for its "religious tolerance."[63] On CBS's *60 Minutes,* Jerry Falwell labeled the prophet Muhammad "a terrorist." Pat Robertson called Islam a violent scam.[64]

In the *National Review,* Ann Coulter had a simple solution. "We should invade Muslim countries, kill their leaders and convert them to Christianity," she wrote.[65]

Politicians who counseled restraint and pointed out that Saddam wasn't behind 9/11 did so at their own political risk. "I am deeply concerned that the course of action that we are presently embarked upon with respect to Iraq has the potential to seriously damage our ability to win the war against terrorism," said Al Gore. "I don't think we should allow anything to diminish our focus on the necessity for avenging the 3,000 Americans who were murdered and dismantling that network of terrorists that we know were responsible for it."[66]

But, with the exception of Gore, and eloquent antiwar speeches by Senator Robert Byrd (D-W. Va.) and Ted Kennedy (D-Mass.), the Democrats were largely mute. As the midterm congressional elections approached, they risked being tarred as soft on terror, or, worse, traitorous liberals who aided and abetted the enemy. Senator Max Cleland (D-Ga.) was a particularly poignant case in point. Up for reelection in Georgia, he advised against rushing into war without first forcing Hussein to agree to unfettered arms inspections. "We should and must give it one final chance before considering military options," he said.[67]

For his efforts, Cleland, a Vietnam War veteran who had lost three limbs in battle for his country, was rewarded with an ad campaign that attacked him by featuring his likeness on the same screen as Osama bin Laden's and Saddam Hussein's and accusing him of being soft on terror. In the context of the war lust and feverish hysteria that gripped the country, Cleland's moderate sentiments were deemed traitorous. Few Americans realized these views were similar to those held by the president's father and Brent Scowcroft.

* * *

Effective as the media campaign was, sooner or later the White House needed specifics that proved Saddam was a grave threat. After all, a Senate-ordered National Intelligence Estimate was in the works and Congress had to authorize the war. As it happened, on September 9, the day after the Judith Miller–Michael Gordon story appeared in the *New York Times*, the White House received a visitor who should have known exactly what the aluminum tubes in question were for: SISMI director Nicolò Pollari. The Italians had used precisely the same tubes that Iraq was seeking in their Medusa air-to-ground missile systems, so Pollari presumably knew that Iraq was not trying to enrich uranium but merely attempting to reproduce obsolete weaponry dating back to an era of military trade between Rome and Baghdad.[68] As *La Repubblica* pointed out, however, he did not set the record straight.

At the time Pollari met with National Security Adviser Stephen Hadley, the Niger documents had already been discredited repeatedly, and, perhaps because the documents originated in Italy, it has been widely speculated that they were the topic of the meeting. Hadley subsequently confirmed that the meeting with Pollari took place, but he declined to say what was discussed. "It was a courtesy call," Hadley told reporters.[69] "Nobody participating in that meeting or asked about that meeting has any recollection of a discussion of natural uranium, or any recollection of any documents being passed."

But there was no need to pass documents. It was significant enough for Pollari to have met with Hadley, a White House official allied with Cheney's hard-liners, rather than with Pollari's American counterpart, George Tenet. "It is completely out of protocol for the head of a foreign intelligence service to circumvent the CIA," said former CIA officer Philip Giraldi.[70] "It is uniquely unusual. In spite of lots of people having seen these documents, and having said they were not right, they went around them."

"To me there is no benign interpretation of this," said Melvin Goodman, a former CIA and State Department analyst. "At the highest level it was known the documents were forgeries. Stephen Hadley knew it. Condi Rice knew it. Everyone at the highest level knew."[71]

Michael Ledeen was one of very few people who had access to both Pollari and Hadley, but he categorically denied setting up the meeting. "I don't even know anything about a meeting between Pollari and Hadley," he said.[72] "And if it happened I had nothing to do with it."[73] Speaking on behalf of George Tenet, a former senior intelligence official said that the former CIA chief had no information suggesting that Pollari or elements of SISMI may have been trying to circumvent the CIA and go directly to the White House.

But the Niger deal had been resurrected once again. Two days later, on September 11, 2002, the first anniversary of the terrorist attacks, Hadley's office asked the CIA to clear language so that President Bush could issue a statement saying, "Within the past few years, Iraq has resumed efforts to purchase large quantities of a type of uranium oxide known as yellowcake. . . . The regime was caught trying to purchase 500 metric tons of this material. It takes about 10 tons to produce enough enriched uranium for a single nuclear weapon."[74] In addition, in a new paper that month, the DIA issued an assessment claiming that "Iraq has been vigorously trying to procure uranium ore and yellowcake."

Later that month, the British published a fifty-page, fourteen-point report on Iraq's pursuit of weapons that said, "There is intelligence that Iraq has sought the supply of significant quantities of uranium from Africa."[75]

"When you are playing a disinformation operation," said former CIA official Milt Bearden, "you're like a conductor who can single out one note in the symphony and say, 'Let the Brits have that.'"[76]

Now it was time for the international media to chime in with independent corroboration. That way it would appear as if there were multiple sources for the story about the Niger-Iraq yellowcake deal, when in fact there was merely an echo chamber of corroboration. On September 24, Prime Minister Tony Blair cited that "dossier of death" and asserted again that Iraq had tried to acquire uranium from Africa. "The reports in [the Niger file] were going around the world, and Bush and Blair were talking about the documents without actually mentioning them," Rocco Martino, the freelance Italian intelligence operative who helped distribute the reports, told Milan's *Il Giornale.* "I turned the television on and I did not believe my ears."[77]

In early October 2002, Martino approached Elisabetta Burba, a journalist at *Panorama,* a Milan-based newsmagazine. Burba and Martino had worked together in the past, but there may have been other reasons he went to her again. Owned by Silvio Berlusconi, *Panorama* was edited by Carlo Rossella, a close ally of the prime minister's. It also counted among its contributors Michael Ledeen.

Martino told Burba he had something truly explosive—documents that proved Saddam was buying yellowcake from Niger. Burba was intrigued, but skeptical. She agreed to pay Martino just over 10,000 euros

APIGEST

REPUBLIQUE DU NIGER

CONSEIL DE RECONCILIATION NATIONALE

MINISTERE DES AFFAIRES ETRANGERES
ET DE L'INTEGRATION AFRICAINE

DIRECTION DES AFFAIRES JURIDIQUES
ET CONSULAIRES

Niamey, le 30 JUIL 1999

N° --05055 /MAE/IA/DAJC/DIR

URGENT

HONNEUR VOUS DEMANDER BIEN VOULOIR CONTACTER S.E. L'AMBASSADEUR D'IRAQ M. WISSAM AL ZAHAWIE POUR CONNAITRE REPONSE DE SON PAYS CONCERNANT FOURNITURE D'URANIUM SELON DERNIERS ACCORDS ETABLIS A NIAMEY LE 29 JUIN 2000.

PRIERE SUIVRE CE DOSSIER TRES CONFIDENTIEL AVEC TOUTE DISCRETION ET DILIGENCE.

NASSIROU SABO

REPUBLIQUE DU NIGER

The Niger documents (above) suggested that Wissam al-Zahawie, the Iraqi ambassador to the Vatican, had traveled to Niger to buy Uranium. But Joseph Wilson, who personally knew al-Zahawie, said, "The idea that he would be entrusted with this super-secret mission to buy 500 tons of uranium from Niger is out of the question." (*La Repubblica*/Reuters/Corbis)

for the documents—then about $12,500—on one condition: Martino would receive his fee only after his dossier had been corroborated by independent authorities. Martino gave her the documents. When Burba told Rossella of her concerns about the authenticity of the Niger documents, he sent her to Africa to investigate. But he also insisted that she give copies to the U.S. embassy. "I think the Americans are very interested in this problem of unconventional weapons," Rossella told her.

On October 17, Burba flew to Niger. Not wanting to give herself away to authorities, she traveled undercover and claimed to be investigating a newly discovered dinosaur fossil in Niger's Tenere desert. Once there, she discovered for herself how difficult it would be to ship five hundred tons of uranium out of Africa. By the time she returned, she believed the real story was not about Saddam's secret nuclear weapons program at all, but about whether someone had forged the documents to fabricate a rationale for invading Iraq. But when she reported her findings to Rossella, he called her off. "I told her to forget the documents," he said. "From my point of view, the story was over."[78]

Now, however, thanks to *Panorama,* the United States had finally received actual hard copies of the Niger documents—rather than just reports about their contents. They were quickly disseminated to the CIA station chief in Rome, who recognized them as the same old story the Italians had been pushing months before, and to nuclear experts at the DIA, the Department of Energy, and the National Security Agency.

The State Department had already twice cast doubt on the reports of the sale of uranium to Iraq. In the fall, Wayne White, who served as the deputy director of INR, the State Department's intelligence unit, and was its principal Iraq analyst, reviewed the papers himself. According to the *Boston Globe,* within fifteen minutes of reviewing the documents he doubted their authenticity.[79]

On occasion, the CIA was successful in thwarting plans by the Bush White House to give credibility to the forged documents. In the middle of October, Bush was scheduled to give a major speech on Iraq in Cincinnati, and the NSC had already forwarded several drafts to the CIA for vetting. About ten days before the speech, according to a report by the Senate Intelligence Committee, the NSC sent the sixth draft of the speech to the CIA, a draft containing a line saying that Saddam "has been caught attempting to purchase up to 500 metric tons of uranium oxide from Africa—an essential ingredient in the enrichment process."[80]

In response, on October 5, the CIA faxed back a memo to Stephen Hadley and chief White House speechwriter Michael Gerson telling them to delete the sentence on uranium "because the amount is in dispute and it is debatable whether it can be acquired from the source. We told Congress that the Brits have exaggerated this issue. Finally, the Iraqis already have 550 metric tons of uranium oxide in their inventory.' "[81]

But the White House refused to make substantive changes. Later that day, Hadley's staff sent over the seventh draft of the Cincinnati speech. It contained a line saying, "the regime has been caught attempting to purchase substantial amounts of uranium oxide from sources in Africa."[82] They had changed the language slightly, but effectively they had inserted the supposed Niger yellowcake deal back into the president's speech.

This time, CIA director George Tenet himself interceded to keep the president from lying. According to his Senate testimony, he told Hadley that the "President should not be a fact witness on this issue," because his analysts had told him the "reporting was weak."[83] The CIA even put the memo in writing, addressed to Hadley and Rice, and faxed it to the NSC.

On October 7, Bush gave the speech at the Cincinnati Museum Center and made it clear why the United States had to finish off the Iraqi dictator. "Facing clear evidence of peril," he told the audience, "we cannot wait for the final proof—the smoking gun—that could come in the form of a mushroom cloud."[84]

There was no specific reference to obtaining uranium from Africa. The CIA had won this round, but the neocons were not done. "That was their favorite technique," said Colonel Larry Wilkerson, who was Colin Powell's chief of staff. "Stick that baby in there forty-seven times and on the forty-seventh time it will stay. I'm serious. It was interesting to watch them do this. At every level of the decision-making process you had to have your ax out, ready to chop their fingers off. Sooner or later you would miss one and it would get in there."[85]

Meanwhile, one by one, those who got in their way were purged. According to Patrick Lang, the former head of Human Intelligence at the DIA, Bruce Hardcastle, defense intelligence officer for the Middle East, South Asia, and Counterterrorism, "told them that the way they were handling evidence was wrong."[86]

As a result, Lang said, they not only removed Hardcastle from his position, they did away with his job entirely. "They wanted just liaison officers who were junior," Lang told *Salon.* "They didn't want a senior intelligence person who argued with them. Hardcastle said, 'I couldn't deal with these people.' They are such ideologues that they knew what the outcome should be. . . . They start with an almost pseudoreligious faith. They wanted the intelligence agencies to produce material to show a threat, particularly an imminent threat. Then they worked back to prove their case. It was the opposite of what the process should have been like, that the evidence should prove the case."[87]

Together, Cheney and Scooter Libby made about ten visits to CIA headquarters in Langley—a highly irregular occurrence. "I was at the CIA for 24 years," said former CIA analyst Melvin Goodman. The only time a vice president came to the CIA building was for a ceremony, to cut a ribbon, to stand on the stage, but not to harangue analysts about finished intelligence."[88]

"Many, many of them have told me they were pressured," said Patrick Lang. "And there are a lot of ways. Pressure takes a lot of forms."[89]

"For the vice president to be meeting with analysts, that was a real red flag," said one State Department official. "It was so unusual. It was clear that people were being leaned on. Usually, if a high-ranking official wants information, it gets tasked out through appropriate channels. It was highly unusual to lock these people in a room and keep pressing. It crossed the line between intellectual inquiry and not accepting the real answer."

The same thing was going on at the State Department, where Under Secretary of State John Bolton barred INR director Greg Thielmann from attending meetings because he had dared question intelligence such as the Niger documents. "Bolton seemed to be troubled because INR was not telling him what he wanted to hear," Thielmann told *The New Yorker*'s Seymour Hersh. "I was intercepted at the door of his office and told, 'The under secretary doesn't need you to attend this meeting anymore. The under secretary wants to keep this in the family.' "[90]

In the mainstream press, virtually no one understood or reported what was going on. Knight Ridder journalists Warren P. Strobel, Jonathan S. Landay, and John Walcott were among the few exceptions. On October 8, 2002, they wrote:

> A growing number of military officers, intelligence professionals and diplomats in [Bush's] own government privately have deep misgivings about the administration's double-time march toward war.
> These officials charge that administration hawks have exaggerated evidence of the threat that Iraqi leader Saddam Hussein poses—including

> distorting his links to the al-Qaida terrorist network—have overstated the amount of international support for attacking Iraq and have downplayed the potential repercussions of a new war in the Middle East.[91]

But with no outlets in either Washington or New York, Knight Ridder had little clout in terms of shaping the national conversation. Editors at the *New York Times* and the *Washington Post* paid them no heed. Strobel, Landay, and Walcott were not powerful enough to stop the onslaught.

Meanwhile, the Bush administration and its allies behaved as if war were a foregone conclusion. In October, Ahmed Chalabi met executives of three unnamed multinational oil companies to negotiate the carving up of Iraq's massive oil reserves after Saddam was gone. "The oil people are naturally nervous," INC spokesman Zaab Sethna told *The Guardian* (England). "We've had discussions with them, but they're not in the habit of going around talking about them."[92] On October 3, the administration announced plans to deploy as many as four aircraft carrier groups to be within striking distance of Iraq in December.[93] The rhetoric escalated. Defense Secretary Donald Rumsfeld said the ties between Al-Qaeda terrorists and Iraq were "accurate and not debatable."[94]

Bush asserted that Saddam "could launch a biological or chemical attack in as little as 45 minutes after the order is given."[95] The need to go to war had become personal for him, it appeared. "There's no doubt [Saddam's] hatred is mainly directed at us," Bush said. "There's no doubt he can't stand us. After all, this is a guy that tried to kill my dad at one time."[96]

On October 1, 2002, the CIA finally delivered the National Intelligence Estimate to Congress. The man in charge was weapons expert Robert Walpole, who, according to the London *Independent*, "had a track record of going back over old intelligence assessments and reworking them in accordance with the wishes of a specific political interest group." In 1998, while acting for a congressional commission chaired by Donald Rumsfeld, Walpole had come up with a particularly alarming estimate of "the missile capabilities of various rogue states," the *Independent* said.[97]

Walpole's role was not the only troubling aspect of the ninety-six-page NIE.[98] It strongly suggested that Saddam had huge supplies of WMDs. But if one read the text carefully, there was "vigorous dissent," as Bob Graham put it, about crucial parts of the program, particularly the aluminum tubes. When Graham personally questioned George Tenet about that, Tenet admitted that the information had not been verified by any operative responsible to the United States. Realizing that the intelligence came from Iraqi exiles or other parties, all of whom had an interest in Saddam's removal, Graham decided the American people needed to understand these caveats, and requested that the CIA prepare an unclassified version of the NIE for public consumption.[99]

In response, on October 4, Tenet presented a twenty-five-page document called "Iraq's Weapons of Mass Destruction Programs." But it was exactly the opposite of what Graham had requested. According to Graham, "It represented an unqualified case that Hussein possessed [WMDs] . . . and omitted the dissenting opinions contained in the classified version. Its conclusions, such as 'If Baghdad acquired sufficient weapons-grade fissile material from abroad, it could make a nuclear weapon within a year,' underscored the White House's claim that exactly such material was being provided from Africa to Iraq." In other words, the Niger documents were still very much alive.

Meanwhile, copies of the complete NIE were kept in two vaults in the secure Hart Senate Office Building, where the Senate's classified material is kept. According to the *Washington Post,* each vault was protected by armed guards. Members of both the Senate and the House of Representatives were allowed to see the NIE if they showed up in person and *without* an accompanying staff member.[100] Even though war was at stake, only six senators and fewer House members bothered to make the trek. Senator John D. Rockefeller IV (D-W. Va.) explained that logistically such a visit was very difficult for busy members of the august legislative body. "Everyone in the world wants to come to see you" in your office, he explained. Rockefeller added that members were not allowed to take notes and that "it's extremely dense reading."[101]

As a result, many senators and congressmen relied only on intelligence briefings from high-level administration officials on the intelligence findings before casting their votes. However, they were not told of the doubts about the Niger documents, the aluminum tubes, or Curveball. Nor were they informed that Foreign Minister Sabri was on the CIA payroll and said Saddam had killed his WMD program.

For the few who actually bothered to read the document, the complete NIE was filled with various caveats and dissenting opinions that filled a total of sixteen pages—much of which was relegated to "annexes" toward the end of the ninety-six-page document.* But caveats aside, ultimately, the document asserted that the United States had a "high confidence" that:

- Iraq is continuing, and in some areas expanding, its chemical, biological, nuclear, and missile programs contrary to U.N. resolutions.
- We are not detecting portions of these weapons programs.
- Iraq possesses proscribed chemical and biological weapons and missiles.
- Iraq could make a nuclear weapon in months to a year once it acquires sufficient weapons-grade fissile material.

More specifically, the NIE stated that Iraq is "vigorously trying" to obtain uranium and "reportedly" is working on a deal to purchase "up to 500 tons" of uranium from Niger.[102] It added that "Baghdad has mobile facilities for producing bacterial and toxin BW agents; these facilities can evade detection and are highly survivable." And, with some qualifications, it asserted that "Iraq's aggressive attempts to obtain high-strength aluminum tubes for centrifuge rotors—as well as Iraq's attempts to acquire magnets, high-speed balancing machines, and machine tools—provide compelling evidence that Saddam is reconstituting a uranium enrichment effort for Baghdad's nuclear weapons program."

At the time, to virtually everyone in Congress, the NIE was still sacrosanct. It was still the last word in American intelligence. Yet even it had been distorted thanks to political pressures from the neocons and the White House. If one took it seriously, the Niger documents were real. Curveball had credibility. And the aluminum tubes were part of Saddam's nuclear program. Only one conclusion could be drawn: Saddam Hussein posed an extraordinarily grave threat. On October 10, the Senate voted 77–23 and the House 296–133 to give President Bush the power and the military mandate he had requested against Saddam Hussein.

Even those who voted against Bush, however, believed the allegations that Saddam had WMDs. "I agree that Iraq presents a genuine threat, especially in the form of weapons of mass destruction: chemical, biological and potentially nuclear weapons," said Russ Feingold (D-Wis.), one of the most forceful opponents of the administration.[103]

*INR's assessments that Iraq's pursuit of uranium in Africa was "highly dubious" and that the aluminum tubes were "not clearly linked" to a nuclear program, for example, were on page eighty-seven of the ninety-six-page document.

The broad language in the resolution gave Bush the authority to declare unilateral, preemptive war. But Bush continued to frame the issue in terms of getting Saddam to disarm. After the congressional vote, the president said it "sends a clear message to the Iraqi regime: It must disarm and comply with all existing U.N. resolutions or it will be forced to comply."[104]

As the Senate vote approached, Bush had repeatedly stated that he had not decided to go to war, but that his aims were to get rid of Iraq's WMDs. "I haven't made up my mind we're going to war with Iraq," he said at an October press conference. "I've made up my mind we need to disarm the man. . . . There needs to be a strong new [United Nations] resolution in order for us to make it clear to the world—and to Saddam Hussein, more importantly—that you must disarm."[105]

Upon signing the congressional resolution on October 16, Bush assuaged the fears of those who thought it was a declaration of war. "I have not ordered the use of force," he said.[106] "I hope the use of force will not become necessary. Hopefully this can be done peacefully. Hopefully we can do this without any military action." He added that he has "carefully weighed the human cost of every option before us," and that he will only send troops "as a last resort."

But Bush's statements didn't mesh with his administration's actions on the ground. On October 23, the CIA set up two field stations in Kurdish-controlled Northern Iraq to prepare for the coming war. Six days later, the BBC reported that the USS *Kitty Hawk* had indeed departed from its base in Japan, and that the USS *Constellation* and her battle group were scheduled to leave for the Persian Gulf in a few days to be on station there by early December.[107] Meanwhile, according to Reuters, the navy sought out more merchant ships to carry huge quantities of ammunition and armor to the Gulf. One order showed that the Military Sealift Command (MSC) was seeking to ship 550 containers of ammunition and explosives—approximately 10,000 tons—from the United States to the Red Sea and the Persian Gulf.[108]

On November 5, the midterm congressional elections took place, and, thanks to the war on terror and Bush's high approval ratings, the Republicans won a narrow majority in the Senate, thereby seizing control from the Democrats. The most notable victim was Democratic senator Max Cleland, the triple amputee who was attacked as weak on terrorism. In the House, the Republicans picked up eight seats and increased their majority over the Democrats to twenty-five.

On November 15, Defense Secretary Donald Rumsfeld assured Americans that if there were a war with Iraq, it would be a short one. "The idea that it's going to be a long, long, long battle of some kind I think is belied by the fact of what happened in 1990," he said on a radio call-in program. "Five days or five weeks or five months, but it certainly isn't going to last any longer than that," he said.[109]

On December 6, in a dramatic shake-up of Bush's economic team, Treasury Secretary Paul O'Neill and Lawrence Lindsay, Bush's top economic adviser, resigned under pressure, Lindsay after having said the cost of the war could be as much as $200 billion, far above the administration's official estimate of $50 billion to $60 billion. Indeed, some in the administration, according to a report drafted by the Center for Strategic and Budgetary Assessment, even argued that all the postwar costs, "the cost of the occupation, the cost for the military administration and providing for a provisional administration, all of that would come out of Iraqi oil."[110]

On the same day, Iraq submitted a 12,200-page document that purportedly itemized all its unconventional weapons.[111] But Colin Powell rejected the document, in part because it did not account for the alleged Niger yellowcake and the aluminum tubes.[112] To back up Powell's statement, the State Department issued a fact sheet saying that "the [Iraqi] Declaration ignores efforts to procure uranium from Niger."[113]

Throughout December, references to the uranium deal resurfaced again and again in "fact sheets," talking-point memos, and speeches. No single component of the administration's case against Saddam was more persuasive than the argument that a brutal dictator armed with nuclear weapons could inflict massive casualties on the United States. Bush, Cheney, Rumsfeld, Powell, and Rice all declared publicly that Iraq had been caught trying to buy uranium from Niger. On December 19, the claim reappeared on a fact sheet published by the State Department.

For all the signs that the nation was headed to war, as 2002 came to an end, President Bush still insisted that the powers given to him by Congress were merely to be used to disarm Saddam. Asked at a New Year's Eve press conference from Crawford, Texas, about "a possible war with Iraq looming," Bush replied, "You said we're headed to war in Iraq—I don't know why you say that. I hope we're not headed to war in Iraq. I'm the person who gets to decide, not you. I hope this can be done peacefully. We've got a military presence there to remind Saddam Hussein, however, that when I say we will lead a coalition of the willing to disarm him if he chooses not to disarm, I mean it."[114]

Not long afterward, in January 2003, Tyler Drumheller, the CIA's head of covert operations in Europe, was asked by CIA headquarters to contact German intelligence again and make yet another stab at authenticating Curveball as a source. The stated reason was that an unnamed U.S. official was about to make a political speech and wanted to make sure German intelligence could vouch for his stories. Drumheller dutifully went back to the Germans, but still could not verify Curveball's claims.

Meanwhile, in light of the many objections to the Niger documents, the Pentagon asked the National Intelligence Council, the body that oversees the fifteen agencies in the U.S. intelligence community, to resolve the matter. According to the *Washington Post*, in a January 2003 memo the council replied unequivocally that "the Niger story was baseless and should be laid to rest."[115] The memo went immediately to Bush and his advisers.

Nevertheless, on January 20, just eight days before the State of the Union address, President Bush submitted a report to Congress citing Iraq's attempts "to acquire uranium and the means to enrich it."[116]

At an NSC meeting on January 27, 2003, George Tenet was given a hard-copy draft of the State of the Union address Bush was to deliver the next day. That day, Tenet returned to CIA headquarters and, without even reading the speech, gave a copy to an assistant who was told to deliver it to the deputy director for intelligence, Jami Miscik. But according to the Senate Intelligence Committee report, no one in the DDI's office recalled receiving the speech or if anyone was ever assigned to review it.

How was it possible that a State of the Union address that was a call for war, that desperately needed to be vetted by the director of Central Intelligence (DCI), had seemingly been misplaced and gone unread? "It is inconceivable to me that George Tenet didn't read that speech," said former CIA officer Milt Bearden. "At that point, he was effectively no longer DCI. He was part of that cabal, and no longer able to carry an honest message."[117] In an e-mail, a former intelligence official close to Tenet said the charge that Tenet was "part of a 'cabal' is absurd." The official added, "Mr. Tenet was unaware of attempts to put the Niger information in the State of the Union speech. Had he been aware, he would have vigorously tried to have it removed."

At 9:01 p.m. on January 28, 2003, George W. Bush delivered the State of the Union address before a joint session of Congress. It is often the custom for special guests to attend. On this occasion, an empty seat was left to symbolize the lives lost in the September 11 attacks. Wearing a dark blue suit and a light blue tie, Bush was composed and somber. His delivery was more effective than it had been on many other occasions, so much so that his speech was interrupted by applause more than seventy times. With Vice President Dick Cheney and Speaker of the House Dennis Hastert behind him, Bush began with a recitation of domestic issues he intended to address—Medicare, the environment, energy independence, faith-based initiatives, AIDS, and the like.[118]

Less than halfway through his speech, however, Bush turned to his true mission—"leading the world in confronting and defeating the man-made evil of international terrorism." His delivery slowed and became more deliberate and insistent. "The war goes on, and we are winning," he said. Both houses of Congress, Democrat and Republican alike, stood and cheered.[119]

Now Bush's tone changed—and so did the text. If the symbolic empty chair lent a portentous note to the proceedings, what followed was even more ominous. Americans had fought in Bosnia under Clinton, but the conflict had not risen to the level of a mission that united the entire nation. Indeed, not since the Gulf War, when Bush's father had been president, had the nation's leader looked his countrymen in the eye and exhorted the American people to join him in battle.

Now, to millions of American families in their suburban homes, and to reservists across the country, wondering if they might be called to war, Bush depicted the state of the union as the state of fear. There was "anthrax, botulinum toxin, Ebola, and plague."[120] There were "outlaw regimes that seek and possess nuclear, chemical, and biological weapons." There was "blackmail, terror, and mass murder." There was Iran, Iraq, and North Korea.

More specifically, there was Saddam Hussein, who "had biological weapons sufficient to produce over 25,000 liters of anthrax—enough doses to kill several million people," and enough botulinum toxin "to subject millions of people to death by respiratory failure," not to mention "the materials to produce as much as 500 tons of sarin, mustard, and VX nerve agent."[121] Saddam's torture techniques, Bush said, included "electric shock, burning with hot irons, dripping acid on the skin, mutilation with electric drills, cutting out tongues, and rape."[122]

Asserting that Saddam aided and protected Al-Qaeda, Bush suggested that we "imagine those 19 hijackers with other weapons and other plans—this time armed by Saddam Hussein. It would take one vial, one canister, one crate slipped into this country to bring a day of horror like none we have ever known."[123]

Bush invited "all free nations" to join America in preventing such attacks. But he added that "the course of this nation does not depend on the decisions of others." There was applause. "Whatever action is required, whenever action is necessary, I will defend the freedom and security of the American people." There was more applause. Bush would act preemptively and unilaterally if necessary.

Bush assured the country that attacks by Saddam were not merely remote possibilities. After all, he said, the United States had learned that Iraq had "mobile biological weapons labs . . . designed to produce germ warfare agents, and can be moved from place to a place to evade inspectors." From Iraqi defectors we had learned that Saddam had "aluminum tubes suitable for nuclear weapons production."[124]

Then came sixteen words that would prove to be the casus belli—but were attributed to one of America's allies. "The British government has learned," Bush said, "that Saddam Hussein recently sought significant quantities of uranium from Africa."[125]

At the time, even the most connected of Capitol Hill insiders were unaware of the machinations behind the intelligence that was being presented. To virtually everyone in the country, Saddam appeared to be a truly grave threat. If Saddam's acquisition of nuclear weapons was not a cause for war, what was?

Bush addressed the men and women of America's armed forces. "Many of you are assembling in or near the Middle East," he said, "and some crucial hours may lay ahead. In those hours, the success of our cause will depend on you. Your training has prepared you. Your honor will guide you. You believe in America, and America believes in you."

As he concluded, cheering him on were Marine Corps general Michael Hagee; Admiral Vern Clark; General Richard B. Myers, the chairman of the Joint Chiefs of Staff; and General Eric Shinseki.[126] A few years earlier, Bush had confided that he thought to be a great president meant being a great commander in chief. Now George W. Bush was leading his nation into war.

CHAPTER SIXTEEN

The Good Soldier

It was not until President Bush's State of the Union address that CIA officer Tyler Drumheller figured out what had just happened—and even then he was utterly dumbfounded. It was now clear to Drumheller that when he had been asked again to try to authenticate Curveball because an unidentified official was giving a speech, the official in question was George W. Bush. The speech was the State of the Union. Drumheller had sent out a report to George Tenet's office saying that the Germans still could not verify Curveball's claims. And yet, his report notwithstanding, Bush had still referred to "several mobile biological weapons labs . . . designed to produce germ warfare agents."[1]

That was hardly the only egregious error in the speech. On at least fourteen different occasions prior to the 2003 State of the Union, analysts at the CIA, the State Department, and other government agencies who had examined the Niger documents or reports about them raised serious doubts about their legitimacy. Yet somehow a reference to the Niger yellowcake deal—a complete fiction, the product of a forgery—had ended up as a vital part of one of the most important State of the Union Addresses in modern history.

How did this happen? Last-minute negotiations between the White House and the CIA led to a decision to attribute reports about the Niger uranium deal to British intelligence rather than classified American intelligence.*[2] Not only had the president of the United States taken

*Stephen Hadley later accepted responsibility for allowing the sixteen-word sentence about the Niger deal to remain in the speech. "I should have recalled at the time of the State of the Union speech that there was controversy associated with the uranium issue," he told reporters at a press conference in July 2003. "When the language in the drafts of the State of the Union referred to efforts to acquire natural uranium, I should have either asked that they—the 16 words given to that subject—be stricken, or I should have alerted DCI Tenet."

a statement that many in his administration knew to be a lie and used it as a cause for war, he had taken the cowardly way out and attributed it to a third party.

Now that President Bush had made the case for war, there was one crucial step left before the United States could invade Iraq. On the morning of Wednesday, January 29, 2003, only hours after Bush's speech, Secretary of State Colin Powell marched into the office next door to his on the seventh floor of the State Department to meet with his longtime aide, Colonel Lawrence B. Wilkerson.[3] Few colleagues were closer to Powell than Wilkerson, who had served as his top aide in the army, in the private sector, and in the State Department for more than a decade.

"I have to present our case to the United Nations Security Council on February 5," Powell told Wilkerson.[4] "I want you to assemble a team. Go out to the Central Intelligence Agency. Deputy Secretary Armitage has prepared the way for you. Mr. Tenet is expecting you. Put together whatever team you need, and I'll be out there as soon as you tell me you are ready. We will begin rehearsal, and on the fifth of February we will be at the United Nations at nine in the morning to present our case."

Powell didn't relish the assignment. Long known as a reluctant war-

---

More to the point, as reported in *The Italian Letter: How the Bush Administration Used a Fake Letter to Build a Case for War in Iraq* by Peter Eisner and Knut Royce, the White House didn't dare go back to George Tenet because he had already rejected the same material for Bush's speech in Cincinnati. Instead, sometime in January—the exact date is unclear—Robert Joseph, special assistant to the president for nonproliferation, called someone else in the CIA who would be even more amenable to White House requests—Alan Foley, the director of the CIA's Weapons Intelligence Non Proliferation and Arms Control Center (WINPAC). Foley, according to Eisner and Royce, had already delivered a clear message to his staff: "If the president wants to go to war, our job is to find the intelligence to allow him to do so." Far more interested in not violating the rules of classification than in whether or not the information was credible, Foley accepted the idea of attributing the information to the British because their report was not classified—even though the CIA had told the White House again and again that it didn't trust the British reports.

As for the British, they have repeatedly claimed to have other sources, but have refused to identify them. According to Joseph Wilson, that refusal is a violation of the U.N. resolution stipulating that member states must share with the International Atomic Energy Agency all information they have on prohibited nuclear programs in Iraq. "The British say they cannot share the information because it comes from a third-country intelligence source," said Wilson. "But that third country is presumably a member of the United Nations, and it, too, should comply with Article 10 of United Nations Resolution 1441." No evidence of a third country has come to light.

rior, he had argued for years against invading Iraq for the same reasons Scowcroft and Bush 41 had put forth. His devoted admirers—and they were many—saw him as an increasingly lonely and isolated voice of moderation in a militaristic administration. Even though he didn't buy all the claims made by the administration, in public he had begun supporting its tough talk on Iraq—initially in a more measured way than his fellow cabinet members.[5]

On occasions like this, when he was forced to implement policies he questioned, Powell sometimes withdrew and became opaque to the point of impenetrability. "He can be the most endearing person you'd ever want to meet in your life," Wilkerson said. "The next minute he can be colder than a fish."[6]

At rare moments, however, Powell shared his qualms with colleagues. The previous November, as the administration marched inexorably to war, Powell had come into Wilkerson's office and looked out the window toward National Airport. "I wonder what will happen," he mused, "if we put half a million troops on the ground and scour the country from one end to the other, and don't find a single WMD."[7]

Such misgivings notwithstanding, as the run-up to war continued, resigning was not a real possibility for Powell. That would mean abandoning both his commander in chief and the soldiers he had fought with for many years. Always the good soldier who never questioned the chain of command, Powell chose to walk a fine line that became ever finer until, at times, it seemed to disappear.

And once Powell accepted that the path to war was inevitable, he became a team player with the White House—to such an extent that insiders felt he had lost control of his own department.[8] Why did Powell allow this to happen? "Powell was completely aware of the machinations going on," said one State Department official. "Presumably he was trying to make peace with Cheney, but things went south pretty quickly. If you were the boss and had John Bolton [a Cheney ally], who was off the reservation, why would you keep him there? One reason is to avoid a crisis inside the administration. By getting rid of the guy you would make headlines and raise all manner of questions."

Regardless of what he *really* believed, Powell ultimately accommodated the White House to such an extent that he became the most articulate spokesman for the war effort. What's more, he played as fast and loose with the facts as the neocons.[9] Just two days before the State of the Union address, on January 26, 2006, Powell addressed the World Economic Forum in Davos, Switzerland, and asked, "Why is Iraq still trying to procure uranium and the special equipment needed to transform it into material for nuclear weapons?"[10] In referring to the Niger deal and the aluminum tubes, Powell was actually betraying his own State Department analysts who had rejected these two key pieces of "evidence" against Saddam.

Now, the day after Bush's speech, Powell handed Wilkerson a forty-eight-page dossier on Iraq's weapons of mass destruction. He said that other files on Iraq's ties to terrorism and human rights violations would arrive shortly. Wilkerson would be working around the clock for the next six days.[11] At the time, he was, relatively speaking, an innocent in terms of doing battle in the intelligence wars.

The next morning, Wilkerson met with Lynne Davidson, Powell's chief speechwriter; Carl Ford, the head of the State Department's Bureau of Intelligence and Research (INR); and Barry Lowenkron, Principal deputy director of policy planning for State. A staffer from the United States Mission to the U.N. was also brought in to help familiarize the team with the logistics of the presentation in New York.

After getting the files from Powell, Wilkerson drove out to the CIA's headquarters, a 258-acre campus in the woods of Langley, Virginia, 10 miles from Washington, where he was joined by George Tenet and his deputy, John McLaughlin. They started working in the National Intelligence Council's offices and later moved to Tenet's conference room for rehearsals and discussions.[12] Tenet had planned a trip to the Middle East, but Powell wouldn't let him go.[13]

At night the next week, Secretary Powell reviewed their work and rehearsed his speech. At various times, National Security Adviser Condoleezza Rice and a host of CIA analysts and National Intelligence officers came and went. Deputy Director McLaughlin was there virtually around the clock. Among the others who attended meetings over the next few days were Deputy National Security Adviser Stephen Hadley;[14] Bob Joseph, the National Security Council official who had made sure mention of the Niger documents got into Bush's speech; Will Tobey, also from the NSC;[15] John Hannah and Scooter Libby from Cheney's staff;[16] Robert Walpole, the author of the NIE; and Lawrence Gershwin, a CIA top adviser on technical intelligence.[17]

Notable in their absence, however, were intelligence analysts from the INR. They had been instructed to fact-check various drafts of Powell's speech from their offices in the State Department. But their physical absence was another striking indication that Powell had capitulated and was trying to avoid a showdown with the White House. "I can only assume that he was under the gun to do this for the president, and he had to find a way to do this and live with it," said a State Department colleague. In other words, the hard-nosed analysts in INR, who had not bowed to White House pressure, would be a political liability for Powell.

Wilkerson had been given three dossiers, about ninety pages of material on Iraq's WMDs, on Iraq's sponsoring of terrorism, and on its violation of human rights. He had not been privy to all the machinations surrounding the WMD issue, but as soon as he examined the documents, he realized something was wrong. The CIA had scores of analysts who had spent years studying Iraq's weapons programs—WMDs, nuclear arms, and other relevant specialties. It had rigorous peer review procedures to eliminate dubious intelligence from fabricators, forgers, double agents, and sources with ulterior motives, and a large budget devoted to Middle East intelligence.[18]

But the dossiers in front of Wilkerson had been prepared by Scooter Libby in Cheney's office, not by the CIA. That was a highly unusual breach of protocol, especially for such a vital matter of state. Explained Richard Clarke, former director of counterterrorism in the National Security Council, "It's very strange for the Vice President's senior adviser to be . . . saying to the Secretary of State, 'This is what you should be saying.' "[19]

As Wilkerson vetted the files he had been given, however, it became increasingly apparent exactly how aggressively Cheney and his men had stacked the deck. The first dossier Wilkerson examined was the forty-eight-page script on WMDs. "It was anything but an intelligence document," he said.[20] "It was, as some people characterized it later, sort of a Chinese menu from which you could pick and choose.

"Every time we had a question, which was virtually every line, John Hannah from the vice president's office would consult a huge clipboard he had," said Wilkerson.[21] Hannah was a former official at the Washington Institute for Near East Policy (WINEP), a pro-Israel think tank. In addition to having coauthored the forty-eight-page document, Hannah had worked closely with Libby on the White House Iraq Group, the team Cheney had put together to cull information about Iraq's WMDs.[22]

Hannah responded to Wilkerson's queries by citing the appropriate source—the CIA, the DIA (Defense Intelligence Agency), the *New York Times,* and others. Then Wilkerson's team tracked down each of the original documents he cited. They pored over satellite photographs, intercepts of Iraqi military communications, briefings of Iraqi defectors, intel from allied services in Europe and the Middle East.

But the more deeply they probed, the more dubious Wilkerson became about Hannah's work. "It was not sourced the way intelligence documents should be sourced. . . ." Wilkerson said. "Once we read the entirety of those documents, we'd find that the context was not quite what the cherry-picked item imported."[23]

Hannah had served as the key liaison between Cheney's office and Ahmed Chalabi.[24] Wilkerson thought that much of the bogus intelligence probably came from Chalabi and the INC—sources who clearly had their own agendas.[25]

That question aside, it would take days to get through the scripts. Within a few hours, both Wilkerson and Tenet were so fed up they decided to junk the entire forty-eight-page dossier on WMDs. "There was no way the secretary of state was going to read off a script about serious matters of intelligence that could lead to war when the script was basically unsourced," Wilkerson said.[26] He turned to Tenet and said that what they were trying to do was not going to work.

Tenet agreed. "Let's go to the NIE," he said.[27] By that Tenet was suggesting that the 2002 National Intelligence Estimate (NIE), the document that had been produced by the CIA in October at the request of the Senate, be used as the basis for Powell's presentation.[28] Relying on the NIE was enormously reassuring to Wilkerson, who did not realize that it, too, had been tampered with. Powell, however, should have known better. After all, key elements of the NIE—the Niger documents, the aluminum tubes—had been disputed by his own State Department analysts.

Later that night, Wilkerson and his colleagues watched a film they had gotten out of the State Department archive of Adlai Stevenson's historic speech before the United Nations Security Council during the 1962 Cuban missile crisis, punctuated by the dramatic unveiling of irrefutable proof of the Soviet Union's perfidy that crystallized world sentiment. They were seeking a similar confluence of evidence and rhetoric, Wilkerson said, so that Powell could have his own historic "Stevenson moment" before the U.N.[29]

While Wilkerson and his team worked at Langley, the State Department's INR fact-checked an early draft of Powell's speech. On January 31, according to *Hubris,* by Michael Isikoff and David Corn, they sent a memo to Powell asserting that thirty-eight allegations in the speech were "unsubstantiated" or "weak."[30] As a result of the memo, twenty-eight of the charges were removed from the draft.

Meanwhile, one of the biggest battles Wilkerson's team faced against Cheney's men was over Iraq's links to terrorism. Acutely aware of the administration's desire to link 9/11 to Iraq, George Tenet had ordered his analysts to dig through vast piles of old intelligence in hopes of finding evidence of such ties. Michael Scheuer, then head of the CIA's bin Laden unit and later the author of *Imperial Hubris,* led the effort. "We went back 10 years," he said. "We examined about 20,000 documents, probably something along the line of 75,000 pages of information. And there was no connection between Iraq and Osama [bin Laden]."[31]

Tenet had personally relayed that message to President Bush and added that if anything were ever found it would be peripheral.[32] But Cheney's men had insisted Tenet was wrong. Among the most explosive allegations in their dossier was the charge that 9/11 hijacker Mohammed Atta had met with an Iraqi intelligence agent in Prague in April 2001. The material had originally been funneled to Cheney via the Counterterrorism Evaluation Group (CTEG), the precursor to the OSP in the Pentagon.[33]

Vice President Cheney had made the claim himself repeatedly on national TV. "It's been pretty well confirmed that [Atta] did go to Prague and he did meet with a senior official of the Iraqi intelligence service," he said on *Meet the Press* not long after 9/11.[34]

But by February 2002, the allegation had already been discredited independently by both the CIA and the FBI. They had determined that at the time of the supposed meeting in Prague, Atta was traveling from Florida to Virginia Beach, Virginia. The FBI even had his rental car and hotel receipts.[35] Nevertheless, Cheney had continued to repeat the charge and his subordinates kept putting it back in play.*[36]

This time, Colin Powell personally threw out the Atta allegation. "He was trying to get rid of everything that didn't have a credible intelligence community–backed source," said Wilkerson.[37]

*One person who knew that the Prague meeting was bogus was NSC counterterrorism head Richard Clarke. One day Clarke was outside the West Wing of the White House when Libby came out of Cheney's office. "I hear you don't believe this report that Mohammed Atta was talking to Iraqi people in Prague," Libby said.

"I don't believe it because it's not true," Clarke replied.

"You're wrong," Libby told him. "You know you're wrong. Go back and find out. Look at the rest of the reports and find out that you're wrong."

Clarke later told PBS *Frontline,* "I understood what he was saying, which was, 'This is a report that we want to believe, and stop saying it's not true. It's a real problem for the vice president's office that you, the counterterrorism coordinator, are walking around saying that this isn't a true report. Shut up.' That's what I was being told."

But Cheney found out—and wanted it back in. A day or two later, Deputy National Security Adviser Stephen Hadley, a Cheney loyalist, raised the issue.

"What happened to the meeting about Mohammed Atta in Prague?" he asked.

"We took that out, Steve," Powell said. "Don't you remember?"

Hadley hung his head sheepishly. "Yes, Mr. Secretary, I do remember."[38]

"They were just relentless," said Colonel Lawrence Wilkerson. "You would take it out and they would stick it back in. That was their favorite bureaucratic technique—ruthless relentlessness."[39]

Normally an even-tempered man, Powell got so angry at such tactics that at times he threw the dossier up in the air. "This is bullshit," he said.[40] "I'm not doing this."

But he continued anyway. One morning that week, as the U.N. speech drew closer, Cheney saw Powell just outside the Oval Office and reassured him. "Your poll numbers are in the seventies," Cheney said. "You can afford to lose a few points."[41]

In light of the immense pressures they faced, Wilkerson was happy that George Tenet served as the CIA's director, for he initially regarded Tenet as a reassuring presence who would fight to keep the facts straight. A Georgetown University alumnus who was infected with the basketball fever common among fans of the Hoyas, Tenet was a likable and gregarious fellow who had won Wilkerson's trust. "Tenet had been there for a number of years," said Wilkerson. "He was a professional. So was his deputy. It didn't seem possible that [the neocons] could exercise much influence."[42]

Working late one night in the Langley pressure cooker, however, Wilkerson heard tales that made him wary about even Tenet. "We had enough instances in the wee small hours of the morning that by the end of that intense, short period I had suspicions about some of the material."[43]

Those doubts were reinforced by a conversation with one of Langley's top analysts. At the time, Iraq had a large network of front companies that it had created to circumvent the U.N.'s economic sanctions. One of those front companies was doing business with an Australian firm that happened to sell mapping software. "This got reported back through intelligence channels as the Iraqis attempted to buy mapping software for the eastern United States," Wilkerson said.[44]

Presumably, that meant Iraq was planning a military attack on the

United States. It was exactly the kind of nugget of information Cheney had sought. "So Mr. Tenet runs over to the vice president's office and briefs Cheney that the Iraqis are buying mapping software for the eastern United States," said Wilkerson. Two weeks later, however, the analyst at Langley discovered the truth: no software had been transferred to Iraq, nor had Saddam expressed any interest in it. He immediately told Tenet.

But Tenet never went back to the vice president to correct the report. "So I began to get the idea that there was political pressure," said Wilkerson. "Why did George Tenet not go back? Because he didn't want to."[45]

At the time, however, Wilkerson put aside his doubts. "Having been so close to George Tenet during those days, it was clear to me that he believed what we were presenting was accurate."[46]

Or so it seemed. He was finally getting a small taste of the extraordinary political pressures at play.

A few days later, Tyler Drumheller was given a draft of Powell's speech[47] in which Powell was to explain that the mobile weapons labs could concoct enough microbes "in a single month to kill thousands upon thousands of people."[48] To Drumheller's astonishment, that meant they were still using the discredited Curveball story. Drumheller immediately met with CIA deputy director John McLaughlin and an aide in a small conference room.[49] According to the *Washington Post,* Drumheller said that McLaughlin promised to investigate immediately.

But Colin Powell was not told about Tyler Drumheller's concerns.[50] He was not told that the United States never had direct access to Curveball or that the Germans had warned the U.S. that he was an unreliable alcoholic. "So all the fine-grained stuff that might have caused [Powell] not to use it, he wasn't given an opportunity to hear firsthand," said David Kay, a former weapons inspector in Iraq.[51]

Meanwhile, on February 2, the State Department's INR was given another look at Powell's speech. This time they found seven items to which they objected. Three of them were removed from the speech. But four items remained.[52]

At about the same time, outside the narrow confines of Colin Powell's speech, the administration was standing firm against resistance it was meeting in the international arena. The International Atomic Energy Agency (IAEA) had just issued a report taking issue with administration claims that Iraq had an active nuclear program. "We have to date found no evidence that Iraq has revived its nuclear weapon program since the elimination of the program in the 1990's," Mohamed ElBaradei, the head of the agency, told the United Nations Security Council.[53]

Likewise, Hans Blix, the chief U.N. weapons inspector, had also told the Security Council that "Iraq has decided in principle" to cooperate on process when it came to granting access to sites and providing support services for U.N. inspectors. "Information provided by member states tells us about the movement and concealment of missiles and chemical weapons and mobile units for biological weapons production," Blix said. "We shall certainly follow up any credible leads given to us and report what we might find as well as any denial of access."[54]

There were many unanswered questions, serious questions, Blix said. But Iraq's concessions to the U.N. inspectors and the IAEA report would make it more difficult for the U.S. to win support for war.

Wilkerson's team worked through the weekend at Langley and early the next week moved its operations to New York. On the night of Tuesday, February 4, the eve of Powell's historic speech, Wilkerson put together a dress rehearsal on the top floor of the U.S. Mission to the U.N. He had the furniture laid out to look exactly like the U.N. Security Council. This was his last chance to refine the presentation, eliminate anything they weren't sure of, to make certain they were getting everything right.

One key piece of Powell's presentation concerned intercepts of conversations between two officers in Iraq's elite military unit, the Republican Guard. Wilkerson was not certain they provided compelling proof. "They were very classy, rat-tat-tat-tat, hitting you fast, like all the TV crap Americans are used to these days, nine-second sound bites. But look at the intercepts. You have this guy at a chemical factory saying, 'Get rid of it.' Suppose he's actually trying to get rid of [the weapons]. . . . [But] all the intercepts could have been interpreted two or three or even more ways. Believe me, I looked at it fifty times."

Of greater concern to Wilkerson was the sourcing for the mobile weapons vans that allegedly manufactured biological and chemical weapons. These facilities were seen to be particularly insidious precisely because their mobility enabled them to escape detection by weapons inspectors. When he spoke about the vans in the Security Council, Powell was to present illustrations of them based on accounts from an undisclosed intelligence source. "Powell and I were both suspicious because there were no pictures of the mobile labs," said Wilkerson.[55]

To make sure the evidence was solid, Wilkerson asked Tenet and John

McLaughlin, the CIA deputy director of intelligence, about the sourcing once again. According to the *Washington Post,* the two men replied that the evidence was "exceptionally strong," and that it was based on several sources whose stories had been independently corroborated.[56] McLaughlin gave no hint that Drumheller had warned him about Curveball.

Wilkerson recalled, "I sat in the room, looking into George Tenet's eyes, as did the Secretary of State, and heard with the firmness that only George could give, . . . I mean eyeball-to-eyeball contact between two of the most powerful men in the administration, Colin Powell and George Tenet, and George Tenet assuring Colin Powell that the information he was presenting to the U.N. was ironclad."[57]

At the end of the rehearsal, Powell turned to Tenet. "Do you stand by this?"

"Absolutely, Mr. Secretary," Tenet replied.[58] He was resolute. As Wilkerson saw it, Tenet had no doubts that the material was irrefutable.

"Good," Powell said, "because you are going to be in camera beside me at the U.N. Security Council tomorrow."

Later that evening, Tenet and Powell retired to their hotel rooms to rest up for the presentation the next morning. At about midnight, Tenet called Tyler Drumheller to check some details about the speech. Well aware of how high the stakes were, Drumheller persisted in making sure that Curveball was not used as a source. "Hey, boss, you're not going to use that stuff in the speech . . . ?" Drumheller told Tenet. "There are real problems with that."[59] Tenet, Drumheller recalled, seemed tired and told him not to worry.[60]

(Both McLaughlin and Tenet said they did not recall having any such conversations with Drumheller. "If someone had made these doubts clear to me, I would not have permitted the reporting to be used in Secretary Powell's speech," McLaughlin said in a prepared statement. Likewise, in *At the Center of the Storm: My Years at the CIA,* Tenet claimed, "I remember no such midnight call or warning. . . . Drumheller had dozens of opportunities before and after the Powell speech to raise the alarm with me [about Curveball], yet he failed to do so.")

Meanwhile, Wilkerson and his team had stayed on to finish up. One lengthy part of the presentation concerned ties between Al-Qaeda and Baghdad. "I was trying to shape it to where it was reasonably believable," Wilkerson said. "There was a hugely lengthy genealogy section that went all the way back to [Osama] bin Laden's and Al-Qaeda's presence in the Sudan and tried to link that somehow to Iraqi intelligence agents going to Khartoum. Talk about connect the dots! It's like Australia is a dot and over there in Moscow is a dot, let's draw a line. It was silly as shit and the secretary got sick and tired of it."

Wilkerson and Powell had deleted it before. "So when I saw that somehow it sneaked its ugly head back in, I just cut it out," said Wilkerson.

At about midnight, Phil Mudd, a CIA terrorism specialist, dropped by, bringing Italian food with him, in hopes of getting a progress report. When Barry Lowenkron, a senior Powell aide, told him that they had tightened the terrorism part, Mudd reviewed the changes.[61]

"Looks fine," he said. Then he left. Not long afterward, Wilkerson returned to the Waldorf-Astoria, where Powell's operations were based.

But at two a.m., Wilkerson got a call. It was one of Tenet's aides.[62] Cheney's office had been insistent that Powell tie Saddam into 9/11. Given that Powell had previously thrown out the allegations that Mohammed Atta had met with Iraqi agents in Prague, Phil Mudd had woken up George Tenet and given him an update. Now Tenet, who was staying at another Manhattan hotel, wanted one last look at the speech.[63]

Wilkerson finally got a few hours' sleep. But the next morning, Cheney's office resumed pressure for him to tie Saddam to Al-Qaeda. Just before Powell's speech began at nine a.m., Wilkerson's phone rang again and again. He looked at the phone's digital screen. The caller was Scooter Libby, Cheney's chief of staff, presumably trying one more time to put the Prague story back in Powell's speech.[64] Wilkerson refused to take it.

At nine a.m. on February 5, 2003, Wilkerson finally took his seat at the United Nations Security Council about ten or twelve feet behind Colin Powell and to the left. As promised, Powell insisted that George Tenet be in camera, just behind him.

Wilkerson sat there opposite the Iraqis, viewing his adversaries intently.[65] "I knew they probably didn't know squat. Saddam Hussein wouldn't have told those guys anything. But I'm sitting there watching them. Are they going to say, 'Oh, God, they've got us!'"

Wilkerson's team had thrown out the entire forty-eight-page WMD dossier given them by Cheney's office. Three-quarters of their file on Saddam's ties to terrorists was in the trash as well. The remaining material was enough for a presentation that would last about an hour and twenty minutes.

Both the audio and the visuals would be dramatic, with fast cutting and fancy editing on big-screen displays. Wilkerson was pleased that they had been so rigorous in eliminating dubious intelligence, but he feared there wasn't much left in the way of evidence. The sought-after "Stevenson moment" was unlikely. "Because we had cut it so severely, I felt that the presentation was fairly solid [in terms of accuracy]," said Wilkerson. "But my biggest concern was the efficacy of the presentation. We had thrown out so much crap—rightfully so. But now the presentation wasn't very effective.

"I felt much the same way the president must have when he said to George Tenet, 'Is this all you've got, George?' That's the way I felt. Is this all we've got? I felt like I'd failed."

But Wilkerson was overlooking one vital factor. The presentation was being made by a man who had risen from the South Bronx to become the most admired and respected person in America. He had served as national security adviser and chairman of the Joint Chiefs of Staff before being appointed secretary of state and becoming the highest-ranking African-American government official in the history of the United States. Blessed with a spectacularly imposing yet reassuring physical presence and an irresistible blend of gravitas and authenticity, poise, and self-assurance, Powell appealed to people at both ends of the political spectrum. At the *New York Times,* liberal columnist Anthony Lewis wrote that Powell had "the dignity, the presence, the directness [that Americans] long for in a president." Neoconservative *Times* columnist A. M. Rosenthal asserted that Powell was "graceful, decisive, courteous, [and] warm."

Dick Cheney had made only one slight miscalculation regarding Powell. He had said Powell's approval ratings were in the seventies. But in fact, at the time, no less than 86 percent of the American people approved of the way Colin Powell was doing his job.[66] He came now before the U.N as a voice of reason, as a man who appeared to have unassailable moral authority, and who had reluctantly concluded that the United States must take up arms against a brutal dictator. All of which made him the most effective salesman on the planet. "There's no question in my mind that Vice President Cheney knew that," said Wilkerson. "That's why he had Powell do it."[67]

Powell did not enjoy the task, but at least Wilkerson and he had weeded out much of the misleading intelligence that Cheney's men had tried to put in his presentation. At least he could go before the world's most important international peacekeeping body on this grave occasion and present intelligence that could be attributed to the NIE or to George Tenet and the CIA.

"My colleagues," he told the United Nations Security Council, "every statement I make today is backed up by sources, solid sources. These are not assertions. What we are giving you are facts and conclusions based on solid intelligence."[68]

The administration's most powerful spokesman, Powell challenged the rest of the world to join the United States. His presentation was masterful. If Saddam Hussein was a villain straight out of central casting, Powell was the hero in the white hat coming in to save the world. Iraq had, he said, seven mobile labs to make germ weapons, unmanned drones, hidden missiles, and terrorist cells.[69] He held up a small vial filled with white powder. "Less than a teaspoon of dry anthrax, a little bit—about this amount . . . shut down the United States Senate in the fall of 2001," Powell said.

"We know that Saddam Hussein is determined to keep his weapons of mass destruction and make more," Powell said. "Given Saddam Hussein's history of aggression, given what we know of his grandiose plans, given what we know of his terrorist associations, and given his determination to exact revenge on those who have opposed him, should we take the risk that he will not someday use these weapons at a time and a place and in a manner of his choosing—at a time when the world is in a much weaker position to respond? . . . Leaving Saddam Hussein in possession of weapons of mass destruction for a few more months or years is not an option. Not in a post–September 11th world."[70]

As soon as his speech was done, it became clear that Powell had hit a home run. The press loved it. More than a hundred media outlets compared his moment to Adlai Stevenson's.[71] That Powell was still thought of as a dove made his credibility unassailable. In the aftermath of Powell's speech, a *Washington Post*/ABC News poll showed that no fewer than 67 percent of Americans felt the United States was justified in going to war because of Iraq's WMDs.[72] As Greg Mitchell of *Editor & Publisher*[73] later reported in a comprehensive critique of the media's response, the *San Francisco Chronicle* called the speech "impressive in its breadth and eloquence." The *Denver Post* romanticized him as "Marshal Dillon facing down a gunslinger in Dodge City," and added that he had shown the world "not just one 'smoking gun' but a battery of them." In Tampa, Florida; Portland, Oregon; and Hartford, Connecticut, the local dailies deemed Powell's presentation "overwhelming," "devastating," and "masterful."[74]

But it was among liberals in the media, especially those who had dismissed George W. Bush as a trigger-happy cowboy, that Colin Powell had the greatest impact. The *Washington Post* asserted that Powell's evidence was "irrefutable." The paper's liberal columnist, Mary McGrory, announced that she was "as tough as France to convince," but Powell had won her over.[75] For Richard Cohen, another *Post* columnist, some of Powell's evidence was "absolutely bone-chilling in its detail . . . [and] had to prove to anyone that Iraq not only hasn't accounted for its weapons of mass destruction but without a doubt still retains them. Only a fool, or possibly a Frenchman, could conclude otherwise."

Finally, there was the *New York Times,* which offered no fewer than three separate news stories praising Powell as "powerful," "sober," "factual," and "nearly encyclopedic." Columnist William Safire saw "half a dozen smoking guns" in Powell's "irrefutable and undeniable" case.[76] At last, concluded the *Times*'s Michael Gordon, "It will be difficult for skeptics to argue that Washington's case against Iraq is based on groundless suspicions and not intelligence information."[77]

However, at least one person watching Powell had a profoundly different reaction, especially as Powell made his presentation about the mobile weapons vans. Somewhere a German intelligence officer was watching—the very same BND officer who had supervised Curveball—and he was absolutely agog that the Americans were knowingly presenting something that was utterly uncorroborated. "We were shocked," he said. "Mein Gott! We had always told them it was not proven. . . . It was not hard intelligence."[78]

As Powell made his presentation, the world instantly knew that it marked the point of no return in launching the Iraq War. How could he have lent his credibility to such an enterprise? When it came to warfare, Powell famously insisted on the use of overwhelming force (a key tenet of what had become known as the Powell Doctrine), yet now, in bureaucratic warfare, he had made the decision to enter battle without being fully armed. His team of intelligence analysts from INR was available whenever necessary,[79] but Powell had not summoned them to Langley or to the U.N.

Why? One of Powell's colleagues told *The Guardian* that this decision was a sign of resignation and defeat. "He didn't have anyone from INR near him," said Greg Thielmann, who served as director of the Strategic, Proliferation and Military Affairs office at INR until September 2002. "Powell wanted to sell a rotten fish. He had decided there was no way to avoid war. His job was to go to war with as much legitimacy as we could scrape up."[80]

Powell has long been reticent about sharing his deepest thoughts, but in an interview posted on the State Department's website in 2003, and cited by *The Guardian* and the *Washington Post*, he demonstrated a revealing historical awareness of his position. In the interview he referred to two of his predecessors, Thomas Jefferson, America's first secretary of state, and General George C. Marshall, who served as secretary of state under Harry Truman. When Dwight Eisenhower was named D-Day commander over Marshall, "whatever disappointment [Marshall] felt over that, he simply ate it," Powell said. In 1948, Marshall had argued bitterly with Truman over America's recognition of the state of Israel, but did not resign. Powell quoted him as saying, "No, gentlemen, you don't take a post of this sort and then resign when the man who has the constitutional responsibility to make decisions makes one you don't like."[81]

Powell was most poignant of all when he came to Jefferson: "He said something along the lines of, 'I go now to the task that you have put before me, in the certain knowledge that I will come out of it diminished.'"[82]

"Diminished" would prove to be a word that applied to Powell, but it would be several years before that became clear. Over time, the meaning of Powell's speech deepened to reveal multiple layers of reality behind the events leading up to the war. In light of his sterling reputation, and what was widely seen as a rigorous vetting of the presentation, it was hard to believe that he had knowingly lied or had presented material that had not been thoroughly checked. But the lies were there nonetheless. There had been a deliberate campaign of disinformation, or black propaganda, which led the White House, the Pentagon, the State Department, and thousands of outlets in the American media to promote the falsehood that Saddam Hussein's WMD program posed a grave risk to the United States. Powell had allowed himself to become a key part of it.

As Wilkerson himself later put it, "My participation in that presentation at the U.N. constitutes the lowest point in my professional life. I participated in a hoax on the American people, the international community, and the United Nations Security Council."

In the end, far from impeding the neocon takeover, Powell had gone so far as to present their case before the entire world and in doing so had given their enterprise credibility and a noble human face. In the wake of his speech, an astounding 90 percent of the American people now believed that Saddam Hussein had a vigorous WMD program that posed a grave threat to the United States.

* * *

Meanwhile, U.N. weapons inspectors frantically continued their work in Iraq—finding nothing.[83] On February 8, the United Nations Team Bravo, led by American biological weapons experts, spent three and a half hours inspecting Curveball's former work site and determined that Curveball had lied.[84]

At the same time, Jacques Baute, head of the International Atomic Energy Agency's Iraq nuclear verification office, examined electronic copies of the Niger documents that had finally been forwarded to the IAEA by the United States. Astonishingly, the Bush administration had attached a note to the documents. "We cannot confirm these reports and have questions regarding some specific claims," it said.[85]

Baute used Google to check a reference in the papers and within minutes discovered the erroneous reference to the Niger's Constitution. "At that point, I completely changed the focus of my search to 'Are these documents real?' rather than 'How can I catch the Iraqis?'" Baute told the *Los Angeles Times.*[86]

By February 10, more than 100,000 U.S. troops were in the Gulf region, including three aircraft carrier groups, with a fourth, the USS *Theodore Roosevelt,* on the way, and a fifth, the USS *Kitty Hawk,* scheduled to be deployed shortly.[87] Three days later, the U.S. activated 39,000 more reservists.[88]

As war fever built, the big oil companies eagerly eyed the prospects of the world's second largest oil reserves opening up for foreign investment. Archie Dunham, chairman of ConocoPhillips, got right to the point. "We know where the best reserves are [and] we covet the opportunity to get those some day," he said.[89] Administration spokesmen and neocon pundits asserted that Iraq's oil riches would make the war a cost-effective, even profitable, affair. "Iraq has tremendous resources. . . ." said White House press secretary Ari Fleischer. "So there are a variety of means that Iraq has to be able to shoulder much of the burden for their own reconstruction."[90] Promised neocon columnist Charles Krauthammer on WUSA-TV: "We will have a bonanza, a financial one, at the other end, if the war is successful."[91]

By February 14, after having combed through more than three hundred sites in Iraq unimpeded for eleven weeks, U.N. inspectors had still found nothing,[92] which only proved, said the neocons, that they were dupes. "UN weapons inspectors are being seriously deceived," declared Richard Perle. "It reminds me of the way the Nazis hoodwinked Red Cross officials inspecting the concentration camp at Theresienstadt in 1944."[93]

The neoconservative moment of glory had finally arrived. Empowered at last, one after another the neocons fanned out across the media and took over the airwaves, intoxicated by the grandiose thrill that comes with juggling the fate of nations. As Ledeen had put it, now was the time to "destroy [our enemies] to advance our historic mission."[94] "It may turn out to be a war to remake the world," he added.[95]

"This is total war . . ." said Richard Perle. "If we just let our vision of the world go forth and we embrace it entirely and we don't try to piece together clever diplomacy but just wage a total war, our children will sing great songs about us years from now."[96]

Iraq was merely the beginning. To Iran, Syria, Saudi Arabia, Lebanon, and the PLO, Perle added, "We could deliver a short message, a two-word message: 'You're next.' "[97] John Bolton assured Israeli officials that after invading Iraq, the United States would take care of Iran, Syria, and North Korea.[98]

On February 15, there were international protests against the coming war, with a million people gathering in Rome, a million in London, and millions more in more than six hundred cities around the globe: New York; Washington; L.A.; Boston; San Francisco; Chicago; Madison, Wisconsin; Eugene, Oregon, and many, many more.

But in the corridors of power in Washington, or, for that matter, in the media culture of America, the protesters had little impact. After all, this was the first great war for a new generation. In the Gulf War of 1991, the United States had pushed back Saddam; in Kosovo, it had fought a repressive dictator. But this was different. The United States was invading a foreign country in order to topple its leader. American prestige and power were on the line. Countless lives were at risk. Hundreds of billions, even trillions of dollars. The geostrategic balance of power.

The chords of patriotism, of romantic nationalism, had been plucked. A giddy triumphalism swept the nation. America's enemies—the perfidious French!—were denounced as "cheese-eating surrender monkeys."[99] Right-wing talk-show hosts—Rush Limbaugh, Bill O'Reilly, Sean Hannity, Glenn Beck, and more—had a field day. Liberals were members of the Al-Qaeda booster club. To O'Reilly, those who opposed war, such as journalist Peter Arnett and actor Sean Penn, were traitors, nothing less.[100] "It's beyond me how anybody can look at these protesters and call them anything other than what they are: anti-American, anti-capitalist, pro-Marxists and communists," Rush Limbaugh told his 20 million listeners.[101]

In the Vietnam era, independent FM radio had served as a counterculture media outlet for disaffected youth to protest the war. But no longer. By 2003, Clear Channel Communications owned more than eleven hundred radio stations in the United States, and sponsored mass pro-war rallies throughout the country. After Natalie Maines, lead singer of the Dixie Chicks, criticized Bush, the bestselling band suddenly got no airplay. The group was denounced as traitors. Sales plummeted. As for the press, the supposedly liberal *Washington Post* ran more than three times as many hawkish op-ed pieces as dovish ones. Its editorials were 100 percent pro-war.[102]

Unseen in the excitement was a purge in the highest reaches of the national security apparatus. Anyone not with the program was pushed overboard. Secretary Treasury Paul O'Neill and counterterrorism czar Richard Clarke had already gone. Elliott Abrams, who had been promoted to special assistant to the president and senior director on the NSC for Near East and North African Affairs, and through Condoleezza Rice had a direct line into President Bush, consolidated his power. Ben Miller, a CIA analyst at the NSC, had dared question whether Ahmed Chalabi was fit to succeed Saddam—and was forced out immediately. Two other officials, Flynt Leverett and Hillary Mann, were out.[103] Rand Beers, the National Security Council's senior director for combating terrorism, resigned after feeling that the war on terror was being set aside for the war in Iraq.[104] Richard N. Haass, director of policy planning at the State Department, another moderate from the Bush 41 era who had been marginalized, was also gone.[105] General Anthony Zinni, former commander in chief of the U.S. Central Command in the Middle East, failed to win reappointment. His sin was to suggest publicly that the aftermath to the war might be prolonged and problematic.[106] There were many others.

On February 25, General Eric Shinseki told a Senate panel that he thought "several hundred thousand" troops would be needed to stabilize postwar Iraq—much to the ire of Donald Rumsfeld, who had pushed to use as small a force as possible. After dismissing Shinseki's estimates as "far off the mark,"[107] Rumsfeld, the *Houston Chronicle* reported, "took the unusual step of announcing that Shinseki would be leaving when his term as army chief of staff end[ed]."[108]

Meanwhile, former senator Daniel Patrick Moynihan (D-N.Y.) was at home, suffering from complications after an emergency appendectomy. A hero of sorts to neocons, who incessantly sought his favor, Moynihan had inherited Scoop Jackson staffers Elliott Abrams and Abram Shulsky, but had always been a maverick who was not averse to pushing over sacred cows. "The war drums were beating as he was going in and out of the hospital," recalled his daughter Maura. "But he took phone calls all the time from Cheney and Rumsfeld."[109]

To their dismay, Moynihan told them there were far better ways to contain Saddam than going to war. "Don't worry about the war, my dear," Moynihan later told his daughter. "The war will last two weeks. It's the ten-year occupation you have to worry about."[110]

On March 6, the foreign ministers of France, Germany, and Russia issued a joint declaration asserting that in view of "encouraging results" that have come out of the renewed U.N. inspections, they could not approve of a "resolution that would authorize the use of force."[111]

In light of the Bush administration's decision to proceed almost unilaterally, with "a coalition of the willing," Brent Scowcroft, more of an outsider than ever, but still a member of the President's Foreign Intelligence Advisory Board, spoke out. "During the campaign [President Bush] made some strong statements about putting more stock in [coalitions]. Clearly, that hasn't happened," he told the *National Journal.* Ultimately, Scowcroft said, such a doctrine "is fundamentally, fatally flawed. . . . [I]t's already given us an image of arrogance and unilateralism, and we're paying a very high price for that image. If we get to the point where everyone secretly hopes the United States gets a black eye because we're so obnoxious, then we'll be totally hamstrung in the war on terror. We'll be like Gulliver with the Lilliputians."[112]

On March 7, the IAEA publicly exposed the Niger documents as forgeries. Not long afterward, on *Meet the Press,* Cheney said they were wrong, that they had "consistently underestimated or missed what it was Saddam Hussein was doing," and went even further. He added, "We know [Saddam] has been absolutely devoted to trying to acquire nuclear weapons. And we believe he has, in fact, reconstituted nuclear weapons."[113]

On March 14, Senator Jay Rockefeller IV, the ranking Democrat on the Senate Intelligence Committee, wrote a letter to FBI chief Robert Mueller asking for an investigation because "the fabrication of these documents may be part of a larger deception campaign aimed at manipulating public opinion and foreign policy regarding Iraq."*

Meanwhile, the Pentagon's PR juggernaut was humming like a finely

*Senator Pat Roberts, of Kansas, the Republican chair of the committee and a Cheney ally, declined to cosign the letter.

tuned machine. Having devised the brilliant strategy of "embedding" reporters as part of the dazzling high-tech war machine, it enlisted hundreds of eager and often overly credulous journalists who had signed on for the adventure of a lifetime, not to mention a plum, career-making assignment.[114] While forces massed in the Middle East, reporters raced to get credentialed by the Pentagon. As the *Columbia Journalism Review* later pointed out, they could write anything they wanted, just so long as they refrained from reporting "about ongoing missions (unless cleared to do so by the on-site commander)"; refrained from "reporting on the specific results of completed missions, or on future, postponed, or canceled missions"; refrained from "breaking embargoes imposed on stories for 'operational security' reasons"; refrained from "traveling in their own vehicles," and a few other practices.[115]

On March 11, Bush continued to talk to world leaders about Iraq.[116] But hopes for another U.N. Security Council agreement were dissipating. The Bush administration stood by its assertion that Saddam had violated U.N. Resolution 1441 by refusing to disarm.

On *Meet the Press,* on March 16, Cheney asserted that "I really do believe that we will be greeted as liberators."

The next day, the Department of Homeland Security raised the level of risk to orange, or "High," after telling Americans to stock up on duct tape as protection against biological or radiological attacks. Fear was everywhere. Knowing the war was about to begin, the U.N. withdrew its staff from Iraq.[117]

That evening, Bush addressed the American people and gave Saddam and his family forty-eight hours to get out of Iraq. "Should Saddam Hussein choose confrontation," Bush said, "the American people can know that every measure has been taken to avoid war, and every measure will be taken to win it."[118] There were about 211,000 military personnel and 1,100 aircraft in position, ready to strike.[119]

The next day, March 18, a letter went out from Bush to Speaker of the House Dennis Hastert and President Pro Tempore of the Senate Ted Stevens declaring that having determined that further diplomacy would not "adequately protect the national security of the United States," he was acting to take the "the necessary actions against international terrorists and terrorist organizations, including those nations, organizations, or persons who planned, authorized, committed, or aided the terrorist attacks that occurred on September 11, 2001."[120]

That morning, Bush's parents appeared on ABC's *Good Morning America.* The president had called them the night before, Barbara Bush said, "just to see how we were and tell us he was fine. He was tired, but he was wonderful.

"I'm not sure that we're any different than any other parents in a lot of ways. . . ." she told Diane Sawyer. "It's very easy for us to be supportive, totally supportive. But it's better to stay, you know, stay off the stage."[121]

With his son leading his country into battle, and the entire nation cheering him on, George H.W. Bush was hardly in a position to express doubts about his son's policies. "Yeah. That's the . . . that's the key point," he chimed in. "What I say is we support the President without reservation."[122]

But Diane Sawyer, well aware of Bush 41's close friendship with Brent Scowcroft, put the latter's recent remarks about "ad hoc coalitions" on the screen.

Bush senior didn't take the bait. "I haven't seen that quotation," he said. "And I'd want to read it and talk to him about it. But I have great respect for him. But . . . I don't, I don't obviously agree with what you've just read."[123]

On the morning of March 19, 2003, Bush started his workday with a 7:40 a.m. phone call to British prime minister Tony Blair, his greatest ally in the "coalition of the willing."[124] Later he went down to the Situation Room in the White House and, through a secure video link, got on line with his field commanders. All of them said they were ready. According to Bob Woodward's *Plan of Attack*, Bush delivered a brief statement: "For the peace of the world and the benefit and freedom of the Iraqi people, I hereby give the order to execute Operation Iraqi Freedom. May God bless our troops."[125]

It was not the first time President Bush had thought of God in relation to the Iraq War. Toward the end of his State of the Union address seven weeks earlier, he had come up with an unusual formulation. "The liberty we prize is not America's gift to the world," Bush declared. "It is God's gift to humanity."[126] That was fundamentally different from asking God to bless America, as presidents often do. Bush was asserting that in bringing freedom to Iraq, America was doing God's work, that America's path was ordained by God.

After leaving the Situation Room, Bush went for a walk outside the Oval Office. "I prayed as I walked around the circle. I prayed that our troops be safe, be protected by the Almighty, that there be minimal loss of life. . . . I was praying for strength to do the Lord's will. . . . I'm surely not going to justify war based upon God. Understand that. Nevertheless, in my case I pray that I be as good a messenger of His will as possible."[127]

Meanwhile, in Iraq, U.S. intelligence intercepted Iraqi communications suggesting that Saddam Hussein was staying at a Tigris River complex referred to as Dora Farms. Reconnaissance revealed that guards and vehicles were hidden in the tree lines around the farm.

The next day, March 20, U.S. forces fired more than forty Tomahawk missiles at Dora Farms, destroying every structure in the compound except the main palace.[128] There were just two mistakes. They missed the main target. And Saddam wasn't there. Nevertheless, the war in Iraq had begun.

CHAPTER SEVENTEEN

Season of Mirth

While the entire world was transfixed by the dazzling pyrotechnics of American military might, one of the most important moments of the entire Bush-Cheney administration took place completely unnoticed. On March 25, 2003, in the immediate wake of Operation Shock and Awe, when thousands of bombs and missiles had been raining a massive firestorm upon the Iraqi leadership, and as the nation held its collective breath hoping for the safe return of Jessica Lynch, a spunky nineteen-year-old supply clerk*[1] who was missing in action,[2] President Bush quietly signed Executive Order 13292.

The event was not secret. The White House even issued a press release about it, the title of which—"Further Amendment to Executive Order

*Lynch became famous after being injured, captured, and listed as missing in action. Nearly every other aspect of what happened was in dispute. The subject of an NBC television movie, *Saving Jessica Lynch,* she later said that the Pentagon fabricated the widely publicized story of her rescue as part of its war propaganda, which the BBC later termed "one of the most stunning pieces of news management ever conceived." An article in the *Washington Post* by Susan Schmidt and Vernon Loeb, among hundreds of other similar stories, portrayed Lynch as heroically fighting on against the Iraqis after being stabbed and "even after she sustained multiple gunshot wounds and watched several other soldiers in her unit die."

Likewise, a biography of her by Rick Bragg, *I Am a Soldier, Too,* says, "The records do not tell whether her captors assaulted her almost lifeless, broken body after she was lifted from the wreckage, or if they assaulted her and then broke her bones into splinters until she was almost dead."

In an interview with ABC's Diane Sawyer, Lynch said, "They used me to symbolize all this stuff. It's wrong. I don't know why they filmed [my rescue] or why they say these things. . . . I did not shoot, not a round, nothing. I went down praying to my knees. And that's the last I remember." She had no memory of being raped. She added that she had been well treated as a captive in Iraq, and that while she was in the Iraqi hospital one person even sang to her to help her feel at home.

12958"*—somehow failed to lure the media away from America's biggest military venture since Vietnam. As a result, not a single major media outlet of any kind reported that President George W. Bush had given Cheney the same powers as the president to classify and declassify intelligence.†

But what had just happened was unprecedented. Bush had granted Cheney the greatest expansion of powers in the history of the vice presidency. Since Cheney had scores of loyalists throughout the Pentagon, the State Department, and the National Security Council who reported to him, in operational terms, he was the man in charge of foreign policy.‡[3] If Cheney wanted to keep something secret, he could classify it. If he wanted to leak information, or disinformation, to the *New York Times* or *Washington Post,* he could declassify it. Or if he wanted to leak the name of an undercover CIA agent, that was fine, too.§[4]

Bush's executive order also effectively granted a measure of legitimacy

*Executive Order 13292 is sometimes referred to officially as "Amendment to Executive Order 12958."

†There is perhaps no more striking example of the chasm between the so-called mainstream media and the work of Internet bloggers than in the coverage of Executive Order 13292. In the entire LexisNexis database, with its many thousands of publications and broadcast outlets, as of August 2007, there are only twenty-two references to Cheney's new powers—none of them in the news pages of the *New York Times, Washington Post, Wall Street Journal,* or on ABC, NBC, CBS. By contrast, the term *Executive Order 13292* appeared on more than twenty thousand websites.

‡Specifically, the key part of the order reads: "Classification Authority. (a) The authority to classify information originally may be exercised only by: (1) the President and, in the performance of executive duties, the Vice President; (2) agency heads and officials designated by the President in the Federal Register. . . ." The complete text can be found at the White House website, where it is labeled "Further Amendment to Executive Order 12958"; http://www.whitehouse.gov/news/releases/2003/03/20030325-11.html.

§Cheney's new powers first came to light in 2006 during the Valerie Plame Wilson CIA–leak case, when Fox News anchor Brit Hume asked Cheney what his aide Scooter Libby may have meant when he testified he had been authorized "by his superiors" to disclose information about the classified National Intelligence Estimate to members of the press. "Is it your view that a Vice President has the authority to declassify information?" Hume asked.

"There is an executive order to that effect," Cheney said.

"There is?"

"Yes."

"Have you done it?"

"Well, I've certainly advocated declassification and participated in declassification decisions. The executive order . . ."

"You ever done it unilaterally?"

"I don't want to get into that."

to Cheney's previous machinations with the national security apparatus, and in doing so it consolidated the totality of his victories. What in fact had taken place might be called, for want of a better term, "an executive coup," a highly unusual but not unprecedented act of state in which a ruling executive, in this case Cheney, seizes extraordinary powers from within the government. Because the executive branch of government is charged with running the day-to-day affairs of the state, the phrase is indeed paradoxical. But in America's constitutional democracy, the executive branch is restricted by checks and balances designed to divide power among competing branches of government through mechanisms such as congressional oversight and judicial restraints, thereby protecting the electorate from tyranny.

By this time, however, a number of components of the complex system of checks and balances that held executive power in restraint had been rendered inoperative. Peer review procedures had been disabled; congressional oversight eliminated; veteran Middle East intelligence analysts barred from crucial meetings;[5] the National Intelligence Estimate subverted. The entire national security apparatus had effectively been disabled and replaced by a parallel national security apparatus that created a disinformation pipeline, which expressed the wished-for reality of Cheney, Rumsfeld, and the neocons.

"It was a stealth organization," said former State Department analyst Greg Thielmann.[6] "They didn't play in the intelligence community proceedings that our office participated in. When the intelligence community met as an intelligence community, there was no OSP [Office of Special Plans, the unit in the Pentagon that "cherry-picked" intelligence for Cheney] represented in these sessions. Because then, if [the OSP] did that, they would have had to subject their views to peer review and they didn't want to do that. Why do that when you can send stuff right in to the Vice-President?"

"We used to say about both [Rumsfeld's office] and the vice president's office that they were going to win nine out of ten battles, because they are ruthless, because they have a strategy, and because they never, ever deviate from that strategy," said Larry Wilkerson. "They make a decision, and they make it in secret, and they make it in a different way than the rest of the bureaucracy makes it, and then suddenly foist it on the government—and the rest of the government is all confused."[7]

"When I say 'secret cabal,' I mean 'secret cabal,'" said Wilkerson, comparing Cheney, Rumsfeld, and the neocons to the Jacobins, the radical zealots of the French Revolution who produced the Reign of Terror of Jean-Paul Marat and Robespierre.* "I see them as messianic advocates of American power from one end of the globe to the other, much as the Jacobins in France were messianic advocates of the French revolution," he said. "I don't care whether utopians are Vladimir Lenin on a sealed train to Moscow or Paul Wolfowitz. You're never going to bring utopia, and you're going to hurt a lot of people in the process."[8]

In addition to having seized control of America's national security apparatus, the administration had, thanks to David Addington and Cheney's legal team, also won vastly expanded powers for the executive branch that had enormous implications both domestically and in terms of the conduct of the wars in Afghanistan and now Iraq. "[Addington] believes that in time of war, there is total authority for the president to waive any rules to carry out his objectives," Congresswoman Jane Harman, the intelligence committee's ranking Democrat, told David Ignatius of the *Washington Post.*[9] "Those views have extremely dangerous implications."

Acting according to a "war paradigm," Cheney's legal team had secretly instituted a warrantless surveillance program through the National Security Agency. It had set up a detainment camp in Guantánamo Bay Naval Base for "enemy combatants," who it said were not entitled to the protections of the Geneva Conventions, which set the widely accepted standards for the humanitarian treatment of prisoners of war. And, according to an August 2002 memo written by Justice Department official John Yoo, based on the radical notion of "the unitary executive," the president's policies on torture, like his policies on detainees and domestic surveillance, could be made by the president alone, unencumbered by such hindrances as the Geneva Conventions, Congress, or other government agencies.†[10]

*Wilkerson's remarks were an apparent reference to the work of Claes Ryn, a professor at Catholic University of America, who christened the neoconservatives "The New Jacobins" in *The Ideology of American Empire.*

†A 2005 debate between John Yoo and Notre Dame law professor Doug Cassel gives a sense of how broadly Yoo defined the president's powers:

CASSEL: If the President deems that he's got to torture somebody, including by crushing the testicles of the person's child, there is no law that can stop him?

YOO: No treaty.

CASSEL: Also no law by Congress. That is what you wrote in the August 2002 memo.

YOO: I think it depends on why the President thinks he needs to do that.

The unitary executive. Torture. Warrantless surveillance. The seizure of the national security apparatus. And now a brand-new war, designed by the neocons to be the first of many that would result in the making of the entire Middle East, a reconfiguration of the geostrategic balance of power that would ensure American supremacy for decades to come.

Patrick Lang, the former DIA defense intelligence officer for the Middle East, South Asia, and Counterterrorism, and the Agency's first director of HUMINT, asserted that what had happened was fundamentally different from anything that had gone on in previous administrations.[11] "[Cheney's men] behaved as though they had seized control of the government in a 'silent coup,'" he wrote in "Drinking the Kool-Aid," an article that appeared in *Middle East Policy*. The result, he added, was "a highly corrupted system of intelligence and policymaking, one twisted to serve specific group goals, ends and beliefs held to the point of religious faith."[12]

At the time, however, few people in America were aware of the extraordinary accumulation of power in the executive branch and almost everyone believed that Saddam had WMDs and posed a real threat. Seven out of ten Americans even blamed him for the atrocities of 9/11.[13] And the stated reasons for the war struck responsive chords among the vast majority of Americans. A Gallup/*USA Today* poll showed that 73 percent believed going to war "was worth it." Eighty-four percent were "very" or "somewhat" confident the United States would find weapons of mass destruction.[14] To everyone, Saddam was indeed a brutal dictator, and the notion of freeing the Iraqi people appealed to America's best humanitarian impulses. Moreover, many Americans, loathe to differentiate between the various Arab peoples, felt the United States was doling out a well-deserved dose of bloody retribution to its Arab foes, taking the battle to them so the U.S. would not have to fight Islamist terrorists on American soil. To neocons and Christian evangelicals, it was America's mission—or God's—to spread democracy throughout the world, and at last that mission was being served.

In addition, initially the war appeared to be going well. It was thrilling. There was the military's death-defying race to Baghdad, marred only by outbreaks of unexpected resistance, and then the heroic taking of Baghdad. For a generation of Americans accustomed to movies and television, the war was one exciting real-life narrative spectacle after another.

Not that it was free of mishaps. In early April, coalition forces* killed a journalist when they fired on al-Jazeera's office,[15] and then hit the Palestine Hotel, killing two more.[16] Looting was rampant.†[17] Mortars, bombs, rocket-propelled grenades (RPGs), and tons of ammunition lined the streets. Seventeen ministries were destroyed. Thousands of archaeological pieces were stolen from Baghdad's National Museum of Antiquities.[18] Baghdad's National Library went up in flames. The Baghdad zoo was looted, its animals stolen.

Then, on April 9, 2003, a huge statue of Saddam Hussein in Baghdad's al-Fardus Square was toppled, with dozens of people stamping on the fallen figure, shouting "Death to Saddam!"‡[19] The symbolism of the moment played well on the TV screens of America, and seemed to augur a rapid conquest. As April wore on, U.S. troops captured one after another of the top fifty-five most wanted men in Iraq, villains whose names and photos were famously printed on the "personality identification playing cards" produced by the military. By April 11 the battle had shifted north. Mosul was on the verge of collapsing.[20] With the help of the Kurds, Kirkuk fell without a shot. The nearby oil fields were secure. There were scenes of genuine jubilation and joy. Saddam had fled, a hunted fugitive, a man without a country.

Now was the neocons' moment in the sun—more than that, their sea-

*The initial invasion force consisted of 250,000 American troops, 45,000 from the United Kingdom, and small numbers of troops from Poland, Australia, and Denmark. Over time, 33 other countries contributed troops ranging from Italy and Spain, with peaks of 3,200 and 1,300 troops, respectively, to Iceland with 2. In August 2007, Coalition forces consisted of 168,786 regular American troops and 182,000 private military contractors. The United Kingdom's forces had drawn down to 5,500 by May 2007.

†Of the looting, Donald Rumsfeld famously said, "Think what's happened in our cities when we've had riots, and problems, and looting. Stuff happens! . . . Freedom's untidy." But, according to *Mother Jones,* a later Coalition Provisional Authority estimate put at $12 billion the cost of the looting that went uncontrolled as U.S. troops sat at the Baghdad airport. *The New Yorker*'s George Packer added that the looting canceled out the "projected revenues of Iraq for the first year after the war. The gutted buildings, the lost equipment, the destroyed records, the damaged infrastructure, would continue to haunt almost every aspect of the reconstruction."

‡Even this historic photo appears to have been a staged media event. The widely distributed photos were tightly cropped to show what appeared to be a large crowd of jubilant Iraqis. But long-distance shots show that the vast square was almost empty. Moreover, in the documentary movie *Control Room,* al-Jazeera producer Samir Khader comments, "[The Americans] brought with them some people—supposedly Iraqis cheering. These people were not Iraqis. I lived in Iraq, I was born there, I was raised there. I can recognize an Iraqi accent . . ."

son of mirth. It was time to celebrate—and gloat. Bush's approval rating soared to 77 percent, according to an ABC News/*Washington Post* poll, his highest rating since 9/11.[21] The media lionized the deputy secretary of defense as "Wolfowitz of Arabia." Richard Perle "took a victory lap" at the American Enterprise Institute.[22] The *Weekly Standard*'s Stephen Schwartz derided liberal critics who had foolishly prophesized that the war would "result in Iraq's being torn apart, as Shias, Sunnis, and Kurds fight each other for power."[23] Likewise, after paying homage to Wolfowitz, Libby, and Cheney, *The Standard*'s David Brooks expressed dismay at liberals who had criticized the war. "I suspect [war opponents] will not even now admit their errors. . . ." he wrote. "[They] may perhaps grow more bitter, lost in the cul-de-sac of their own alienation."[24]

On April 10, neocon Kenneth Adelman, the former director of the U.S. Arms Control & Disarmament Agency who had famously written that "liberating Iraq would be a cakewalk,"[25] and had been chided for it when U.S. troops briefly encountered unexpected resistance, figured it was time to celebrate and said so in a *Washington Post* column, taking aim at neocon critics along the way. "Taking first prize among the many frightful forecasters was the respected former national security adviser Brent Scowcroft," he wrote.[26]

No one got a bigger kick out of Adelman's column than Dick Cheney, who immediately phoned and invited him and his wife to a small dinner party to celebrate, as Bob Woodward reported in *Plan of Attack*. The only other guests would be Scooter Libby and Paul Wolfowitz.[27]

When Adelman arrived, he was so happy he broke into tears and hugged Cheney. They reminisced briefly abut the 1991 Gulf War, then Adelman interrupted, "Hold it! Hold it! Let's talk about *this* Gulf War. It's so wonderful to celebrate. . . . Paul and Scooter, you give advice inside and the president listens. Dick, your advice is the most important, the Cadillac." The war, Adelman said, had been awesome. "So I just want to make a toast without getting too cheesy. To the president of the United States."[28]

As it happened, the neocons' allies on the Christian Right were equally thrilled. No fewer than 87 percent of all white evangelical Christians in the United States supported the president's decision in April 2003.[29] Both Franklin Graham, the son of Billy Graham, and Marvin Olasky, Bush's adviser on faith-based policy, asserted that the American invasion created exciting new prospects for converting Muslims to

Christianity. And for Left Behind prophet Tim LaHaye, the stakes were even higher. Iraq was "a focal point of end-time events," he said. Its special role in the End of Days would become clear only with time.

About two weeks after Cheney's party, on May 1, 2003, came Bush's big moment. At the time, his critics were carping that no WMDs had been found in Iraq and that Shi'ite clerics were fomenting trouble. But as Karl Rove saw it, Bush was now a war hero, and with the 2004 election coming up, it was time to exploit the war's popularity to the hilt. "The big event is over," said Michael O'Hanlon, a military specialist at the Brookings Institution.[30] "Why not take a victory lap, and what politico would advise against it?"

Now Bush had his chance to star in the White House's version of the big show. As Frank Rich of the *New York Times* noted in *The Greatest Story Ever Sold: The Decline and Fall of Truth,* it was a double feature of sorts, opening with George W. Bush in a pilot's suit à la Tom Cruise in *Top Gun,* complete with parachute, as a copilot of a navy jet—specially renamed for the occasion, Navy One, and painted with the legend GEORGE W. BUSH, COMMANDER IN CHIEF—landing on the USS *Abraham Lincoln* off the coast of southern California.[31]

Even though the carrier was just thirty-nine miles from San Diego, Bush, instead of taking a helicopter, opted for the heroic spectacle of a fighter jet making a dramatic tail-hook landing on the carrier.[32] Only after the fact was it revealed that the *Abraham Lincoln* was so close to land that its navigators had to position the ship precisely to make sure that views of the shoreline did not mar images of Bush at sea. The president returned the short distance to land via helicopter.[33]

Emerging from the cockpit in a flight suit as the conquering hero, Bush tucked his helmet under his arm and greeted the crew. "Yes, I flew it!" Bush shouted.[34] "Of course I liked it!" As absurd as it might be that the president would actually land a jet on an aircraft carrier, an extremely difficult and dangerous task, the press, for the most part, embraced the heroic imagery. He was bringing his "daring mission to a manly end," said the *Detroit Free Press.*[35] And the *Washington Post*'s David Broder, proverbial dean of the White House press corps, told his fellow panelists on *Meet the Press* that the "president has learned to move in a way that just conveys a great sense of authority and command."[36]

Three hours after the landing, Bush emerged in a business suit. The time was dusk, also known as the magic hour[37] to Hollywood cinematographers because of the special glow low light from the horizon casts upon its subjects. Scott Sforza, a former ABC producer[38] who was working with the White House, had boarded the carrier several days earlier to attend to key production values and made sure that a banner reading MISSION ACCOMPLISHED would be stretched high above the president's head as he gave his historic address.[39] "Officers and sailors of the USS *Abraham Lincoln,* my fellow Americans, major combat operations have ended," Bush announced. "In the Battle of Iraq, the United States and our allies have prevailed."*[40]

The neocons joined in the celebration. "This was a war worth fighting," Richard Perle wrote in a *USA Today* article that day titled "Relax, Celebrate Victory." "It ended quickly with few civilian casualties and with little damage to Iraq's cities, towns or infrastructure. It ended without the Arab world rising up against us, as the war's critics feared, without the quagmire they predicted, without the heavy losses in house-to-house fighting they warned us to expect."[41]

In reality, the war had just begun. Initially, the administration had planned for American troops to leave Iraq shortly after deposing Saddam. "The idea was we'd go in, get rid of Saddam; the government could function with new direction coming from the top; the economy would be revitalized by oil revenue," said Anthony Cordesman of the Center for Strategic and International Studies, in an interview on PBS's *Frontline.* "There would not be major ethnic or sectarian struggles; there wouldn't be any resistance or resurgence of pro-Saddam movements."[42]

But there was an insurgency—and the administration had no plans for stopping it. One reason they were not prepared was because the neocons had little genuine understanding of Iraq's history and culture. "In all the literature they've written about Iraq, you will not see anything by them that has anything about the nature of Iraqi society," said a State Department official. "It was as if that stuff did not exist. Likewise, they did not think about Iraq in the context of the region. They viewed those things as distractions from their larger objectives. That is what was so frustrating to [Brent] Scowcroft and others who spent a lifetime developing

*Later the White House tried to blame the prematurely celebratory banner on the military. But, Bob Woodward reported, in *State of Denial,* the two offending words—"mission accomplished"—had actually been in Bush's speech until Donald Rumsfeld insisted they be deleted. "I took 'Mission Accomplished' out," said Rumsfeld. "I was in Baghdad and I was given a draft of that thing and I just died. And I said, it's too inclusive. . . . They fixed the speech but not the sign."

knowledge and expertise to examine the unintended consequences of various policies."

In fact, even as Bush gave his "Mission Accomplished" speech, looting swept across Iraq. To make matters worse, the United States had disbanded the Iraqi police and military, more than half a million strong, and many were still armed. Further, the U.S. had alienated the Sunnis by banning anyone affiliated with Saddam Hussein's Ba'athist Party from participating in the government.[43] It was a recipe for chaos, fear, random murder, and the rise of local strongmen and warlords. Throughout the summer of 2003, Washington dismissed the bombings, assassinations, and lootings as the work of a few dead-enders, while the insurgency grew. In August, the U.N.'s headquarters in Baghdad was destroyed by a huge truck bomb. Seventeen were killed when the Jordanian embassy was blown up, and riots shook the city of Basra.[44]

Meanwhile, officials at the highest levels of the administration were frantically trying to figure out what had happened to Saddam's WMDs. Now that combat operations in Iraq had ostensibly ended and weapons inspectors were finally coming back from the field with definitive reports, George Tenet called Colin Powell with some bad news. "I'm really sorry to have to tell you," Tenet told the secretary of state. "We don't believe there were any mobile labs for making biological weapons."

According to Wilkerson, this was the third or fourth time Tenet had called Powell with such a message;[45] virtually every important allegation Colin Powell made about Saddam's WMDs was turning out to have been wrong. Powell had held up a small vial filled with white powder at the U.N. But Saddam had no active program to manufacture anthrax. The last known batch had been produced in 1991, and its shelf life had long since expired.

Powell had also said that Iraq had repeatedly tried to acquire aluminum tubes that "can be adapted for centrifuge use" in a nuclear weapons program.[46] These assertions were false as well.

For Tenet, the calls to Powell were not easy. "George recognized that not only was he, George, out front on this issue . . . but the CIA and the intelligence community was going to be called to task if it didn't pan out," said John Brennan, who served as Tenet's chief of staff before becoming deputy executive director of the CIA.[47]

"It's fair to say the secretary and Mr. Tenet, at that point, ceased being close," said Wilkerson. "I mean, you can be sincere and you can be honest and you can believe what you're telling the secretary. But three or four times on substantive issues like that?"[48]

At the same time, the ongoing chaos in Iraq notwithstanding, the neocons, intoxicated with the prospect of implementing their grand design for the Middle East, dreamed about moving on to the ultimate prize that had been in their sights for more than a decade: the Islamic Republic of Iran. With a population of roughly 70 million people—triple Iraq's—Iran, with an area bigger than the United Kingdom, Spain, France, and Germany combined, was the eight-hundred-pound gorilla in the Middle East. Iran had the second largest natural gas reserves in the world and the third largest oil reserves, making it a grand prize on the global chessboard in which great powers vied for vital strategic resources. In the aftermath of its Islamic revolution in 1979, Iran had been the bane of Israel's existence because of its support of Hamas and Hezbollah, both of which consisted of militant, anti-Israeli Islamists.* In terms of the balance of power in the volatile Middle East, Iran vied with Saudi Arabia for dominance of the region. In addition, Iraq had long been a counterweight to Iran, so Saddam's fall raised the inevitable question of what would happen to Iran now that he was gone.

It was Iran's support of Hamas and Hezbollah that made the Islamic Republic of vital concern to Israel and the neocons. High-level Israelis had repeatedly made it clear to the Bush administration that taking on Iran was their highest priority. "If you look at President Bush's 'axis of evil' list, all of us said North Korea and Iran are more urgent," said former Mossad director of intelligence Uzi Arad, who served as Benjamin Netanyahu's foreign policy adviser.[49] "Iraq was already semi-controlled because there were [U.N.-imposed economic] sanctions. It was outlawed. Sometimes the answer [from the neocons] was 'Let's do first things first. Once we do Iraq, we'll have a military presence in Iraq,

*Hamas, which is an acronym for "Islamic Resistance Movement," is a militant Palestinian Sunni organization that was elected as the government of the Palestinian people in January 2006, but was ousted by rival Fatah after using force to take over the Gaza Strip. Because of its frequent use of suicide bombings and other violent tactics, it is widely considered a terrorist organization. Hezbollah, which means Party of God, is a radical Shi'ite Islamist political and paramilitary group based in Lebanon with widespread political support in Lebanon. The United States and the United Kingdom regard Hezbollah as a terrorist organization, but most people in the Arab world and in the Muslim world regard it as a legitimate resistance group.

which would enable us to handle the Iranians from closer quarters, would give us more leverage.' "*[50]

"It's time to bring down the other terror masters," wrote Michael Ledeen just after the Iraq invasion. He added that "The Iranian, Syrian, and Saudi tyrants know that if we win a quick victory in Iraq and then establish a free government in Baghdad, their doom is sealed."[51]

Meanwhile, the invasion of Iraq appeared to have had a real impact on Tehran and the Iranian leadership. Shortly afterward, when U.S. success in Iraq seemed assured, Iran came to fear that it would be next in America's crosshairs. To stave off that possibility, Iran's leadership, including Supreme Leader Ayatollah Ali Khamenei, began to assemble a negotiating package. Suddenly, everything was on the table—Iran's nuclear program, Iran's policy toward Israel, its support of Hamas and Hezbollah, and control over Al-Qaeda operatives captured since the United States went to war in Afghanistan.

This comprehensive proposal, which diplomats referred to as "the grand bargain," was sent to Washington on May 2, 2003, just before a meeting in Geneva between Iran's U.N. ambassador, Javad Zarif, and Zalmay Khalilzad,† then a senior director at the National Security Council.[52] According to a report by Gareth Porter in the *American Prospect,* Iran offered to take "decisive action against any terrorists (above all, al-Qaeda) in Iranian territory." In exchange, Iran wanted the U.S. to pursue "anti-Iranian terrorists"—such as the Mujahideen e-Khalq, or MEK.[53] Specifically, Iran offered to share the names of senior Al-Qaeda operatives in its custody in return for the names of MEK cadres captured by the United States in Iraq.

Well aware that the United States was concerned about its nuclear program, Iran proclaimed its right to "full access to peaceful nuclear technology," but offered to submit to much stricter inspections by the International Atomic Energy Agency (IAEA).[54] On the subject of Israel, Iran offered to join with moderate Arab regimes such as Egypt and Jordan in accepting the 2002 Arab League Beirut declaration calling for peace with Israel in return for Israel's withdrawal to its pre-1967 borders.

*To make absolutely certain the White House understood Israel's priorities, in February 2002, Israeli prime minister Ariel Sharon and defense minister Benjamin Ben-Eliezer visited Washington to make the case to President Bush that Iran was a much greater threat than Saddam Hussein. "Today, everybody is busy with Iraq," Ben-Eliezer said. "Iraq is a problem . . . But you should understand, if you ask me, today Iran is more dangerous than Iraq."

†Khalilzad went on to become the U.S. ambassador to Iraq and later to the U.N.

The negotiating package also included proposals to normalize Hezbollah into a mere "political organization within Lebanon," to bring about a "stop of any material support to Palestinian opposition groups (Hamas, Jihad, etc.) from Iranian territory," and to apply "pressure on these organizations to stop violent actions against civilians within borders of 1967."[55]

To be sure, Iran's proposal was only a first step. There were countless unanswered questions, and many reasons not to trust the Islamic Republic. But given the initiative's historic scope, it was striking that the Bush administration simply declined to respond. There was not even an interagency meeting to discuss the proposal. "The State Department knew it had no chance at the interagency level of arguing the case for it successfully," former NSC staffer Flynt Leverett told the *American Prospect.* "They weren't going to waste [Colin] Powell's rapidly diminishing capital on something that unlikely."

Iran had sent its initiative through an intermediary, Tim Guldimann, the Swiss ambassador to Iran. A few days later, Leverett said, the White House had the State Department send Guldimann a message. "We're not interested in any grand bargain," said Under Secretary of State for Arms Control and International Security John Bolton, who later became interim ambassador to the U.N.[56]

The grand bargain was dead. Flush with a false sense of victory, Bush, Cheney, and Rumsfeld felt no need to negotiate with the enormous oil-rich country that shared a border with the country America had just invaded.

By the spring of 2003, with America's 2004 presidential election more than a year away, observant citizens began to see a footrace between two competing realities, a day-by-day battle for control of the national narrative. On the one hand, the Bush administration's proposition was being tested using America's blood and treasure. Would democracy take hold in Iraq, as advertised, then sweep through the Middle East? Would the terrorists be hunted down and killed, as promised?

Or, as the United States suffered one setback after another in Iraq, were the contradictions and fallacies in the neoconservative argument emerging, and Iraq's deeper historic and political realities rising up through the layers of ideological and theological fog? Would the Democrats challenge Bush, Cheney, and the neocons over the decision to go to war?

Meanwhile, information describing the discredited intelligence reports about the aluminum tubes, Curveball, and the mobile weapons labs began to surface, each story eroding the rationale for war served so convincingly to the American people. On May 6, without mentioning former ambassador Joseph Wilson by name, *New York Times* columnist Nicholas Kristof published the first account of Wilson's trip to Africa. "I'm told by a person involved in the Niger caper that more than a year ago the vice president's office asked for an investigation of the uranium deal, so a former U.S. ambassador to Africa was dispatched to Niger," he wrote. "In February 2002, according to someone present at the meetings, that envoy reported to the C.I.A. and State Department that the information was unequivocally wrong and that the documents had been forged."[57]

Now the secret was out with regard to the Niger documents. Not only had the IAEA determined they were forgeries, but it was clear that the administration *knew* the Niger deal was phony even before Bush cited them in his State of the Union address.

With the election season starting to heat up, the Democrats could have a field day if everything unraveled. But the Bush-Cheney administration struck back fast and hard, leaking information they thought would protect them on the Niger question. In mid-June, Deputy Secretary of State Richard Armitage told Bob Woodward that Wilson's wife worked for the CIA on weapons of mass destruction.[58] At the time, Woodward did not write a story about the revelation. But, apparently for the first time, an administration official had revealed the identity of Valerie Plame Wilson, an undercover CIA agent.

On June 23, Scooter Libby told *Times* reporter Judith Miller that Wilson's wife had something to do with the CIA.[59] But nothing came of that either. Then, in late June, conservative syndicated columnist Bob Novak got word that Deputy Secretary of State Richard Armitage, whom Novak had been trying to get an appointment with for some time, had finally agreed to see him. They set up a meeting for just after the July 4 holiday.[60]

But on July 6, 2003, before Armitage and Novak met, Joe Wilson published his now famous op-ed piece in the *New York Times,* "What I Didn't Find in Africa." "Based on my experience with the administration in the months leading up to the war, I have little choice but to conclude that some of the intelligence related to Iraq's nuclear weapons program was twisted to exaggerate the Iraqi threat," Wilson wrote. After spending eight days with dozens of people associated with the country's uranium business, "It did not take long to conclude that it was highly doubtful that any such transaction had ever taken place."[61]

When evaluating Wilson's role in the political firestorm that resulted from his piece, it is crucial to remember that he was a veteran diplomat who had served in the administration of Bush 41 and ended up as deputy chief of mission at the U.S. embassy in Baghdad between 1988 and 1991. In the wake of Iraq's 1990 invasion of Kuwait, Wilson was the last American diplomat to meet with Saddam Hussein. As tensions built between the United States and Iraq, he had been housing some sixty Americans at the ambassador's residence when he received a threatening note saying that anyone who harbored foreigners would be executed. In response, Wilson staged a press conference at which he wore a noose around his neck. "If the choice is to allow American citizens to be taken hostage or to be executed, I will bring my own fucking rope," he said.[62] Wilson, it was clear, was tough enough to stand up to Saddam Hussein personally, a point rarely recalled in subsequent descriptions of him.

President George H.W. Bush was so impressed with the way Wilson handled himself that he cabled him after one of his appearances. "I recently saw you on CNN saying what you thought of Saddam's latest attempt to derive political gain out of understandable concern for the hostages," Bush wrote. "I could not have said it better. . . . What you are doing day in and day out under the most trying conditions is truly inspiring. Keep fighting the good fight; you and your stalwart colleagues are always in our thoughts and prayers."[63]

As the administration marched to war in late 2002, Wilson continued to stay in touch with both Brent Scowcroft and Bush 41. But now, in the *Times,* Wilson was telling the country, for the first time, that the administration of Bush 43 was using phony intelligence to lead the country into war.

Wilson's story was big news. Within hours, the Bush administration came under fire as never before. President Bush, Colin Powell, Condoleezza Rice all were questioned or criticized over the Niger yellowcake deal.

To orchestrate damage control and cast off doubt of Wilson's findings, Karl Rove went into action. On July 11, he spoke to Matt Cooper, a reporter at *Time.* According to Cooper's e-mail notes about the conversation as later reported by *Newsweek,* Rove had said it was Wilson's wife "who apparently works at the agency on wmd issues who authorized the trip . . . not only [*sic*] the genesis of the trip is flawed an[d] suspect but so is the report. he [Rove] implied strongly there's still plenty to implicate iraqi interest in acquiring uranium fro[m] Niger. . . ."[64]

The entire conversation was on background, but the White House was getting its message across: something about Joe Wilson's trip was suspect. And a White House press corps that relied heavily on access to high-level administration officials was listening intently and was holding its fire.

Fortunately for the Bush administration, Cheney also had friends in Congress. On July 11, 2003, the same day that Rove spoke to Matt Cooper, Senator Pat Roberts, chairman of the Senate Select Committee on Intelligence and a Cheney ally, faced with public pressure to investigate the Niger forgeries, issued a statement defending the White House and blaming the CIA. "So far, I am very disturbed by what appears to be extremely sloppy handling of the issue from the outset by the CIA," he said. "What now concerns me most, however, is what appears to be a campaign of press leaks by the CIA in an effort to discredit the President."*[65] The vast majority of Americans, of course, were unfamiliar with the arcanae of the Niger forgeries. Tens of millions of people now read about CIA intelligence failures—which was very different than learning about the administration's use of forgeries to lead the nation to war.

Slowly, a counternarrative emerged in opposition to the official version put forth by the White House. But by and large, it was confined to outlets such as *The Daily Show with Jon Stewart,* HBO's *Real Time with Bill Maher,* MSNBC's *Countdown with Keith Olbermann,* and hundreds of liberal blogs that sprang up on the Internet. For the most part, however, as the battle to control the narrative was played out in the media, the Washington press corps continued to promote the Bush administration's narrative—and on occasion, did special favors for them. Columnist Bob Novak was a case in point. Karl Rove confirmed Valerie Plame Wilson's connection to the CIA for Novak,[66] and eight days after Wilson's column appeared, on July 14, Novak wrote, "Wilson never worked for the CIA, but his wife, Valerie Plame, is an Agency operative on weapons of mass destruction. Two senior administration officials told me Wilson's wife suggested sending

*Under Roberts's aegis, the Senate Intelligence Committee that investigated the Niger affair issued a report that came to some extraordinary conclusions. "At the time the President delivered the State of the Union address, no one in the IC [intelligence community] had asked anyone in the White House to remove the sentence from the speech," read the report. It added that "CIA Iraq nuclear analysts told Committee staff that at the time of the State of the Union, they still believed that Iraq was probably seeking uranium from Africa."

him to Niger to investigate the Italian report. The CIA says its counter-proliferation officials selected Wilson and asked his wife to contact him."[67]

The implication from the administration was that the CIA's selection of Wilson to go to Niger was somehow tainted because his wife was at the CIA. But, more important, the administration had put out a message to any and all potential whistle-blowers: if you dare speak out, we will strike back. To that end, the cover of Valerie Plame Wilson, a CIA operative specializing in WMDs, had been blown by a White House that was supposedly orchestrating a worldwide war against terror.

Meanwhile, the news from Iraq got worse. The insurgency grew. In December 2003, Saddam Hussein was captured, but that did not change the fundamental realities. Sectarian violence increased. There were snipers. Assassinations. Car bombings. The expected oil bonanza failed to materialize. Casualties mounted, but the administration refused to allow photos of coffins to be shown. As George Packer reported in *The New Yorker*, the neocon scenario had envisioned handing off the responsibility of managing Iraq to Ahmed Chalabi and other pro-West exiles, without the United States having to commit more troops or money. "There was a desire by some in the Vice-President's office and the Pentagon to cut and run from Iraq and leave it up to Chalabi to run it," said one senior administration official. "The idea was to put our guy in there and he was going to be so compliant that he'd recognize Israel and all the problems in the Middle East would be solved. He would be our man in Baghdad. Everything would be hunky-dory. . . . It isn't pragmatism, it isn't Realpolitik, it isn't conservatism, it isn't liberalism. It's theology."[68]

But in June, U.S. intelligence officials told the *New York Times* that Chalabi had given American state secrets to Iran, including news that "the United States had broken the secret communications code of Iran's intelligence service," thereby destroying Washington's most valuable source of information about Iran.[69] Chalabi consistently denied having given any intelligence to Iran, maintaining that the allegations were a CIA smear. Regardless of the truth of the allegations, within Iraq, Chalabi was widely seen as a stooge of the United States. According to a February 2004 survey of nearly three thousand Iraqis, only .2 percent of respondents saw Chalabi as Iraq's most trustworthy leader. The Chalabi option had vaporized.

As the war continued, hundreds of millions of dollars' worth of federal contracts in Iraq were awarded without competitive bidding, including $425 million to Halliburton, Dick Cheney's old firm.[70] The White House continued to insist that U.S. troops would find weapons of mass destruction, but fewer and fewer people believed them. With a year to go before the elections, the administration issued one denial after another. "The American people were not told lie after lie after lie," said Colin Powell.[71] "No one said that [they had nuclear weapons]," said Rumsfeld.[72] "We will find [Osama bin Laden and Saddam Hussein," Bush told a group of reporters in November.[73] "Okay?"

Internationally, the astonishing goodwill that the United States enjoyed after 9/11 had been thrown away. As for the cryptic, sphinxlike Cheney, as the revelations and unanswered questions tumbled forth, he quietly embraced the behind-the-scenes intrigue and his enigmatic image. "Am I the evil genius in the corner that nobody ever sees come out of his hole?" he asked a reporter.[74] "It's a nice way to operate, actually."

By the spring of 2004, the only jubilation in the streets of Iraq was of the macabre variety, in the Sunni Triangle, near Fallujah, when an enraged mob killed four American contractors, triumphantly dragged their charred bodies through the streets, and then hung some of the bodies high, jeering in delight.[75] Unable to set up a truly stable representative Iraqi government, constantly changing its strategy, alienating the local populace, the Bush administration decided to try to build up the Iraqi army. Now the new policy was that as the new Iraqi army stood up, America would stand down.

In late April and May 2004, CBS's *60 Minutes II* and *The New Yorker* magazine's Seymour Hersh reported the systematic abuse, torture, and rape of Iraqi prisoners at the Abu Ghraib prison in Iraq. Graphic and disturbing images of torture and humiliation instantly circled the world. There were photos of hooded men with electrodes attached to their bodies, naked prisoners who had been smeared with feces, and, later, reports of detainees who had been dragged across the floor pulled by ropes tied to their penises, of young Iraqi boys sodomized by American soldiers, and of phosphoric acid being poured on detainees.

One oft-stated reason for overthrowing Saddam had been his systematic and brutal policies of torture. But now Americans had become the torturers.

In the wake of the Abu Ghraib revelations, violence in Iraq escalated throughout the spring of 2004. The Sunni insurgency grew. A year earlier, there had been merely a few hundred attacks per month against Iraqi and coalition forces. Now, there were roughly two thousand attacks per month.[76] Iraq's oil and electricity infrastructure had been destroyed. America retreated to a "light footprint," meaning that the United States had elected to engage in what is sometimes called "war tourism," going out on mounted missions occasionally from gigantic bases stocked with McDonald's and Burger King. But essentially, the United States had no strategy to defeat the insurgency. Moqtada al-Sadr's Shi'ite militia, the Mahdi Army, grew into a powerful force with which occupation forces had to contend. To people on the ground, the liberation of Iraq had become a distant notion, the idea of spreading democracy there a quaint but hollow dream.

But the media in America still aggressively promoted the idea of democratizing the Middle East. Even America's most respected pundits continued to accept the official narrative—or at least hedge their bets. In November 2003, *New York Times* foreign affairs columnist Thomas Friedman predicted that "The next six months in Iraq—which will determine the prospects for democracy-building there—are the most important six months in U.S. foreign policy in a long, long time."[77]

By June 2004, however, as the media watchdog group FAIR (Fairness and Accuracy in Reporting) pointed out,[78] Friedman, in an interview on National Public Radio's *Fresh Air,* revised his prediction. "It might be over in a week, it might be over in a month, it might be over in six months, but what's the rush? Can we let this play out please?"[79] On October 3, 2004, in an appearance on CBS's *Face the Nation,* Friedman extended the deadline again: "What we're gonna find out, Bob [Schieffer], in the next six to nine months is whether we have liberated a country or uncorked a civil war."

Less than two months later, on November 30, Friedman was confident that the United States was in the homestretch. "This is crunch time," he wrote. "Iraq will be won or lost in the next few months. But it won't be won with high rhetoric. It will be won on the ground in a war over the last mile." Ten months later, Friedman decided to extend the deadline once again. "I think we're in the end game now . . ." he said on NBC's *Meet the Press.* "I think we're in a six-month window here where it's going to become very clear."

And so it went, again and again, in his *Times* column, on *Face the Nation, Meet the Press, Charlie Rose, The Oprah Winfrey Show,* MSNBC's *Hardball,* Friedman made essentially the same prediction no fewer than fourteen different times, ultimately extending the deadline by more than three years. (For Friedman's other predictions, see Note 79 in Chapter 17.)

Perhaps the best indication of what was really happening on the ground came not from news reports, but from a widely circulated e-mail filed by Farnaz Fassihi, a Middle East correspondent for the *Wall Street Journal*,[80] who wrote her friends:

> Being a foreign correspondent in Baghdad these days is like being under virtual house arrest . . . I am house bound. . . . I avoid going to people's homes and never walk in the streets. I can't go grocery shopping any more, can't eat [in] restaurants, can't strike a conversation with strangers, can't look for stories, can't drive in any thing but a full armored car, can't go . . . And can't and can't. . . . There has been one too many close calls, including a car bomb so near our house that it blew out all the windows. So now my most pressing concern every day is not to write a kick-ass story but to stay alive and make sure our Iraqi employees stay alive. . . . It's hard to pinpoint when the "turning point" exactly began. Was it April when Fallujah fell out of the grasp of the Americans? Was it when Moqtada and Jish Mahdi declared war on the U.S. military? Was it when Sadr City, home to ten percent of Iraq's population, became a nightly battlefield for the Americans? Or was it when the insurgency began spreading from isolated pockets in the Sunni triangle to include most of Iraq? Despite President Bush's rosy assessments, Iraq remains a disaster. If under Saddam it was a "potential" threat, under the Americans it has been transformed to "imminent and active threat," a foreign policy failure bound to haunt the United States for decades to come.
>
> Iraqis like to call this mess "the situation." . . . What they mean by situation is this: the Iraqi government doesn't control most Iraqi cities, there are several car bombs going off each day around the country killing and injuring scores of innocent people, the country's roads are becoming impassable and littered by hundreds of landmines and explosive devices aimed to kill American soldiers, there are assassinations, kidnappings and beheadings. The situation, basically, means a raging barbaric guerilla war. In four days, 110 people died and over 300 got injured in Baghdad alone.

Worse, highly placed military figures such as General William Odom, former head of the National Security Agency under Ronald Reagan, said that the disaster that Fassihi described on a personal level was true in geostrategic terms as well. "Bush hasn't found the WMD. Al-Qaeda, it's worse, he's lost on that front," Odom told the *Guardian*. "That he's going to achieve a democracy there? That goal is lost, too. It's lost.... Right now, the course we're on, we're achieving Bin Laden's ends."[81]

Retired general Joseph Hoare, the former marine commandant and head of U.S. Central Command, added: "The idea that this is going to go the way these guys planned is ludicrous. There are no good options...."

By the summer of 2004, Bush's numbers had dropped nearly 30 points from their "Mission Accomplished" high.[82] As a result, when the Democratic National Convention came to an end in Boston on July 29, nobody knew who would be the next president.*[83] Most polls gave the John Kerry–John Edwards ticket a lead of from 2 to 7 points over their Republican rivals.[84] But when one took into consideration the usual postconvention spike, the presidential race was essentially a dead heat.

A liberal three-term senator from Massachusetts at a time when the disastrous 1988 candidacy of former Massachusetts governor Michael Dukakis was still a painful memory for Democrats, Kerry was not a likely nominee to oppose a conservative incumbent president from Texas in a profoundly conservative era. A blue-blooded, blue-state candidate in a red-state world, he had a patrician pedigree that included a Yale degree, relatives from France, and an heiress for a wife. In the seventies, as a leader of the Vietnam Veterans Against the War, he had testified before Congress as an antiwar activist. Although now critical of the war effort, Kerry had voted for the war powers authorization, which left him open to charges that, as Bush put it, he had "entered the flip-flop hall of fame."[85] On the other hand, unlike Bush or Cheney, Kerry had faced actual combat and had a sterling war record in Vietnam.

As antiwar passions grew, the Democrats and their supporters spent tens of millions of dollars on sophisticated get-out-the vote efforts. So-

*On the same day that the Democratic National Convention ended, David Kay of the Iraq Survey Group briefed the president, Cheney, Rumsfeld, Tenet, Rice, Wolfowitz, and other top aides on the fact that his group had found no evidence of WMDs in Iraq. Describing the president's reaction, in *Hubris: The Inside Story of Spin, Scandal, and the Selling of the Iraq War* by Michael Isikoff and David Corn, Kay said: "I'm not sure I've spoken to anyone at that level who seemed less inquisitive. He was interested but not pressing any questions.... I cannot stress too much that the president was the one in the room who was the least unhappy and the least disappointed about the lack of WMDs. I came out of the Oval Office uncertain as to how to read the president. Here was an individual who was oblivious to the problems created by the failure to find WMDs. Or was this an individual who was completely at peace with himself on the decision to go to war, who didn't question that, and who was totally focused on the here and now of what was to come?"

called 527 groups* such as America Coming Together, The Media Fund, and Moveon.org targeted millions of antiwar Democrats who had been outraged by the lies about WMDs, the abuses of Abu Ghraib, and Bush's policies. For a time, it appeared that these grassroots organizations, many linked to websites, might be rewiring the American political edifice; the organizations were instrumental in drawing younger, idealistic, Internet-centric voters to become involved in the election, had helped momentarily elevate the unlikely prospects of Howard Dean, former governor of Vermont, and seemingly energized the political discourse.

But the Republicans continued to control the narrative as it played on Main Street. In the immediate aftermath of the Democratic Convention, on August 1, Secretary of Homeland Security Tom Ridge raised the threat level to Code Orange—high—asserting that the government had "new and unusually specific information about where al-Qaeda would like to attack. And as a result, today, the United States Government is raising the threat level to Code Orange for the financial services sector in New York City, Northern New Jersey and Washington, DC."[†86] The timing was opportune, for it distracted attention from Kerry's official nomination as the Democratic candidate and reminded voters that they saw the Republicans as strong on national security. And if Democrats railed about torture at Abu Ghraib, right-wing talk-show host Rush

*A 527 group is a tax-exempt advocacy group named after a section the U.S. tax code, 26 U.S.C. § 527.

†Another key factor in the 2004 election was fear itself. As John Judis explained in "Death Grip," an August 2007 piece in *The New Republic,* studies by psychologists Sheldon Solomon, Jeff Greenberg, and Tom Pyszczynski proved "that the mere thought of one's mortality can trigger a range of emotions—from disdain for other races, religions, and nations, to a preference for charismatic over pragmatic leaders, to a heightened attraction to traditional mores."

According to Judis, beginning in 1989, the three psychologists began a series of experiments to test their hypothesis that "the recognition of mortality evokes 'world-wide defense'"—their term for a series of emotional responses, including intolerance, religiosity, a preference for law and order, the protection of traditional values against social experimentation and individual prerogatives. As Judis explains, "For many conservatives, this means opposition to abortion and gay marriage."

In September 2004, the three psychologists tested their theories on 131 Rutgers University undergraduates who were registered and planned to vote. Predictably enough, a control group of students at the liberal university favored John Kerry by a margin of four to 1. However, the study showed that among those who were subjected to "mortality reminders" in the form of having "911" or "WTC" flashed subliminally between word associations, the students favored Bush two to one.

Limbaugh dismissed the abuses as "sort of like hazing, a fraternity prank. Sort of like that kind of fun."*[87]

For most Americans, what was really going on in Iraq had not yet coalesced into a coherent reality. Many of the essential facts of Nigergate had been revealed—that the documents had been forged, that the White House knew the information was untrue before the war, that the White House leaked Valerie Plame Wilson's identity to the press in an effort to discredit Joe Wilson. But no scandal ensued, no investigation linked disparate events into one clear story line. With Senator Pat Roberts, a Cheney ally, running the Senate Intelligence Committee, congressional oversight had essentially been nullified.

As for the supposedly liberal Washington press corps, many key figures simply liked Bush, Rove, and company too much to act as watchdogs. "Let me disclose my own bias . . . ," wrote David Broder, the lead political columnist for the *Washington Post.* "I like Karl Rove. In the days when he was operating from Austin, we had many long and rewarding conversations. I have eaten quail at his table and admired the splendid Hill Country landscape from the porch of the historic cabin Karl and his wife Darby found miles away and had carted to its present site on their land."[88]

Moreover, many Americans were simply resistant to the notion that a miscalculation of historic proportions had taken place; few wanted to believe that the rising toll of blood lost and citizens' tax money spent would be in vain; it was both emotionally and intellectually easier to have faith in the presidency and the American military. That the country's leaders, many of whom had come of age in the Vietnam era, had actually induced a catastrophic geopolitical blunder verged on an unthinkable cultural taboo, one hard to sustain while going about the everyday business of earning a living, raising a family, and being subjected to the infinite distractions and entertainments of American culture.

To the vast majority of Americans, in fact, there was nothing wrong with our Iraq policy. According to a Harris Poll, in October 2004, 76

*It did not help that a photograph of Kerry appeared in the national media in late August showing him in an expensive, tight-fitting wet suit and windsurfing, an image that seemed both ill-advised in that it both showed him at leisure and pursuing a flaky, nonmainstream diversion. Even for those who wanted to see Kerry victorious, and who were perhaps frustrated with his unwillingness to defend himself against the "Swift-boating" attacks on his war record, the photo suggested that Kerry did not understand his own national political persona.

percent of Americans believed the United States had made life better for most Iraqis, and 63 percent believed Iraq had been a serious threat to the United States. In the eyes of most Americans, the enemy was embodied by the villainous Saddam Hussein and his brutal henchman, who were often conflated with Osama bin Laden.[89] According to the same Harris Poll, 62 percent of Americans still believed Saddam "had strong links to al Qaeda," and 41 percent believed he "helped plan and support the hijackers who attacked the U.S. on September 11, 2001."

In the context of 9/11, how could one possibly question which side to be on? There were enormous stirrings of patriotism and national pride over the achievements of the country's men in uniform who had put their lives on the line to make life better for Iraqis under Saddam's thumb and to eliminate a serious threat to U.S. security. Again and again, Bush told them, "In times of war and in hours of crisis, Senator Kerry has turned his back on 'pay any price' and 'bear any burden,' and he's replaced those commitments with 'wait and see' and 'cut and run.' "[90]

Calling attention to presidential elections scheduled in Iraq in January, Bush told voters, "Think how far that country has come from the days of torture chambers and mass graves. Freedom is on the march, and we're more secure for it." Echoing his State of the Union address, Bush added, "Freedom is not America's gift to the world. Freedom is the Almighty God's gift to each man and woman in this world."[91]

Bush's invocation of God when discussing foreign policy served as a clever rhetorical splice to the other issues that mattered to Americans—issues that the leaders of the Christian Right, working in concert with the White House, intended to use to leverage the election. According to Jerry Falwell, the Arlington Group, a powerful group of evangelical leaders within the much larger Council for National Policy, was regularly in contact with Karl Rove to strategize. Among their highest priorities was the so-called marriage amendment initiative—to protect "the sanctity of marriage" from gays.

As Falwell described their strategy, the Arlington Group began focusing on crucial swing states such as Ohio. "One of our Arlington Group men, Phil Burress, started the ground work for the marriage amendment initiative to be on the ballot in 2004 in Ohio," said Falwell.[92] "We had two things in mind. First, to save the family, but also to turn out the pro-family vote for Bush in 2004."

Burress, a self-described former pornography addict, gathered 575,000 signatures to get the measure on the ballot in fewer than ninety days—and then made sure the signatories turned up to vote for Bush.

"In 21 years of organizing, I've never seen anything like this," Mr. Burress, sixty-two, told the *New York Times.*[93] "It's a forest fire with a 100-mile-per-hour wind behind it."

Focusing on swing states such as Ohio, Karl Rove quietly put together a much bigger and better campaign machine than he had in 2000, setting a clear goal that became dogma among religious conservatives. In order to win in November 2004, he said, Bush would have to draw 4 million more evangelical voters than he had in 2000. In June, via a live telecast from the White House, Bush had addressed the Southern Baptist Convention, lending fuel to the first major voter registration drive in history for the 16 million Southern Baptists.[94] Ralph Reed, formerly head of the Christian Coalition, officially joined the Bush-Cheney campaign to bring in the religious vote. As tax-exempt religious institutions, churches are prohibited from endorsing candidates. But at a reception for prominent pastors, Reed told them, "Without advocating on behalf of any candidate or political party, you can make sure that everyone in your circle of influence is registered to vote."[95]

So it was that Dr. Jack Graham, former president of the Southern Baptist Convention, hosted the Bush-Cheney reception, nonetheless asserting that there was a clear distinction between his personal views and his role as a religious figure. "This is Jack Graham the person, not Jack Graham the president of the Southern Baptists," he told the *New York Times.* Declaring that the Bible told Christians to support godly politicians, he added, "You can connect the dots. I don't mind if you connect the dots. You can't separate what you believe from the political process."[96]

Ultimately, the views of Jack Graham went out to thousands of pastors who in turn influenced millions of parishioners. On a Sunday before the election, at least one pastor asked his congregants who were registered Republicans to stand. When about 80 percent of the assembled were standing, he told them, "Now, I want you to pray for the sinners who are present."[97]

The effectiveness of its powerful network notwithstanding, it's important to remember that the Christian Right was not monolithic. Rather, it consisted of tens of millions of ordinary Americans, middle-aged couples, the elderly and young singles, from San Diego, Dallas, Atlanta, and Pittsburgh, and hundreds of towns in between. They were real estate brokers and lawyers, computer consultants and pastors, nurses, and housewives. They drove pickup trucks and American cars, and shopped at Wal-Mart and the mall—and Bush's folksy West Texas mannerisms appealed to them far more than Kerry's sophistication and aloof, almost wooden, persona. And many of them hailed from the same regions that provided the men and women who fought in the all-volunteer army—and therefore felt an especially close kinship to the military. Their basic wish to believe in a right and just America was probably little different from that of the East Coast college students signing up to bus to Ohio on election day to serve as poll watchers; the difference was in what each group thought was wrong with the country and which candidate would have a better chance of fixing it.

As Election Day approached, the situation in Iraq continued to deteriorate. By September 7, the U.S. death toll hit one thousand—in part because the soldiers' "light-skinned" Humvees lacked the requisite protective armor. On September 15, U.N. secretary-general Kofi Annan finally declared the Iraq War illegal.[98] The conflict had taken on the character of a classical guerrilla war. Or was it also a civil war? No matter which analysis held sway, the war was clearly out of America's control. It seemed Kerry had a chance.

With memories of the bitter 2000 presidential election controversy still fresh in their minds, impassioned activists from both parties flooded crucial swing states—such as Florida, Ohio, and Pennsylvania—with organizers and poll watchers. On November 2, more than 122 million voters came to the polls, the highest turnout in history, and, with 6.4 percent more voters than in 2000, the biggest increase in decades.*[99] When the results from the early exit poll came in, Kerry supporters were especially optimistic. In an online chat, *Washington Post* managing editor Steve Coll wrote that "the last wave of national exit polls we received . . . showed Kerry winning the popular vote by 51 to 48 percent—if true, surely enough to carry the Electoral College."[100]

The contest, indeed, came down to Ohio, and when the votes were counted, Bush had won the state by a 119,000 votes—2 percent—amid widespread accusations of fraud. When John Kerry decided not to contest Bush's victory in Ohio, Bush effectively had won reelection by 286 to 251 in the Electoral College and a 3-million-vote margin in the popular vote.

How had Bush won? Ultimately, Iraq had mattered, but, to the astonishment of secular America, not as much as "moral values."

*Voter turnout, which has historically been low in the United States, also improved significantly with nearly 57 percent of the voting-age population going to the polls, up from 51 percent in 2000.

According to *Christianity Today,* exit polls showed that 22 percent of the electorate cited "moral values," such as abortion, gay marriage, and the marriage amendment initiative, as the single most important factor in determining their vote—more than the economy, terrorism, or Iraq.[101] Nationally, for those who cited moral values as their greatest concern, Bush defeated Kerry 79 to 18. But the numbers were even more striking in Ohio, the swing state that ultimately determined the election, where moral values voters supported Bush over Kerry 85 to 14.

To the dismay of secular America and to those citizens concerned about the fabrication of intelligence used to start the war, about America's use of torture and the detainment of prisoners without rights, about the erosion of their own constitutional rights, and about the complex system of checks and balances that held the executive branch in restraint, the Christian Right had narrowly won again.

The day after the election, Bob Jones III, the right-wing president of Bob Jones University, who had once denounced Ronald Reagan as "a traitor to God's people" for choosing as his vice president George H.W. Bush, whom Jones called "a devil," wrote a personal letter to Bush expressing far warmer sentiments than he had expressed about his father. "In your re-election," he wrote, "God has graciously granted America—though she doesn't deserve it—a reprieve from the agenda of paganism. You have been given a mandate. We the people expect your and with it, voice to be like the clear and certain sound of a trumpet. Because you seek the Lord daily, we who know the Lord will follow that kind of voice eagerly. Don't equivocate. Put your agenda on the front burner and let it boil. You owe the liberals nothing. They despise you because they despise your Christ. Honor the Lord, and He will honor you." In addition, the neoconservatives had retained enough power to pursue their agenda.[102]

Now that George W. Bush had been reelected, he had finally bested his father, a one-term president. In historic terms, *he* was now the most potent Bush, *he* had now become the ultimate embodiment of the House of Bush. The election result was not just a point of family pride but a popular ratification that provided him new political powers. "Let me put it to you this way," Bush told reporters after the election. "I earned capital in the campaign, political capital, and now I intend to spend it."[103]

CHAPTER EIGHTEEN

An Angel Directs the Storm

It was not until after George W. Bush and Dick Cheney were narrowly reelected that many Americans began to realize that the Iraq War represented a dangerous moment in American history, a turning point both in terms of America's place on the global chessboard and, domestically, in terms of its fate as a constitutional democracy. Gradually, the horrors of the war, its related scandals, and its ramifications began to reveal themselves.

On November 7, 2004, five days after the election, it was reported that thousands of surface-to-air missiles that had once been under Saddam's control were unaccounted for because the U.S.-led force had not secured all the weapons depots in Iraq.[1] The next day, U.S.-led forces moved in to clear out Fallujah, a stronghold for Sunni insurgents, launching a ferocious ten-day battle that killed at least one thousand insurgents and left fifty-four Americans dead and more than four hundred seriously wounded.[2] Colonel Gary Brandl led his troops into battle with words evocative of a Holy War. "The enemy has got a face," he said.[3] "He's called Satan. He's in Fallujah and we're going to destroy him."

During the assault, a marine deliberately shot and killed an unarmed Iraqi civilian in a mosque, and the videotaped incident was televised across the world. In response, violence raged across Iraq. On November 9, militants kidnapped three members of interim prime minister Ayad Allawi's family. A few days later, in the north, saboteurs set fire to four oil wells northwest of Kirkuk.[4] Astoundingly, despite having the second largest oil reserves in the world, Iraq was forced to *import* oil from nearby Kuwait because of lack of refining capacity and hundreds of terrorist attacks on its facilities.[5]

By now, repercussions from the war were also being felt throughout the entire Middle East. Iraqi authorities had already captured Saudis crossing the Saudi border into Iraq to fight the United States. In response to Fallujah, twenty-six prominent Saudi religious scholars urged their followers to support "jihad" against U.S.-led forces. Militant Islamists from America's oil-rich ally had now taken up arms against the United States.[6]

Paradoxically, even though their policy failures were finally evident, the neocons had become empowered as never before. Just before the election, Bush had quietly dismissed Brent Scowcroft as chairman of PFIAB—without even bothering to speak to him personally. Cheney and Bush had both known that the phenomenally popular Colin Powell was crucial to their reelection chances. But now he, too, was expendable. On November 10, eight days after the election, Powell got the phone call from White House chief of staff Andy Card.[7] He was out.

Doing everything possible to put a good face on his resignation, Powell told reporters at a November 15 press briefing that "it has always been my intention that I would serve one term," and that he and Bush "came to a mutual agreement that it would be appropriate to leave at this time."[8] But Frank Carlucci, a former secretary of defense himself during the Reagan administration, who was close to both Powell and Cheney, and who continued to think highly of Cheney, was more forthright. "Colin has been used," he said.[9]

Bush and Cheney reshuffled the cabinet, strengthening the neocon hand. Condoleezza Rice replaced Powell. Much to Scowcroft's dismay, she had proven to be less a voice for the realists than an enabler and repeater of others' formulations, in effect a neocon fellow traveler. Her deputy, Stephen Hadley, a Cheney ally, in turn took her old job as national security adviser. As for intelligence, Porter J. Goss, a former Republican congressman from Florida, who had become the new CIA director before the election, issued a memo to CIA employees that instantly confirmed his reputation as an administration loyalist: "As agency employees we do not identify with, support or champion opposition to the administration or its policies." The memo added that their job was "to support the administration and its policies in our work."[10]

With Rice, Hadley, and Goss in key positions, Bush, Cheney, and Rumsfeld had consolidated control over national security to an unprecedented degree. The notion that America's $40 billion intelligence apparatus would speak truth to power had become a pipe dream. State Department veterans desperately fantasized that Scowcroft, former Secretary of State James Baker, or even Bush 41 himself would somehow soon ride to the rescue.

* * *

Meanwhile, in both Iraq and Washington, the dream of spreading democracy throughout the Middle East continued to be mocked by the brutal realities of war. On January 30, 2005, 58 percent of the Iraqi electorate defied threats of violence to vote in the first elections since Saddam's ouster. After reaching the polls, Iraqis proudly displayed their ink-dipped purple fingers as indications that they had voted. In Washington, Republican congressmen flaunted purple fingers as a sign of solidarity with Bush and pride at how the United States had brought democracy to Iraq. "Giving Terrorism the Purple Finger," read a headline.[11] "Purple finger" cocktails were concocted, consisting of grenadine, cassis, black currants, and vodka.[12]

After nearly two years of bombings, kidnappings, and assassinations in Iraq, at last the White House had a concrete achievement to celebrate—one that no one could deny. In his 2005 State of the Union address on February 2, President Bush proudly saluted the Iraqi voters and the American soldiers who had made the election possible, introducing as his special guest Iraqi human rights advocate Safia Taleb al-Suhail: "Eleven years ago, Safia's father was assassinated by Saddam's intelligence service. Three days ago in Baghdad, Safia was finally able to vote for the leaders of her country—and we are honored that she is with us tonight."[13] At last, Bush said, Iraq had turned the corner.

The speech also showed that Bush had been reading from the neocon handbook; he proclaimed to the world that his administration's goal was the promotion of "democratic movements and institutions in every nation and culture, with the ultimate goal of ending tyranny in our world."[14]

"This is real neoconservatism," Robert Kagan, a leading neocon, told the *Los Angeles Times.* "It would be hard to express it more clearly. If people were expecting Bush to rein in his ambitions and enthusiasms after the first term, they are discovering that they were wrong."[15]

Dimitri Simes, president of the Nixon Center, a conservative think tank that hewed more closely to realist policies, had a different point of view. "If Bush means it literally, then it means we have an extremist in the White House," he said. "I hope and pray that he didn't mean it."[16]

To anyone who believed in democracy, the sight of Iraqis voting was potentially inspiring. But the political reality on the ground was starkly different. Yes, Shi'ites flocked to the polls in huge numbers. But the Sunnis, alienated by America's de-Ba'athification policies, which removed members of the largely Sunni Ba'athist regime from government, angry because they had lost jobs and security when the United States disbanded the police and the military, and enraged by the American assault on Sunni mosques in Fallujah, boycotted the election in droves. Even before the elections were held, Brent Scowcroft had warned that voting had "great potential for deepening the conflict" in Iraq by exacerbating the divisions between Shi'ite and Sunni Muslims, and that it might lead to a civil war.[17] As the Bush White House basked in the glory of having shown the world it could create a new democracy in the Middle East, it soon became clear that Scowcroft had been prescient.

The Shi'ites took office, but the Sunni insurgency went after new targets. U.S. forces had protected its own bases, including the Green Zone, but not the general population in Baghdad or any of the major cities. "When we did not secure the population," General Jack Keane told PBS's *Frontline*, "the enemy realized that the population was fair game. . . . All through '05 they exploited it. They began to kill people, take them on. . . . In ever-increasing numbers they began to kill more and more of the Iraqis. . . . They were exposed."[18]

Immediately after the election, the Sunnis struck back with a vengeance. On February 3, bombs killed at least 20 people in Baghdad; insurgents stopped a minibus near Kirkuk and gunned down 12 of its occupants; gunmen ambushed and killed 2 Iraqi contractors near Baghdad; others overran a police station in the town of Samawah—not to mention innumerable assassination attempts, car bombs, and the like. On February 17, a string of attacks killed at least 36 people, mostly Shi'ites. The next day, at least 8 suicide bombings and other attacks targeted Shi'ite worshippers observing the religious festival of Ashura. By the end of the month, suicide bombers targeted crowded marketplaces near Baghdad, killing as many as 115 people with one bomb.[19]

Meanwhile, Osama bin Laden had ordered his supporters to attack Iraqi oil facilities—which they had begun to do with considerable success. Terrorists had begun an all-out war against the country's oil facilities, costing it billions in lost revenue.

Having put so much stock in the Iraqi elections, the Bush administration now had another problem. Like it or not, the administration was wedded to a Shi'ite government led by Prime Minister Ibrahim al-Jaafari* of the Islamic Dawa Party, one of two major Shia parties in the ruling coalition, the United Iraqi Alliance. A militant Shi'ite Islamic group

*Al-Jaafari's successor, Nouri al-Maliki was also a leader of the Dawa Party.

that had supported the 1979 Islamic revolution in Iran and that had received support from the Iranian government during the Iran-Iraq War, the Dawa Party had moved its headquarters to Tehran in 1979. There, according to Juan Cole, professor of Middle Eastern history at the University of Michigan, it "spun off a shadowy set of special ops units generically called 'Islamic Jihad,' which operated in places like Kuwait and Lebanon." The party, Cole wrote, was also "at the nexus of splinter groups that later, in 1982, began to coalesce into Hezbollah."[20] Moreover, the party had been founded by Muhammad Baqir al-Sadr, the uncle of Moqtada al-Sadr, the powerful Shi'ite leader of the Mahdi* Army, which has been tied to ethnic cleansing of Sunnis.[21]

One by one the contradictions behind America's Middle East policies emerged—and with them, the enormity of its catastrophic blunder. Gradually America's real agenda was coming to light—not its stated agenda to rid Iraq of WMDs, which had been nonexistent, not regime change, which had already been accomplished, but the neoconservative dream of "democratizing" the region by installing pro-West, pro-Israeli governments, led by the likes of Ahmed Chalabi, in oil-rich Middle East states.

Now that Chalabi had been eliminated as a potential leader amid accusations that he had been secretly working for Iran, and the Sunnis had opted out of the elections entirely, the United States, by default, was backing a democratically elected government that maintained close ties to Iran and was linked to Shi'ite leaders whose powerful Shi'ite militias were battling the Sunnis.

Professing to train Iraqi soldiers to "stand up," so Americans could "stand down," the United States was in fact training soldiers who were loyal to the Shi'ite cause, rather than to any concept of Western democracy. "[T]hey weren't really Iraqi security forces," explained journalist and author Nir Rosen.[22] "They were loyal primarily to Moqtada al-Sadr, to Abdul Aziz al Hakim [the Shia leader of the Supreme Council for the Islamic Revolution in Iraq], but not to the Iraqi state and not to anybody in the Green Zone." As shown in a PBS *Frontline* documentary, "Gangs of Iraq," Iraqi soldiers, even when accompanied by Americans who were training them, intentionally kept the Americans away from large weapons caches that could be used against the Sunnis.[23] Unwittingly, America was spending billions of dollars to fuel a Sunni-Shi'ite civil war.

Even worse, in the larger context of the region, by deposing Saddam

*The Mahdi is a messianic figure who, many Muslims believe, will appear before the Day of Judgment and fight alongside Jesus against the Antichrist.

and supporting the Iran-leaning Shi'ites, the United States had inadvertently empowered Iran, its biggest foe in the Middle East. And Iran's ascendancy posed problems for Israel and Saudi Arabia as well. Potentially, the Sunni-Shi'ite conflict could spread throughout the entire region.

By 2005, for tens of millions of Americans, it was increasingly impossible to ignore the realities of what was happening in Iraq—the absence of WMDs, the escalating sectarian violence, the vast expenditures of blood and treasure in pursuit of a mission that was unclear at best, constantly changing, and had never been accomplished at all. Polarizing the nation more profoundly than at any time since the Vietnam era, the war had become a litmus test issue that defined and linked whole sets of belief systems—red state America versus blue, evangelical Christians, antiabortion activists, NASCAR dads, and other denizens of the Bible Belt versus the secular, post-Enlightenment America that has long been on the cutting edge of science and the embodiment of modernism. Those who questioned U.S. policies in the Middle East, as their foes saw it, were cut-and-run traitors who aided and abetted the enemy. On the other side were Neanderthals waging a holy war in the Middle East, shredding the Constitution, destroying civil liberties, rolling back not just the New Deal but the Enlightenment, all in the name of God.

Hate filled the air, at times evoking the specter of McCarthyism, the hate and fear mongering of Father Coughlin, and even the assault against reason undertaken by the Puritans. Right-wing pundit Ann Coulter expressed her regret that Oklahoma City bomber Timothy McVeigh "did not go to the *New York Times* building." Americans who did not vote for Bush, she said, were "traitors," her critics, members of the "Treason Lobby."[24] To Rush Limbaugh, Democrats "had aligned themselves with the enemy" and were "PR spokespeople for Al Qaeda."[25] To Fox News host Bill O'Reilly, the American Civil Liberties Union were "terrorists" who were almost as dangerous "as Al-Qaeda."*[26] Thanks to

*A study of O'Reilly by the Indiana University Media Relations Department compared him to Father Charles Coughlin, the radio broadcaster in the thirties who sang the praises of Hitler and Mussolini, and found that O'Reilly was three times as likely to be a "name caller" as Coughlin and that he used age-old propaganda techniques, such as "glittering generalities" and "selective use of the facts," far more often than Father Coughlin ever did—roughly thirteen times a minute—in his editorials. The study also showed that 96.3 percent of his references to Democrats cast them as villains.

the neocons and religious conservatives, the radical right was driving America as never before.

With the Republicans still in control of Congress, Bush's critics vested their few remaining hopes for retribution in Patrick Fitzgerald, a newly appointed federal prosecutor who had recently taken charge of the Valerie Plame Wilson–CIA leak investigation. But in many respects, it seemed as if the nation had regressed to the era of the Scopes Monkey Trial. Tens of millions of people in the only country that had put a man on the moon, that had unraveled the human genome, now questioned whether evolution was real. A Creation Museum was under construction near Cincinnati, Ohio, to demonstrate that it wasn't. Tourists to the Grand Canyon were treated to creationist tours assuring them that geologists had been wrong, and that one of America's greatest wonders had not been formed slowly over millions of years, but was God's creation dating "to the early part of Noah's flood."[27] The Kansas State Board of Education held hearings about redefining the word *science* to remove bias toward "naturalistic" (nontheistic) belief systems.[28] Pennsylvania senator Rick Santorum—who believed that states should be able to arrest gay lovers in the privacy of their bedrooms—backed an amendment to allow the teaching of intelligent design as an alternative theory to evolution.[29]

The Bush administration and the religious right declared war on science. Slogans that had once been bumper stickers—"Just a Theory"—became government policy: global warming is a hoax; condoms don't work; intelligent design is legitimate science. The administration's initiative to fund AIDS programs in Africa was hailed by the press, but information about the benefits of condoms was removed from government websites. The global warming section of the Environmental Protection Agency was dropped entirely. In deference to the Christian Right, morning-after contraceptive sales were banned, even after having been approved by the Food and Drug Administration. According to Howard Dean, the former Vermont governor and 2004 Democratic presidential hopeful, a National Cancer Institute fact sheet was "doctored to suggest that abortion increases breast-cancer risk, even though the American Cancer Society concluded that the best study discounts that."[30]

And when it came to dealing with the "liberal" judiciary, Pat Robertson sought help from God during a prayer retreat, and the Lord told him, "I will remove judges from the Supreme Court quickly, and their successors will refuse to sanction the attacks on religious faith." Asking his television audience to pray that three liberal Supreme Court justices retire, Robertson said, "I don't care which three, I mean as long as the three conservatives stay on. . . . There's six liberals, so it's up to the Lord."[31]

If the once powerful Christian Coalition had become moribund—and it had—that was because it had been replaced by a far more powerful institution: the Republican Party. Indeed, in 2004, no fewer than forty-one out of fifty-one Republican senators voted with the Christian Coalition 100 percent of the time.[32] When the new Congress took office in early 2005, it included Tom Coburn, newly elected senator from Oklahoma, who believed that doctors who performed abortions should be executed. Asserting that global warming was a hoax, Senator Jim Inhofe (R-Okla.) compared environmentalists to the Nazis. He argued that American policy in the Middle East should be based on the Bible,* that Israel had a right to the West Bank "because God said so."[33] And on the Senate floor, in a speech about the proposed Federal Marriage Amendment, he displayed an enormous photo of his extended family, and told the august assembly, "We have 20 kids and grandkids. I'm really proud to say that in the recorded history of our family, we've never had a divorce or any kind of homosexual relationship."[34]

Meanwhile, the White House sought extraordinary means to get its message across. In late January 2005, a man named James Guckert showed up at a presidential news conference using Jeff Gannon as a pseudonym, and lobbed softball questions to President Bush. "Senate Democratic leaders have painted a very bleak picture of the U.S. econ-

*Inhofe said: "I believe very strongly that we ought to support Israel; that it has a right to the land. This is the most important reason: Because God said so. As I said a minute ago, look it up in the book of Genesis. It is right up there on the desk. In Genesis 13:14–17, the Bible says:

> The Lord said to Abram, "Lift up now your eyes, and look from the place where you are northward, and southward, and eastward and westward: for all the land which you see, to you will I give it, and to your seed forever. . . . Arise, walk through the land in the length of it and in the breadth of it; for I will give it to thee."

"That is God talking.

The Bible says that Abram removed his tent and came and dwelt in the plain of Mamre, which is in Hebron, and built there an altar before the Lord. Hebron is in the West Bank. It is at this place where God appeared to Abram and said, 'I am giving you this land—the West Bank.' This is not a political battle at all. It is a contest over whether or not the word of God is true."

omy. . . ." he told President Bush. "Yet in the same breath they say that Social Security is rock solid and there's no crisis there. How are you going to work—you've said you are going to reach out to these people—how are you going to work with people who seem to have divorced themselves from reality?"

Gannon's questions were *so* friendly, critics suspected that they might have been planted, and found out that he worked for Talon News, an apparent front for the conservative website GOPUSA. More titillating, Gannon had appeared naked on several gay escort sites such as hotmil itarystud.com and was reported to be a "a $200-an-hour gay prostitute." More titillating yet were reports that Gannon visited the White House regularly, often on days in which there were no press conferences.[35] Was it possible that he might be part of what was known in Washington circles as the Lavender Bund,* the coterie of closeted right-wing gays who helped the religious right and the Republicans advance an agenda that was often explicitly antigay? Later came revelations about Congressman Mark Foley and his suggestive e-mails to young congressional pages, and Ted Haggard, head of the National Association of Evangelicals, who had a relationship with a male prostitute.[†36]

As the culture and political wars continued, they took a toll on the White House's credibility. In March 2005, Republican politicians and the religious right—most of whom, theoretically at least, had been proponents of States' rights—ignited a national controversy when they tried to intervene on behalf of Terri Schiavo, a Florida woman in a persistent vegetative state, to prevent the removal of her feeding tube.

In April, federal prosecutor Patrick Fitzgerald continued to investigate the leak of CIA officer Valerie Plame Wilson's name. But journalists Matthew Cooper of *Time* and Judith Miller of the *New York Times* refused to divulge their sources. The question of who in the Bush administration had leaked her name was both a Washington parlor game and a profound inquiry into what was really going on in the White House.

*The term appears to date back to the 1980s as reference by openly gay Washingtonians to closeted gay Republicans who promoted the social conservative agenda.

†Another technique the Bush administration used to ensure that it got favorable coverage was to pay for it. In January 2005, *USA Today* reported that in return for promoting the Bush administration's No Child Left Behind Act, newspaper columnist and television host Armstrong Williams had been paid $240,000. The payment was part of a $1 million contract between the U.S. Department of Education and a public relations company.

* * *

Meanwhile, two years into the war, America's all-volunteer military force was being drained. With ongoing wars in Afghanistan and Iraq, there were not enough boots on the ground. To replenish their forces, officials raised the age limit for enlistment from thirty-four to forty.[37] Tours of duty for soldiers were extended repeatedly, leaving many of them feeling tricked and demoralized. In particular, the military relied on call-ups from the National Guard, many of whom were "weekend warriors," middle-aged men wrenched away from their families and jobs, at great sacrifice.

And what about Osama bin Laden—the all-but-forgotten villain behind 9/11? "We're on a constant hunt for bin Laden," Bush reassured America. "We're keeping the pressure on him, keeping him in hiding."[38]

But Bush's promises were wearing thin. The administration's practice of transferring prisoners from Guantánamo to other countries where they might be tortured was called into question. There were multiple reports of brutal treatment of detainees by the government. Likewise, attorneys for Guantánamo detainee Salim Ahmed Hamdan, who had been Osama bin Laden's driver, argued in court that their client must be afforded the same legal protections that American citizens have. The numbers of wiretaps and secret searches soared.

By late spring of 2005, approximately $200 billion had been spent on the war in Iraq. Tens of thousands of people had been killed. Countless more were wounded or living as refugees. There were no WMDs. Iraq's oil riches were being destroyed by saboteurs and stolen by terrorists. A report prepared for the U.N. Human Rights Commission showed that malnutrition rates in Iraqi children under five had nearly doubled since the U.S. invasion.[39]

Yet the administration continued to assert that victory was around the corner. "The level of activity that we see today, from a military standpoint, I think will clearly decline," Cheney told Larry King in May 2005. "I think that they're in the last throes, if you will, of the insurgency."[40]

But by the end of June, more than seventeen hundred Americans had been killed in Iraq. Baghdad's mayor decried his city's crumbling infrastructure. The Iraqi capital of more than six million people was now plagued by shortages of electricity and fuel, incessant bombings and suicide attacks, and did not even have adequate drinking water for its residents.[41] With one revelation after another about the Bush administration's secret rendition policies, detention of prisoners with rights at Guantánamo, and Abu Ghraib, America, rather than Saddam, had become known for torture and abuse.

Then, on July 7, 2005, four terrorist explosions rocked London's transport system at the height of rush hour, killing at least thirty-three and wounding roughly a thousand others.[42] A group calling itself the Secret Organization of the Al-Qaeda Jihad in Europe later claimed credit for the attacks, and asserted that the attacks were payback for Britain's involvement in the Iraq and Afghanistan wars.[43] The bombings sent a ripple of dread through Americans, especially New Yorkers. Many people could not help but wonder if the war in Iraq might induce such attacks on American soil rather than prevent them.

President Bush had argued, "If we were not fighting and destroying this enemy in Iraq, they would not be idle. They would be plotting and killing Americans across the world and within our own borders."[44] But the London bombing proved that exactly the opposite was true. According to a study published in *Mother Jones* by Peter Bergen and Paul Cruickshank, research fellows at the Center on Law and Security at the NYU School of Law, the net effect of the Iraq War was that it increased global terrorism by a factor of seven. "The rate of terrorist attacks around the world by jihadist groups and the rate of fatalities in those attacks increased dramatically after the invasion of Iraq. . . ." said the study. "A large part of this rise occurred in Iraq, which accounts for fully half of the global total of jihadist terrorist attacks in the post–Iraq War period. But even excluding Iraq, the average yearly number of jihadist terrorist attacks and resulting fatalities still rose sharply around the world by 265 percent and 58 percent, respectively."

Four days after London, a suicide bomber in Baghdad killed twenty-three people outside an army recruiting center in Baghdad. Among other victims that day were nine members of a Shi'ite family. It was all but official. As Iraq's former interim prime minister Ayad Allawi now asserted, Iraq was facing a civil war, and the consequences would be dire not just for Iraq but for Europe and America. A longtime ally of Washington, Allawi said, "The problem is that the Americans have no vision and no clear policy on how to go about in Iraq."[45]

As if the situation in Iraq were not enough, the neocons still had their eyes on Iran. To that end, in July 2005, House intelligence committee chairman Peter Hoekstra (R-Mich.) and committee member Curt Weldon (R-Pa.)* met secretly in Paris with an Iranian exile known as "Ali." Weldon had just published a book called *Countdown to Terror: The Top-Secret Information That Could Prevent the Next Terrorist Attack on America . . . and How the CIA Has Ignored It,* alleging that the CIA was ignoring intelligence about Iranian-sponsored terror plots against the U.S., and Ali had been one of their main sources. But according to the CIA's former Paris station chief Bill Murray, Ali, whose real name is Fereidoun Mahdavi, fabricated much of the information. "Mahdavi works for [Iranian arms dealer and intelligence fabricator Manucher] Ghorbanifar," Murray told Laura Rozen of the *American Prospect.* "The two are inseparable. Ghorbanifar put Mahdavi out to meet with Weldon."[46]

In a similar vein, in a speech before the National Press Club in late 2005, neocon Raymond Tanter, of the Washington Institute for Near East Policy, recommended that the Bush administration use the MEK [the Mujahideen e-Khalq, the Marxist-Islamic urban guerrilla group of Iranian dissidents who had been designated as a terrorist organization by the United States] and its political arm, the National Council of Resistance of Iran (NCRI), as an insurgent militia against Iran. "The NCRI and MEK are also a possible ally of the West in bringing about regime change in Tehran," he said.[47]

Tanter even suggested that the United States consider using tactical nuclear weapons against Iran. "One military option is the Robust Nuclear Earth Penetrator, which may have the capability to destroy hardened deeply buried targets. That is, bunker-busting bombs could destroy tunnels and other underground facilities." He granted that the Non-Proliferation Treaty bans the use of nuclear weapons against non-nuclear states, such as Iran, but added that "the United States has sold Israel bunker-busting bombs, which keeps the military option on the table."[48] In other words, the United States couldn't nuke Iran, but Israel, which never signed the treaty and maintains an unacknowledged nuclear arsenal, could.

If the MEK was being cast as the Iranian counterpart to the INC, there were more than enough Iranian and Syrian Ahmed Chalabis to go around. Reza Pahlavi, the son of the late shah, who was installed by the United States but had lost power as a result of the Islamic Revolution, was shopped around Washington as a prospective leader of Iran. And Farid Ghadry, a Syrian exile in Virginia who founded the Reform Party

*Weldon was defeated in the 2006 midterm elections.

of Syria, was the neocon favorite to rule Syria. Ghadry has an unusual résumé for a Syrian—he's been a member of the American Israel Public Affairs Committee, the right-wing pro-Israel lobbying group—and has endured so many comparisons to the disgraced leader of the INC that he once sent out a mass e-mail headlined, "I am not Ahmed Chalabi."

Nevertheless, according to a report in the *American Prospect,* Meyrav Wurmser introduced Ghadry to key administration figures, including the vice president's daughter Elizabeth Cheney, who, as principal deputy assistant secretary of state for Near Eastern affairs and coordinator for broader Middle East and North Africa initiatives, played a key role in the Bush administration's policy in the region.*[49]

The biggest blow of all to Bush came on August 29, 2005, when Hurricane Katrina devastated the gulf coast of Louisiana and Mississippi, killing more than 1,836 people and causing more than $81 billion in damage. It was not the storm itself, of course, but the monumental incompetence of the Bush administration and its inability to manage the disaster that devastated New Orleans. Under Michael Brown's aegis, the Federal Emergency Management Agency (FEMA) failed to heed warnings that the city's levees might be breached, failed to evacuate the city, and failed to bring housing and relief to the victims after the storm. Disengaged and ineffective, Bush, most memorably, told the director of FEMA, "Heckuva job, Brownie."†

With Katrina, whatever myths were left about Bush's presidency had been shattered. His approval ratings plummeted to 38 percent.[50] When New Orleans needed the National Guard, the National Guard was in Iraq. Only 34 percent of the public approved of Bush's handling of Iraq—roughly the same percentage who had approved of LBJ's handling of Vietnam in March of 1968.[51]

*According to the *Financial Times,* Elizabeth Cheney had supervised the State Department's Iran-Syria Operations Group, created in 2006 to plot a strategy to democratize those two "rogue" states. One of her responsibilities was to oversee a projected $85 million program to produce anti-Iran propaganda and support dissidents.

†Among other famous lines that came in the wake of the hurricane, one came from rapper Kanye West at a televised Hurricane Katrina relief concert: "George Bush doesn't care about black people."

The president's mother, Barbara Bush, did not help her son's case when she observed, "What I'm hearing which is sort of scary is they all want to stay in Texas. Everyone is so overwhelmed by the hospitality. And so many of the people in the [Houston Astrodome] here, you know, were underprivileged anyway, so this . . . this [she chuckles slightly] is working very well for them."

By this time, any chances that American forces could prevail in Iraq were gone. Less than a year after the marines' horrific siege, Fallujah had morphed into a police state patrolled by thousands of Iraqi and American troops who lived in its bombed-out buildings. But the Sunni insurgency there had somehow survived. In a twelve-day stretch in late summer, forty-eight Americans died.[52] They would not be the last. Bush's fate was sealed. His presidency was an irrevocable failure.

On October 28, Scooter Libby resigned as Dick Cheney's chief of staff after being indicted on five counts of perjury, obstruction of justice, and making false statements. Bush's approval ratings hit 36 percent, a new low. The war and its causes and manifestations now filled the media. In October, Democratic senators Jay Rockefeller and Harry Reid, ranking Democrats on the Senate Intelligence Committee, even staged a dramatic shutdown of the Senate and challenged Republican senator Pat Roberts to get to the bottom of the Niger forgeries.

But with the Republicans still in control of Congress they could not get very far. "The fact is that at any time the Senate Intelligence Committee pursued a line of questioning that brought us close to the White House, our efforts were thwarted," said Senator Rockefeller. The Republican-controlled Senate committee failed to produce a more extensive report.

But now even conservatives began to question the merits of "democratizing" the Middle East. "Everything the advocates of war said would happen hasn't happened," said the president of Americans for Tax Reform, Grover Norquist, an influential conservative who had backed the Iraq invasion.[53] "And all the things the critics said would happen have happened. [The president's neoconservative advisers] are effectively saying, 'Invade Iran. Then everyone will see how smart we are.' But after you've lost X number of times at the roulette wheel, do you double-down?"

In *Tyranny's Ally: America's Failure to Defeat Saddam Hussein,* David Wurmser, the low-profile but powerful policy maker who had become Cheney's Middle East adviser, argued that overthrowing Saddam was the key to toppling the mullahs in Iran. "Any serious display of American determination" will cause "our regional enemies to wilt," he wrote.[54] Razing Saddam's Ba'athism "will send terrifying shock waves into Teheran . . . and will promote pro-American coalitions in the region, unravel hostile coalitions." He concluded that "the Iraqi Shi'ites, if liberated from [Saddam's tyranny], can be expected to present a challenge to Iran's influence and revolution."[55]

But, in fact, exactly the opposite had happened. The strategic fiasco created by the Iraq War had actually emboldened Iran, given it power and influence within Iraq, and increased the danger posed by Iran to Israel.

"Nobody thought going into this war that these guys would screw it up so badly, that Iraq would be taken out of the balance of power, that it would implode, and that Iran would become dominant," said Martin Indyk, who served as U.S. ambassador to Israel under Clinton.[56]

"[Bush's wars] have put Israel in the worst strategic and operational situation she's been in since 1948," said Larry Wilkerson, who was Colin Powell's chief of staff in the State Department.[57] "If you take down Iraq, you eliminate Iran's number one enemy. And, oh, by the way, if you eliminate the Taliban, they might reasonably be assumed to be Iran's number two enemy."[58]

As a result, many Israelis believed that diplomacy was doomed and that Iran would have to be dealt with sooner or later. "Attacking Iraq when it had no WMDs may have been the wrong step," said Uzi Arad, the former Mossad intelligence chief. "But then to ignore Iran would compound the disaster. Israel will be left alone, and American interests will be affected catastrophically."[59]

Writing anonymously in the online magazine *Salon,* a senior State Department official saw the oncoming Bush agenda in stark terms—a key part of which meant eliminating Iran's nuclear facilities.[60]

Meanwhile, Iranian oil minister Bijan Namdar Zanganeh announced Iran's new policy was to give preference in its exports to China,[61] which had just signed a new oil-and-gas accord with Iran valued at $70 billion to $100 billion. Given its veto power on the U.N. Security Council, American observers feared that China's emerging alliance would interfere with U.S.-led attempts to put international pressure on Iran.[62]

Even though their Iraq policy was a fiasco, now the neocons' Iran initiative gathered steam. In August 2006, Peter Hoekstra released a House intelligence committee report titled "Recognizing Iran as a Strategic Threat: An Intelligence Challenge for the United States."[63] Written by Frederick Fleitz, former special assistant to John Bolton, the report asserted that the CIA lacked "the ability to acquire essential information necessary to make judgments" on Tehran's nuclear program.

The House report received widespread national publicity, but critics were quicker to point out its errors. Gary Sick, senior research scholar at the Middle East Institute of Columbia University's School of International and Public Affairs, and an Iran specialist with the NSC under Presidents

Ford, Reagan, and Carter, said the report overstates both the number and range of Iran's missiles and neglected to mention that the IAEA found no evidence of weapons production or activity. "Some people will recall that the IAEA inspectors, in their caution, were closer to the truth about Iraqi WMD than, say, the Vice President's office," Sick remarked.[64]

"This is like pre-war Iraq all over again," David Albright said in the *Washington Post.*[65] "You have an Iranian nuclear threat that is spun up, using bad information that's cherry-picked and a report that trashes the inspectors."

As the midterm congressional elections approached in November 2006, at last the Democrats had an opportunity to strike back. Their odds were enhanced by the fact that the Republican scandals were almost too numerous to list. Disgraced super-lobbyist Jack Abramoff had pleaded guilty to charges of tax fraud and conspiracy to defraud clients and bribe a public official. California congressman Duke Cunningham had been convicted of taking more than $2 million in bribes. Texas congressman Tom DeLay, once the powerful whip of the House, had been forced to drop out of the race because of Texas investigations into the financing of his political action committees. The aforementioned Republican congressman Mark Foley had sent suggestive e-mails and sexually explicit instant messages to teenage boys. Republican senator George Allen was caught on video using a racial slur—"macaca"—to describe an Indian-American, thus dooming him to defeat. The CIA leak investigation proceeded. The aftermath of Katrina remained a festering sore, with tens of thousands of people living in trailers, their fates uncertain, New Orleans still in ruins. A U.N. report on detainees at Guantánamo concluded that the rights of the prisoners were violated and in some cases U.S. treatment constituted torture. And just before the election, Pastor Ted Haggard, head of the National Association of Evangelicals and an outspoken foe of "homosexual activity," was exposed as having had a long relationship with a gay prostitute.

But most important of all, now that the public had turned against the war, everything that had once made Rove, Bush, and the Republicans so strong—their discipline, the insistence of their message that only they knew how to fight the war on terror—now made them seem weak, in denial, out of touch.

Even the *New York Times*'s foreign affairs columnist Thomas Friedman, a longtime supporter of the war who had patiently given the administration more than three years to get it right, had given up. "It is now obvious that we are not midwifing democracy in Iraq," Friedman wrote on August 4, 2006. "We are babysitting a civil war."[66] Even the most unyielding centrists had turned against Bush.

And when the votes were counted on November 7, it turned out Friedman had indeed been an accurate bellwether. According to an analysis by CBS News, fully 80 percent of those who disapproved of the war voted for the Democratic congressional candidate. For the first time in American history, not a single Republican won any House, Senate, or gubernatorial seat previously held by a Democrat.[67] In the end, the Democrats won the House of Representatives by 233 to 202 seats, and the Senate 51 to 49. The Democrats had won everything but the White House.

On November 8, the day after the election, President Bush announced the resignation of Donald Rumsfeld. Though Rumsfeld was the most prominent casualty of the rising antiwar sentiment, in the aftermath of the elections, another shoe was about to drop. Months earlier, a number of congressmen had succeed in putting together the Iraq Study Group (ISG), the blue ribbon commission chaired by former secretary of state James Baker and former congressman Lee Hamilton, to come up with a bipartisan report that offered a face-saving strategy to exit Iraq. Who better than Baker, after all, the Bush family's longtime friend and consigliere, to talk some sense into the president?

Baker's entry into the scene didn't just raise new questions about Bush's openness to pragmatic solutions. It also introduced an Oedipal element into the drama. Baker's and Bush's father, after all, were best friends. More than forty years earlier, when George W. was a sixteen-year-old student at Andover, Baker had given him a summer job as a messenger at Baker Botts, his Houston law firm. Now, along with Brent Scowcroft, Baker was leading a coterie of multilateralists and realists who found themselves aghast at the radical direction the younger Bush was taking American foreign policy, and desperate to reverse it.

Even though he had been let go from the administration, Scowcroft had not given up his lonely battle. In July 2006, after Israel's disastrous attack on Hezbollah in Lebanon, Scowcroft offered the administration more foreign policy advice on the opinion page of the *Washington Post,* arguing that the crisis in Lebanon provided a "historic opportunity" to achieve a comprehensive settlement of the Israeli-Palestinian conflict. Resolving that conflict, Scowcroft argued, was crucial to stabilizing the region—including Iraq. But, as usual, it fell on deaf ears.

Nevertheless, according to an article in *Salon* by Sidney Blumenthal, Scowcroft, with the assent of Baker and the elder Bush, sought and found support for this notion from the rulers of Egypt and Saudi Arabia.[68] Even Secretary of State Condoleezza Rice seemed receptive, so Scowcroft asked her to help open the president's mind to the forthcoming ISG report.

The president's relentless commitment to his failed vision in Iraq notwithstanding, there was reason to be optimistic that Baker might succeed. Key neoconservative architects of the war in Iraq—Paul Wolfowitz, Douglas Feith, and Richard Perle—were no longer part of the Bush foreign policy team. The State Department, all but inoperative during the run-up to the Iraq War, was showing new signs of life. Rumsfeld was out. And Bush owed Baker because he had come to Bush's rescue during the controversial Florida recount battle in 2000. "Here you have Baker coming back trying to pull the president's chestnuts out of the fire," said a former State Department official. "Not only did he help Bush out in Florida, but now he is doing the Baker Hamilton Commission. He and Scowcroft were talking relentlessly during the policy formulation of the Iraq Study Group report. Baker was keeping the president informed the whole time. He is trying to throw him a lifeline and give him an exit."

Scowcroft strategized with Baker throughout the entire period, and talked to Condoleezza Rice to try to get her to intervene with Bush. "There was an assumption, given the debt of gratitude Bush owed Baker, that at least he would get a real hearing," said the official.

On December 6, the Iraq Study Group finally released its report, "The Way Forward—A New Approach." Describing the situation in Iraq as "grave and deteriorating," the ISG report did not shy away from pointing out that the new Iraqi army, the police force, and even Prime Minister Nuri Kamal al-Maliki often showed greater loyalty to their ethnic identities than to the ideal of a nonsectarian, democratic Iraq. The report concluded that sending more American soldiers to Iraq would not resolve what were fundamentally political problems.

The subtext was clear: America's policies in Iraq had failed. It was time for the administration to cut its losses. A Gallup poll from December 12 showed that, among people who had an opinion on the subject, five out of six supported implementing the report's recommendations.

The only American whose opinion mattered, however, was not

impressed. President Bush, *Salon* reported, dismissed the ISG study with an obscenity. "That did not sit well with Scowcroft *at all,*" said an administration official. To Scowcroft's enormous frustration, Condoleezza Rice had never carried forth his entreaty to the president.

Just eight days later, on December 14, Bush found a study that was more to his liking. Not surprisingly, it came from the American Enterprise Institute, the intellectual stronghold of neoconservatism. The author, Frederick Kagan, a resident scholar at the AEI, is the son of Donald Kagan and the brother of Robert Kagan, who signed PNAC's famous 1998 letter to President Bill Clinton urging him to overthrow Saddam Hussein. According to Kagan, the project began in late September or early October at the instigation of his boss, Danielle Pletka, vice president for foreign and defense policy studies at AEI. She decided "it would be helpful to do a realistic evaluation of what would be required to secure Baghdad," Kagan said.[69]

The project culminated in a four-day planning exercise in early December, Kagan said, that just happened to coincide with the release of the Iraq Study Group report. But he rejected the notion that his study had been initiated by the White House as an alternative to the bipartisan assessment. "I'm aware of some of the rumors," Kagan said. "This was not designed to be an anti-ISG report. . . . Any conspiracy theories beyond that are nonsense.

"There was no contact with the Bush administration. We put this together on our own. I did not have any contact with the vice president's office prior to . . . well, I don't want to say that. I have had periodic contact with the vice president's office, but I can't tell you the dates."[70]

Sharply at odds with the consensus forged by the top brass in Iraq, Kagan's study suggested that with a surge of new troops America could finally succeed. Focusing on holding certain areas of Baghdad, it concluded that the deployment of twenty thousand additional troops would be enough to pacify significant sections of the city. Even the title of Kagan's report must have been more appealing to Bush: "Choosing Victory: A Plan for Success in Iraq." The escalation was on.

The year ended with Saddam Hussein's grotesque hanging, marked by shouting between Saddam and his executioners and the secret recording of the event with cell phones. Ultimately, the episode was ugly, pathetic, and disturbing. The grotesque manner of Saddam's death seemed like a Shi'ite revenge fantasy against its Sunni dictator and only fueled the Iraqi civil war.

* * *

By the time Bush made his January 10, 2007, speech on the war in Iraq, millions of listeners feared that American policy in the Middle East was about to enter a new phase.

It wasn't just that Bush was doubling down on an extravagantly costly bet by sending a "surge" of 21,500 more American troops to Iraq; there were also indications that he was upping the ante by an order of magnitude. The most conspicuous clue was a four-letter word that Bush uttered six times in the course of his speech: Iran.

In a clear reference to the Islamic Republic and its sometime ally Syria, Bush vowed to "seek out and destroy the networks providing advanced weaponry and training to our enemies." At about the same time his speech was taking place, U.S. troops stormed an Iranian liaison office in Erbil, a Kurdish-controlled city in northern Iraq, and arrested and detained five Iranians working there.

By this time, many Americans had at least an inkling of how neoconservatives had taken control of American foreign policy. But few were aware that the same tactics were in play with Iran, that once again neocon ideologues were flogging questionable intelligence about WMDs, and dubious Middle East exile groups were making the rounds in Washington—this time urging regime change in Syria and Iran.

At the same time, a series of moves by the military lent credence to widespread reports that the United States may have been secretly preparing for a massive air attack against Iran. (No one suggested a ground invasion.) First came the deployment order of U.S. Navy ships to the Persian Gulf. Then came high-level personnel shifts signaling a new focus on naval and air operations rather than the ground combat that predominates in Iraq. In his January 10 speech, Bush announced that he was sending Patriot missiles to the Middle East to defend U.S. allies—presumably from Iran. And he pointedly asserted that Iran was "providing material support for attacks on American troops," a charge that could easily evolve into a casus belli.

"It is absolutely parallel," said Philip Giraldi, a former CIA counterterrorism specialist.[71] "They're using the same dance steps—demonize the bad guys, the pretext of diplomacy, keep out of negotiations, use proxies. It is Iraq redux."

As the neocons argued, the Iraqi debacle was not the product of their failed policies. Rather, it was the result of America's failure to think big. "It's a mess, isn't it?" said Wurmser, who now serves as director of the

Center for Middle East Policy at the Hudson Institute. "My argument has always been that this war is senseless if you don't give it a regional context."[72]

One neocon after another made similar pleas: Iraq was the beginning, not the end. Former Israeli prime minister Benjamin Netanyahu himself had gone as far as to frame the issue in terms of the Holocaust. "Iran is Germany, and it's 1938," he told CNN. "Except that this Nazi regime that is in Iran . . . wants to dominate the world, annihilate the Jews, but also annihilate America."

Even before Bush's January 10 speech, many inside the military had concluded that the decision to bomb Iran had already been made. "Bush's 'redline' for going to war is Iran having the knowledge to produce nuclear weapons, which is probably what they already have now," said Sam Gardiner, a retired air-force colonel who specializes in staging war games on the Middle East. "The president first said [that was his redline] in December 2005, and he has repeated it four times since then."[73]

According to Gardiner, the most telling sign that a decision to bomb Iran had been made was the deployment order of minesweepers to the Persian Gulf, presumably to counter any attempt by Iran to blockade the Strait of Hormuz. "These have to be towed to the Gulf," said Gardiner. "They are really small ships, the size of cabin cruisers, made of fiberglass and wood. And towing them to the Gulf can take three to four weeks."[74]

Another serious development was the growing role of the U.S. Strategic Command (StratCom), which oversees nuclear weapons, missile defense, and protection against weapons of mass destruction. Bush directed StratCom to draw up plans for a massive strike against Iran, at a time when CentCom had its hands full overseeing operations in Iraq and Afghanistan. "Shifting to StratCom indicates that they are talking about a really punishing air force and naval air attack [on Iran]," said Colonel Patrick Lang.[75]

Horrifying as the Iraq War has been, a conflict with Iran would be likely to have consequences that are even more dire. It is widely believed that Iran might respond to an attack by blockading the Strait of Hormuz, a twenty-mile-wide narrows in the eastern part of the Persian Gulf through which about 40 percent of the world's oil exports are transported. Oil analysts say a blockade could propel the price of oil through the roof, sending the world economy into a tailspin. At a time when the era of "easy oil" was coming to an end, when China's and India's energy consumption was soaring, there was now the possibility of vast international oil wars. Iran could act on its fierce rhetoric against Israel. The Saudis, Russia, China, and other countries might all join the fray.

"I think of war with Iran as ending America's present role in the world," former national security adviser Zbigniew Brzezinski told the *Washington Post.* "Iraq may have been a preview of that, we'll get dragged down for 20 or 30 years. The world will condemn us. We will lose our position in the world."[76]

In the spring and summer of 2007, the ongoing fiascos associated with the Bush White House and its allies were simply too numerous for the media, Congress, or the American people to process or fully comprehend. There had been the scandals of Jack Abramoff, Tom DeLay, and Mark Foley. As the situation in Iraq deteriorated, even longtime loyalists, such as speechwriter Matthew Dowd, became fed up, defected, and went public with the story of how he had fallen out of love with the president. "If the American public says they're done with something, our leaders have to understand what they want," Dowd said. "They're saying, 'Get out of Iraq.' "[77]

In May, Paul Wolfowitz was forced to resign as head of the World Bank, thanks to an overly generous raise he gave to Shaha Ali Riza, his onetime girlfriend. In the Valerie Plame Wilson–CIA leak case, Scooter Libby was convicted of perjury and obstruction of justice, but his sentence was commuted by Bush in June. The army's cover-up about the death of former NFL star Pat Tillman came unraveled—with evidence emerging in July suggesting he may even have been murdered.[78] For much of the summer, Congress investigated the politicization of the Justice Department—and by extension, the federal judiciary—because of the firing of nine U.S. attorneys for political reasons. In the ensuing investigation, there was one subpoena after another. White House officials defied Congress by refusing to testify and deleting millions of e-mails. Attorney General Alberto Gonzales repeatedly stonewalled Congress, apparently perjured himself—and, ultimately, resigned under fire.

In late August, *Roll Call* broke the story that Republican senator Larry Craig had been arrested for lewd conduct in a Minneapolis airport and had pleaded guilty to a lesser charge of disorderly conduct, thereby becoming yet another politician whose life was at odds with the "family values" he piously promoted.* Even Karl Rove jumped ship, leaving the president almost all alone. Bush's most devoted retainers were gone. His approval ratings hovered in the low thirties and sometimes dipped into the twenties. The presidency was in free fall.

In September 2007, Lieutenant General David Petraeus, the commander of coalition forces in Iraq, went before Congress and argued that the surge in troop strength was working as a strategy and should be given more time. But by a number of measures, it had been a failure. Although the surge brought down violence in some sectors of Baghdad, throughout the country the death toll from sectarian violence was averaging sixty-two fatalities per day, nearly double the toll of a year earlier.[79] On one day alone, August 16, 2007, more than four hundred people had been killed in the worst string of suicide bombings of the war. And the Iraqi government had failed to meet most of the benchmarks and timelines it had promised to achieve in terms of providing security.

As the White House evaluated its options, a *New York Times*/CBS News poll showed that the number of Americans who trusted Bush to bring the Iraq War to a successful conclusion had sunk to an astonishingly low figure—5 percent.[80] Leading historians such as Princeton University professor Sean Wilentz asserted that Bush was arguably the worst president in history.[81] A website called "Backwards Bush" sold keychains with "Bush Countdown Clocks," that counted the amount of time he had left in office.[82] Broadly speaking, across the country, there was a sense of an utterly failed presidency. America waited impatiently for his term to end.

Yet through it all, Bush remained unfazed—besieged, but somehow utterly certain, unquestioning, and oddly at ease with himself, even when confronted with overwhelming evidence that he had created the greatest foreign policy disaster in American history. Ultimately, he had a fatalistic approach toward his legacy. "Look, everybody is trying to write the history of this administration even before it's over," Bush said. "I'm reading about George Washington still. My attitude is if they are

*In 1999, speaking on NBC's *Meet the Press,* Senator Craig lashed out at President Bill Clinton for his behavior in the Monica Lewinsky scandal: "The American people already know that Bill Clinton is a bad boy—a naughty boy. I'm going to speak out for the citizens of my state, who in the majority think that Bill Clinton is probably even a nasty, bad, naughty boy."

still analyzing number one, forty-three ought not to worry about it and just do what he thinks is right."[83]

Bush's real concerns were best reflected in a series of exclusive interviews he gave to Robert Draper in *Dead Certain: The Presidency of George W. Bush,* a book that revealed the president's true passion by discussing his regimen of jogging and biking on no fewer than twenty-seven pages. What emerged was a portrait of a president, at a time in which the country was truly in crisis, who had developed an astonishing, single-minded obsession with improving his biking times and creating the perfect "single-track" mountain-biking trail on his ranch in Crawford, trails that challenged his stamina and reflexes. After his interviews with Bush, Draper channeled the president: "Loving how he could sustain a heart rate of 140 to 175 for ninety minutes—something he could never do when he was running—while regularly checking his calorie burn, relishing the metrics: 1,000 calories, 1,200, sometimes 1,500. . . . Loving the pain, seeking it out. . . . Absolutely geared to the *mentality* of the sport: grinding, pushing, meeting resistance head-on. Quitting only when the hurt consumed him. And loving the feeling that, even in this solitary pastime, he was ahead of the curve, among the first Baby Boomers with failing joints to give up the jogging track for the cycle. . . . Even now, *he was setting an example—he was leading.*"[84]

As if that was not an unusual obsession for a man whose presidency was in crisis, when it came to the most crucial policies of his administration, choices that cost thousands of lives, such as the widely criticized decision to disband the Iraqi army shortly after the 2003 invasion, Bush seemed curiously confused and disengaged, asserting that he had wanted to maintain the Iraqi army intact and couldn't remember why it had been disbanded. "The policy was to keep the army intact; didn't happen," he told Draper. When Draper asked Bush why his chief administrator for Iraq, L. Paul Bremer III, issued an order in May 2003 to disband the 400,000-man army, Bush replied, "Yeah, I can't remember; I'm sure I said, 'This is the policy, what happened?' "[85]

In fact, on May 23, 2003, just after Bremer disbanded the Iraqi army, Bush had written Bremer a letter saying, "You have my full support and confidence."[86]

As for what the future might hold, after his administration ended, Bush intended to create the Freedom Institute at Southern Methodist University in Dallas, where young leaders would be given stipends to do research, lecture, and write. In addition, in *Dead Certain,* Bush said he harbored more mundane concerns such as "replenishing the ol' coffers." He was reasonably sure he could make "ridiculous" money by lecturing. "I don't know what my dad gets. But it's more than fifty, seventy-five. . . . Clinton's making a lot of money."[87]

Even if Bush preferred to think that judgment of his administration would be rendered only many years in the future, with more than fourteen months left in his last term, his legacy was largely sealed. By embodying a vision of American exceptionalism shared by the neocons and the Christian Right, and by implementing policies aligned with that vision, Bush, with the help of so many of his father's nemeses—Rumsfeld, the neocons, the Christian Right—had put America on an extraordinarily radical new course. As his father's tearful speech in Tallahassee in 2006 had suggested, George W. Bush had indeed destroyed his father's legacy. The family's political future appeared to be effectively dead as well. In early 2007, in his final days* as governor of Florida, Jeb Bush, once hailed as a prospect to become the third Bush to occupy the White House, explained it simply but dramatically to Spanish-speaking reporters. "*Yo no tengo futuro,*" he said. "I have no future."†

But what George W. Bush had done to his family paled compared to the impact of his policies on the rest of the world. By the fall of 2007, more than 3,700 Americans had died in Iraq, and tens of thousands had been seriously wounded. Tens of thousands of Iraqis had been killed, and four million Iraqis had been displaced. In dollars, the cost to American taxpayers was in the hundreds of billions, at least—and by some estimates, would stretch into the trillions when the future cost of caring for wounded U.S. veterans was tallied.

Bush's war had fueled, not extinguished, the flames of terror, and the repercussions from the Sunni-Shi'ite conflict reverberated throughout the region. The White House announced plans to sell $20 billion of weapons to Saudi Arabia, in large part to pacify its Sunni ally who felt threatened by a newly emboldened Iran. Moreover, the war in Iraq had overshadowed and drawn attention and resources from the conflict in Afghanistan. Initially, American troops had had success against Al-Qaeda and the Taliban in Afghanistan, but both those forces had

*Due to term limits under state law, Jeb Bush was unable to seek a third term.

†A spokesman for the governor later said he had made the comment in jest. But like many jokes, this one had at least a grain of truth to it and even allies conceded that his prospects had dimmed because of "Bush fatigue."

regrouped and begun to attack American forces again.* And the neocons had been successful in thwarting the Israeli-Palestinian peace process by creating "a clean break' from the land-for-peace path carved out in Oslo.

In addition, Bush and Cheney had gone even further than Richard Nixon in pursuit of the imperial presidency, in the process inflicting a brutal assault on America as a constitutional democracy. The intricate system of checks and balances that made America's constitutional democracy so exceptional was in tatters. The home of the free had given birth to the gulag of Guantánamo, the torture of Abu Ghraib. America had abrogated the Geneva Convention, secretly rendering detainees to countries famous for their techniques in torture. And it had begun spying on its own citizens without court authorization.

Perhaps sensing that the die had been irrevocably cast, Americans shifted their attention to the 2008 presidential race. Among the Republican contenders, at least, the neocons were far from dead. A number of them expressed their "neoculpas" in *Vanity Fair,* effectively blaming the Bush administration for incompetently implementing their vision, which they hoped to rectify with the next administration. Norman Podhoretz, one of the godfathers of neoconservatism and author of a May 2007 *Wall Street Journal* piece that called for war with Iran, joined Rudy Giuliani's team as senior foreign policy adviser. Other GOP hopefuls—Mitt Romney, Fred Thompson, and John McCain—were eyed by neocon policy makers as well.

As for the Christian Right, it was still an extraordinarily potent political force but, relative to the 2004 campaign, the role it would play was far less clear. Jerry Falwell had died. Thanks to evangelical leaders such as Jim Wallis and publications such as *Sojourners,* a new religious left had emerged, focusing on peace and justice and reviving the evangelical traditions of the social gospel. Likewise, led by Richard Cizik, the vice president for governmental affairs of the National Association of Evangelicals (NAE), a new breed of evangelical environmentalist, concerned with global warming, called for "creation care," and created new divisions in the Christian Right. Finally, as the presidential primaries approached, the leading candidates for the Republican nomination carefully sculpted their views to win the support of the Christian Right, but none were as natural a fit as Bush had been. Ulti-

*Casualty levels rose dramatically in Afghanistan, and by the end of 2006, it was statistically as dangerous for U.S. soldiers to serve in Afghanistan as it was in Iraq.

mately, Bush's and Karl Rove's attempt to create a permanent Republican majority by wedding it to religious conservatives may have had the opposite of its intended effect.

With more than a year remaining in Bush's term, there was still reason to fear that the worst was yet to come. A large flotilla of U.S. ships had entered the Persian Gulf, just two weeks after Cheney vowed to do what was necessary to prevent Iran from building nuclear weapons and "dominating the region."[88] The United States now had enough forces in place to mount a massive air assault on Iran.

Bush's rhetoric with regard to Iran became increasingly bellicose—evoking memories of the "smoking-gun–mushroom-cloud" campaign against Saddam five years earlier. Referring to Iran as "the world's leading state sponsor of terrorism" in a late August speech, Bush said the Islamic Republic "threatens to put a region already known for instability and violence under the shadow of a nuclear holocaust," and vowed "to confront this danger before it's too late."[89] The administration considered designating Iran's Islamic Revolutionary Guard as a "specially designated global terrorist" organization, an act that arguably could allow the administration to bomb Iran without seeking congressional approval. McClatchy* newspapers' Warren Strobel, John Walcott, and Nancy A. Youssef—part of the same team of Knight Ridder journalists that had seen through many of the phony intelligence reports disseminated by the neocons—reported that Cheney had proposed launching air strikes "at suspected training camps in Iran run by the Quds force, a special unit of the Iranian Revolutionary Guard Corps."[90]

On September 12, 2007, Fox News joined in, reporting that, "according to a well-placed Bush administration source, 'everyone in town' is now participating in a broad discussion about the costs and benefits of military action against Iran, with the likely time frame for any such course of action being over the next eight to ten months, after the presidential primaries have probably been decided, but well before the November 2008 elections."

Fox added that "consideration is being given as to how long it would take to degrade Iranian air defenses [so] U.S. fighter jets could then begin a systematic attack on Iran's known nuclear targets. . . . The Bush administration 'has just about had it with Iran,' said one foreign diplo-

*In 2006, McClatchy bought Knight Ridder, the nation's second-largest chain of daily newspapers, and in the process took over its Washington bureau.

mat. . . . 'There are a number of people in the administration who do not want their legacy to be leaving behind an Iran that is nuclear armed.' "[90]

On September 23, 2007, the *Sunday Times* of London reported that Israel had secretly bombed a Syrian military installation on September 6, after seizing North Korean nuclear material from the base. The article said that President Bush approved the attack "after Washington was shown evidence the material was nuclear related."[92] Both North Korea and Syria denied the charges that an exchange of nuclear materials had taken place, however, and what really happened was shrouded in mystery. Some observers speculated that the incident was yet another disinformation operation, and the the Israeli bombing raid was a dry run for a future attack on Iran.

Whatever the truth, any number of such incidents involving the United States or Israel against Syria or Iran could easily ignite a larger conflagration. The Bush White House had built the fire. Whether it would light the match remained to be seen.

But even if Bush and Cheney did not strike Iran, historians had already amassed enough information with which to assess the damages wrought by the Bush administration. Driven by delusional idealism and religious zeal, Bush, after all, had already made one catastrophic blunder, the true historic dimensions of which have yet to emerge. To fully appreciate its consequences, one cannot overlook the fact that the Iraq War took place in the twilight of the hydrocarbon era, during China's extraordinary ascendancy. Far from safeguarding America as promised, the Iraq War had jeopardized the country's security and with it, potentially, America's vital access to the Middle East oil so crucial to fueling the most powerful economic engine in history.

Who knows how much stronger America's geostrategic position might be if the Bush administration had not squandered the incalculable goodwill the United States had after 9/11? Who knows how much better off America might be, if instead of wasting its time and money on Iraq, it had invested those same resources in education or the health care system that was in crisis, or in developing alternative energy sources and a strategy to free the country from its dependence on Middle East oil? Such losses are truly impossible to calculate.

ACKNOWLEDGMENTS

This book would not have been possible without the help of many people. Once again, I have been privileged to be edited by Colin Harrison at Scribner. His editorial judgment, once again, has been superb. It's a pleasure to work with him, and this time he went above and beyond the call of duty. I am also grateful to Susan Moldow and Nan Graham, who oversaw a terrific team at Scribner that treated the book with the highest level of professionalism. They include Roz Lippel, John Fulbrook, Karen Thompson, Katy Sprinkel, Katie Rizzo, Erich Hobbing, Jane Herman, and Dan Cabrera. My thanks also go to Elisa Rivlin for her legal review, and to Molly Dorozenski, Steve Oppenheim, Brian Belfiglio, and Elizabeth Hayes for their work on the book's publicity.

My agent, Sloan Harris of International Creative Management, has been enormously supportive throughout and has always been there with valuable advice. Kate Jones of ICM in London has my gratitude as well. Cynthia Carris, my photo editor, again performed with grace and professionalism under deadline pressure. My thanks also go to James Hamilton for the author's photo and to Michelle Risley for editorial assistance. Alistair Wandesforde-Smith at AW Systems provided excellent computer support.

Parts of the book have appeared in *Vanity Fair* in three articles between 2005 and 2007—"American Rapture"; "The War They Wanted, The Lies They Needed"; and "From the Wonderful Folks Who Brought You Iraq." I'm deeply indebted to Graydon Carter and Michael Hogan for their editing and to Julian Sancton for his assistance.

The Center on Law and Security at New York University School of Law has proven to be a wonderful resource, and I'd like to thank Karen Greenberg, Stephen Holmes, Colleen Larkin, Nicole Bruno, Daniel Freifeld, Jeff Grossman, Tara McKelvey, and David Tucker for their assistance and collegiality and for providing a lively intellectual community with Larry Wright, Paul Cruickshank, Peter Bergen, Michael Sheehan, Scott Horton, Nir Rosen, and many others.

Among the many people who were either interview subjects or gave me assistance, I'd like to thank Uzi Arad, Karen Armstrong, Sondra Baras, Milt Bearden, John Berger, Chip Berlet, Arthur Blessitt, Max Blumenthal, Carlo Bonini, Frank Brodhead, Vince Cannistraro, Frederick Clarkson, Yechiel Eckstein, Daniel Ellsberg, Benny Elon, the late Jerry Falwell, Yitzhak Fhantich, Sam Gardiner, Philip Giraldi, Melvin Goodman, Gershom Gorenberg, Mickey Herskowitz, Grant Hopkins, Martin Indyk, Larry C. Johnson, Yechiel Kadishai, Frederick Kagan, Fred Kaplan, Jack Keane, Karen Kwiatkowski, Tim and Beverley LaHaye, Pat Lang, Michael Ledeen, Avishai Margalit, Ann McFeatters, Ray McGovern, Marji Mendelson, Guido Moltedo, Grover Norquist, Richard Pipes, Skipp Porteous, Carlo Rosella, Jim Sale, Matt Simmons, Gershom Solomon, Bill Spindle, Bob Strauss, Pete Teeley, Yaroslav Trofimov, Sarah Vail, Donald Wagner, Larry Wilkerson, Joe Wilson, and Meyrav Wurmser. In Jerusalem, I'm indebted to the staff of Mishkenot Sha'ananim. I would also like to thank Steve Clemons of the New America Foundation and Zbigniew Brzezinski.

In addition, I am particularly indebted to the many people in and out of government who spoke to me on the condition of anonymity.

Helpful as such sources have been, this book also relies extensively on declassified government documents, congressional investigations, and news accounts from thousands of newspapers and journals from all over the world. This book is deeply critical of the American media's complicity in spreading disinformation about Iraq's alleged WMDs and other facets of the Bush administration's policies. At the same time, this book, in large measure, is an attempt to craft a counternarrative of the era, and it would not be possible to do so without relying on the dogged persistence of many fine reporters who decided not to take everything the administration said at face value. I am indebted to them and they are credited extensively in the book's notes.

It would have been impossible to research this book without the Internet and I am especially grateful to the people and institutions who have built the Internet research tools that enabled me to search through such vast amounts of material from all over the world so quickly. Wherever possible, I have cited relevant websites in the notes. The reader should be advised, however, that Internet links are not eternal and some web addresses may be out of date.

Specifically, my thanks go to Gary Sick and Columbia University's

Gulf2000, an Internet group that afforded me access to hundreds of scholars, diplomats, and policy makers all over the world who specialize in the Middle East. Gulf2000's vast Internet archives of clippings on the Middle East were of great value and the thousands of e-mails they sent out enabled me to be privy to a dialog with hundreds of specialists in the field. I am particularly indebted to the work of G2K members Jim Lobe, Robert Dreyfuss, and Gareth Porter in covering the neoconservatives.

Because I made a practice of citing original sources, a number of extraordinarily useful resources do not appear in my notes nearly as often as they should. Among them, I'm particularly grateful to the Center for Cooperative Research, *Mother Jones*'s "Lie by Lie" timeline, the Downing Street Memo database, and the excellent websites produced by PBS *Frontline* and *Bill Moyers Journal.* Their timelines about the war in Iraq and other issues often helped me find exactly what I was looking for.

Researching and writing a book like this is by its nature an obsessive undertaking. As a result, I am particularly grateful to my friends and family members who were either in my corner—or tolerated my absences.

I'm especially grateful to Sidney Blumenthal, whose friendship and insights have been invaluable, as have his articles in *Salon* and *The Guardian.* Many other friends and colleagues helped either by contributing in one way or another to the book itself or through much-needed moral support. They include John Anderson, Len Belzer and Emily Squires, Max Blumenthal, Peter Carey, Joe Conason, Susan Ennis, Leon Falk, Will Fulton, Rob Kaufelt and Nina Planck, Martin Kilian, Heidi Larson, Don Leavitt, Robin and Susan Madden, Craig McCord and Staci Strauss, Celia and Henry McGee, Maura Moynihan, Bob Parry, and Cody Shearer. Susan Letteer's support was essential. Pazit Ravina's assistance in Israel was vital to this book. My thanks to John Strahinich for reading the manuscript in galleys and for his acute critique. And finally, my gratitude goes to my family—my mother, Barbara; my father, Roger; Marlise and Romy-Michelle; Chris, Shanti, Thomas, and Marley; and Jimmy, Marie-Claude, Adam, and Matthew.

NOTES

Chapter One: Oedipus Tex

1. Hugh Sidey, "A Former President's Mad Dash to 80," *Time,* June 7, 2004, p. 49.
2. Michael Graczyk, "Ex-president Makes Birthday Skydive," *USA Today,* June 13, 2004, http://www.usatoday.com/news/nation/2004-06-13-bush-jump_x.htm.
3. "Iraq Prisoner Abuse Photos," February 23, 2005, http://news.bbc.co.uk/2/hi/in_pictures/4185719.stm.
4. Author interview with a friend of the Bush family, May 2007.
5. George H.W. Bush and Brent Scowcroft, *A World Transformed,* p. 489.
6. "President Outlines Path for Lasting Prosperity in Wednesday Speech," April 21, 2004, http://www.whitehouse.gov/news/releases/2004/04/20040421-5.html.
7. Steve Campbell, "The Masters," *Houston Chronicle,* April 11, 2004, p. 8.
8. Matthew Cooper, "Cue Parachutes for Bush 41," *Time,* April 12, 2004, p. 21.
9. *Anderson Cooper 360,* CNN, aired April 21, 2004, http://transcripts.cnn.com/TRANSCRIPTS/0404/21/acd.00.html.
10. Gus Tyler, "President Bush Failed to Heed His Father's Warnings," *Forward* 107, no. 31 (April 16, 2004): 5. And William Hamilton, "Behind Diplomacy, Military Plan Set in Motion," *Washington Post,* April 18, 2004, p. A1.
11. Author interview with Bob Strauss, February 2007.
12. William Kristol and David Brooks, "What Ails Conservatism," *Wall Street Journal,* September 15, 1997, referred to by Claes Ryn in "The Ideology of American Empire," in *Neo-conned! Again,* edited by D. Liam O'Huallachain and J. Forrest Sharpe (Light in the Darkness Publications, 2005), p. 67.
13. Interview with Bob Strauss.
14. Brent Scowcroft, "Don't Attack Saddam," OpinionJournal, WSJ.com, August 15, 2002, http://www.opinionjournal.com/editorial/feature.html?id=110002133.
15. CBS *Face the Nation,* August 5, 2002, http://www.cbsnews.com/stories/2002/08/05/ftn/main517523.shtml.
16. *CBS Morning News,* August 26, 2002.
17. Jeffrey Goldberg, "Breaking Ranks: What Turned Brent Scowcroft Against the Bush Administration, *The New Yorker,* October 31, 2005, http://www.newyorker.com/archive/2005/10/31/051031fa_fact2?printable=true.
18. Ibid.
19. Ibid.
20. Peter Schweizer and Rochelle Schweizer, *The Bushes,* p.535.
21. Maureen Dowd, "Family Feud? Scowcroft's Opposition to Iraq War Plan Hints at a Signal from Elder Bush," *Pittsburgh Post-Gazette,* August 19, 2002, p. A13.
22. Bob Woodward, *Plan of Attack* (New York: Simon & Schuster, 2004), pp. 420–21.
23. Interview with Bob Strauss.

24. Sidney Blumenthal, "Shuttle Without Diplomacy," *Salon,* http://www.salon.com/opinion/blumenthal/2007/01/10/condi_rice/.
25. "How Did Condoleeza Rice Became the Most Powerful Woman in the World," January 16, 2005, http://www.guardian.co.uk/usa/story/0,12271,1391579,00 .html.
26. Blumenthal, "Shuttle Without Diplomacy."
27. Keynote address by Philip Zelikow, counselor to the Department of State; Howard Berkowitz, president of the Washington Institute for Near Easy Policy; Robert Satloff, executive director of the Washington Institute for Near East Policy. Weinberg Founders Conference 2006, Federal News Service, State Department Briefing, September 15, 2006. And Eli Lake, "Bush, Rice to Revive Mideast Peace Process," *New York Sun,* September 18, 2006, p. 5.
28. David S. Cloud and Eric Schmitt, "More Retired Generals Call for Rumsfeld's Resignation," *New York Times,* April 14, 2006, http://select.nytimes.com/search/restricted/article?res=F00917FD3D5B0C778DDDAD0894DE404482.
29. Blumenthal, "Shuttle Without Diplomacy."
30. Herb Keinon, "Rice: No Link Between Palestinian Track and Iran," *Jerusalem Post,* September 19, 2006, p. 1.
31. Eli Lake, "Rice: Iran Sanctions Will Not Be Linked to Israel Peace Talks," *New York Sun,* September 19, 2006, p. 7.
32. Ibid.
33. Blumenthal, "Shuttle Without Diplomacy."
34. "The Iraq Study Group Releases Its Report," *The NewsHour with Jim Lehrer,* PBS, December 6, 2006.
35. "George Bush Holds a Media Availability Following a Meeting with the Iraq Study Group," *Congressional Quarterly,* December 6, 2006.
36. David Montgomery, "Footnote to History: Rituals of Delivering the Iraq Report," *Washington Post,* December 7, 2006, p. C1.
37. "Iraq Study Group Releases Report," CNN Newsroom, Transcript #120603CN.V11, December 6, 2006.
38. Blumenthal, "Shuttle Without Diplomacy."
39. http://www.youtube.com/watch?v=s1ceoZHTv1w.
40. Peggy Noonan, "A Father's Tears," *Wall Street Journal,* OpinionJournal, WSJ.com, http://www.opinionjournal.com/columnists/pnoonan/?id=110009355.
41. Ibid.
42. Ibid.
43. Ibid.
44. "Combat Diary," *ABC Nightline,* December 2, 2004.
45. Mark Gregory, "Baghdad's 'Missing' Billions," BBC World Service, http://news.bbc.co.uk/1/hi/business/6129612.stm
46. "Secretary Rumsfeld Town Hall Meeting At Aviano Air Base," February 7, 2003, http://www.defenselink.mil/transcripts/transcript.aspx?transcriptid=1900.
47. Paul Blustein, "Wolfowitz Strives to Quell Criticism," *Washington Post,* March 21, 2005, p. A1, http://www.washingtonpost.com/wp-dyn/articles/A52375-2005Mar20.html.
48. Charles M. Young, "The $2 Trillion War," Rollingstone.com, December 15, 2006, http://www.rollingstone.com/politics/story/12855294/national_affairs_the_2_trillion_dollar_war/print.
49. James Glanz, "Billions in Oil Missing in Iraq, U.S. Study Says," *New York Times,* May 12, 2007.
50. "Oil Prices Medium Term," http://en.wikipedia.org/wiki/Image:Oil_Prices_Medium_Term.png.
51. Woodward, *Plan of Attack,* pp. 420–21.

Chapter Two: Redeemer Nation

1. Gayle White, "LaHaye Makes Mark on America," *Washington Times,* July 26, 2001.
2. The author was present with Gary Frazier and Tim LaHaye at Megiddo and videotaped the events, June 2005. All quotes come from that video.
3. 1 Thessalonians 4:16–18.
4. Revelation 19:15.
5. Revelation 19:20.
6. Revelation 14.
7. Ibid.
8. Videotape at Megiddo.
9. Jane Lampman, "The End of the World," *Christian Science Monitor,* February 18, 2004, http://www.csmonitor.com/2004/0218/p11s01-lire.html.
10. Nancy Gibbs, "Apocalypse Now," *Time,* http://www.time.com/time/covers/1101020710/story.html.
11. "Survey Explores Who Qualifies as an Evangelical," January 18, 2007, http://www.barna.org/FlexPage.aspx?Page=BarnaUpdateNarrowPreview&BarnaUpdateID=263.
12. Ibid.
13. Daniel Wojcik, "Faith, Fatalism, and Apocalypse in America," *Social Science,* 1999, p. 21.
14. Thomas Ice, *Lovers of Zion: A History of Christian Zionism,* http://www.raptureready.com/ice/featured/AHistoryOfChristianZionism.html.
15. Kelly Ingram, "Christian Zionism," The Link 16, no. 4 (November 1983), http://ameu.org/page.asp?iid=74&aid=116&pg=2.
16. There is some debate as to whether Winthrop delivered his sermon on board the *Arbella* or before its departure.
17. Though sometimes synonymous with Jerusalem, in general *Zion* refers to the land of Israel, especially Jerusalem and, more specifically, Solomon's Temple. It derives from Mount Zion, the name of a mountain near Jerusalem.
18. John Winthrop, "A Modell of Christian Charity," also known as "A City Upon a Hill," http://en.wikisource.org/wiki/City_upon_a_Hill.
19. Cotton Mather, *Nehemiah Americanus,* Chapter Four, http://xroads.virginia.edu/~DRBR/cotton1.html.
20. Donald Wagner, "Evangelicals and Israel: Theological Roots of a Political Alliance," *The Christian Century,* November 4, 1998, p. 1020.
21. Daniel Lapin, "Living as Jews in Christian America," *Jewish Action Magazine,* Winter 2004, http://www.towardtradition.org/Jewish%20Action%20Magazine.pdf.
22. Ernest Lee Tuveson, *Redeemer Nation: The Idea of America's Millennial Role,* (Chicago: University of Chicago Press, 1980), p.99.
23. Ibid., p. 106.
24. Lapin, "Living as Jews in Christian America."
25. Michael Northcott, *An Angel Directs the Storm: Apocalyptic Religion and American Empire* (London: I. B. Tauris, 2004), p. 5, and http://www.greatseal.com/committees/firstcomm/reverse.html.
26. Robert Fuller, *Religious Revolutionaries: The Rebels Who Reshaped American Religion* (New York: Palgrave MacMillan, 2004), p. 6.
27. Forrest Church, *The Separation of Church and State: Writings on a Fundamental Freedom by America's Founders* (Boston: Beacon Press, 2004), p. x.
28. Daniel Wojcik, *The End of the World as We Know It: Faith, Fatalism, and Apocalypse in America* (New York: New York University Press, 1997), p. 24.
29. Church, *The Separation of Church and State,* p. x, "The Treaty of Tripoli," Article 11. There is no mention of God or Christianity in the Constitution and constitutional scholars often cite the Treaty of Tripoli as the most explicit evidence that the omission was intentional. The treaty, which was drawn up to protect U.S. merchant

ships from Barbary pirates, was signed by President John Adams in 1796 and ratified by the Senate the following year (p. 123).

30. Rousas John Rushdoony, *The Institutes of Biblical Law* (Phillipsburg, N.J.: P&R Publishers, 1978), pp. 1, 2.
31. Steve Erickson, "George Bush and the Treacherous Country," *LA Weekly,* February 12, 2004, http://www.laweekly.com/general/features/george-bush-and-the-treacherous-country/1987/.
32. Church, *The Separation of Church and State,* p. ix.
33. Ibid., p. x.
34. Karen Armstrong, *The Battle for God* (New York: Ballantine, 2001), p. 217.
35. Paul Boyer, *When Time Shall Be No More: Prophecy Belief in Modern American Culture* (Cambridge, Mass.: Belknap Press, 1994), p. 87.
36. Armstrong, *The Battle for God,* p. 118.
37. Ronald M. Henzel, *Darby, Dualism, and the Decline of Dispensationalism* (Tucson, Ariz.: Fenestra Books, 2003), p. 54.
38. Northcott, *An Angel Directs the Storm,* p. 58.
39. Ibid., p. 15.
40. Boyer, *When Time Shall Be No More,* p. 88.
41. Armstrong, *The Battle for God,* p. 93.
42. Ibid., p. 137.
43. Boyer, *When Time Shall Be No More,* p. 89.
44. http://www2.yale.edu/timeline/1757/index.html.
45. Boyer, *When Time Shall Be No More,* p. 90.
46. George M. Marsden, *Understanding Fundamentalism and Evangelicalism* (Grand Rapids, Mich.: Wm. B. Eerdmans Publishing Co., 1991), p. 21.
47. Boyer, *When Time Shall Be No More,* p. 98.
48. John B. Judis, *The Folly of Empire: What George W. Bush Could Learn from Theodore Roosevelt and Woodrow Wilson* (New York: Oxford University Press, 2004), pp. 41, 42.
49. Ibid., p. 43.
50. *John Thomas Scopes v. The State, Supreme Court of Tennessee,* http://www.law.umkc.edu/faculty/projects/ftrials/scopes/statcase.htm. And "New TV Special Connects Darwin to Hitler," http://www.coralridge.org/imp/impact08061 .aspx.
51. Armstrong, *The Battle for God,* p. 177.
52. Ibid., p. 175.
53. Boyer, *When Time Shall Be No More,* p. 105.
54. Armstrong, *The Battle for God,* p. 178.
55. Chip Berlet and Matthew N. Lyons, *Right-Wing Populism in America: Too Close for Comfort* (New York: Guilford Press, 2000), p. 201.
56. William Martin, *With God on Our Side: The Rise of the Religious Right in America* (New York: Broadway, 2005), p. 37.
57. Armstrong, *The Battle for God,* p. 215.
58. Harvey Cox, *The Secular City,* http://www.smithcreekmusic.com/Hymnology/Theology/Secular.City.html.
59. Michael Hamilton, "The Dissatisfaction of Francis Schaeffer," *Christianity Today,* 1997.
60. "Toward a Hidden God," *Time,* April 8, 1966.
61. Ibid.
62. Ibid.
63. Robert D. Putnam, *Bowling Alone: The Collapse and Revival of American Community* (New York: Simon & Schuster, 2001), p. 76.
64. The Creed of Bob Jones University, which is recited by students four days a week at chapel, gives an idea of its fundamentalists precepts: "I believe in the inspiration of the Bible (both the Old and the New Testaments); the creation of man by the direct act of God; the incarnation and virgin birth of our Lord and Savior, Jesus Christ; His identification as the Son of God; His vicarious atonement for the sins of mankind by the shedding of His blood on the cross; the resurrection of His body from the tomb;

His power to save men from sin; the new birth through the regeneration by the Holy Spirit; and the gift of eternal life by the grace of God."

65. Armstrong, *The Battle for God,* p. 214.
66. The Association of Biblical Higher Education, http://abhe.gospelcom.net/index.html.
67. Armstrong, *The Battle for God,* p. 216.
68. Kevin Phillips, *American Theocracy: The Peril and Politics of Radical Religion, Oil, and Borrowed Money in the 21st Century* (New York: Viking, 2006), p. 113.
69. http://www.cbn.com/CBNTelevision.aspx?WT.svl=menu.
70. "Trinity Broadcasting Network," http://en.wikipedia.org/wiki/Trinity_Broadcasting_Network.
71. Martin, *With God on Our Side,* p. 31.
72. Ibid., document.
73. Armstrong, *The Battle for God,* p. 216.
74. "To Change the World," *Time,* http://www.time.com/time/magazine/article/0,9171,854920-2,00.html.
75. Martin, *With God on Our Side,* p. 45.
76. Max Blumenthal, "Agent of Intolerance," http://www.thenation.com/doc/20070528/blumenthal (posted online on May 16, 2007). And Jonathan A.Wright, *Shapers of the Great Debate on the Freedom of Religion,* p. 227.
77. Author interview with Jerry Falwell, May 2005.
78. Ibid.

Chapter Three: Birth of the Neocons

1. Sidney Blumenthal, *The Rise of the Counter-Establishment: From Conservative Ideology to Political Power* (New York: HarperCollins, 1988), p. 123.
2. Ibid., p. 124.
3. Norman Podhoretz, *Ex-Friends: Falling Out with Allen Ginsberg, Lionel and Diana Trilling, Lillian Hellman, Hannah Arendt, and Norman Mailer* (San Francisco: Encounter Books, 2000).
4. Stefan Halper and Jonathan Clarke, *America Alone: The Neoconservatives and the Global Order* (New York: Cambridge University Press, 2004), p. 46.
5. Podhoretz, *Ex-Friends,* p. 1.
6. Ibid., pp. 27–28.
7. The poem's memorable opening line: "*I saw the best minds of my generation destroyed by madness, starving hysterical naked, dragging themselves through the negro streets at dawn looking for an angry fix.*"
8. Gore Vidal, *Sexually Speaking: Collected Sex Writings* (San Francisco: Cleis Press, 2001), p. 119.
9. Podhoretz, *Ex-Friends,* p. 42.
10. Blumenthal, *The Rise of the Counter-Establishment,* p. 138. Podhoretz has denied that the incident took place, and disputed it in a letter to the *New York Times,* which referred to it in a review of Sidney Blumenthal's *The Rise of the Counter-Establishment,* the original source of the story, which attributed the anecdote to George McGovern. In turn, McGovern told the *Times,* "I'd better not comment on it. It's 25 years ago, it was a very personal matter anyway and I had no idea anyone was going to comment on it."

    Blumenthal, *The Rise of the Counter-Establishment,* p. 127. See also Podhoretz, *Ex-Friends,* p. 207; and ibid., p. 208.

    Alexander Cockburn gives a similar account of the episode in the CounterPunch website, http://www.counterpunch.org/cockburn02122003.html.
11. Alex Fryer, "Scoop Jackson's Protégés Shaping Bush's Foreign Policy," *Seattle Times,* January 12, 2004, http://archives.seattletimes.nwsource.com/cgi-bin/texis.cgi/web/vortex/display?slug=jackson12m&date=20040112.
12. Julian Borger, "Democrat Hawk Whose Ghost Guides Bush," *The Guardian* (Lon-

don), December 6, 2002, http://www.guardian.co.uk/international/story/0,,854703,00.html.

13. Fryer, "Scoop Jackson's Protégés Shaping Bush's Foreign Policy."
14. http://www.sourcewatch.org/index.php?title=Abram_N._Shulsky.
15. Fryer, "Scoop Jackson's Protégés Shaping Bush's Foreign Policy."
16. Sidney Blumenthal, "Richard Perle's Nuclear Legacy; An Acolyte's Education and the Passing of the Torch," *Washington Post,* November 24, 1987, p. D1.
17. Sidney Blumenthal, *Our Long National Daydream: A Political Pageant of the Reagan Era* (New York: HarperCollins, 1990), p. 236.
18. Ibid.
19. Jeffrey Goldberg, "A Little Learning," *The New Yorker,* http://www.newyorker.com/fact/content/articles/050509fa_fact.
20. http://www.palestinecenter.org/palestine/conference_2003_panel2.htm.
21. Author interview with Stefan Halper, September 2006.
22. http://www.socialdemocrats.org/MayDayTranscript.html#kirkpatrick. And George Packer, *The Assassin's Gate,* p. 68.
23. Author interview with Richard Pipes, March 2007.
24. David Brooks, "The Era of Distortion," *New York Times,* January 6, 2004, p. A23.
25. Joe Hagan, "President Bush's Neoconservatives Were Spawned Right Here in N.Y.C., New Home of the Right-Wing Gloat," *New York Observer,* April 28, 2003, p. 1.
26. Robert Lieber, "The Neoconservative-Conspiracy Theory: Pure Myth," May 2, 2003, http://chronicle.com/free/v49/i34/34b01401.htm.
27. Stefan Halper and Jonathan Clarke, *America Alone: The Neoconservatives and the Global Order* (New York: Cambridge University Press, 2004), p. 43.
28. Jeet Heer, "Trotsky's Ghost Wandering the White House: Influence on Bush Aides: Bolshevic's Writings Supported the Idea of Pre-emptive War," *National Post* (Canada), June 7, 2003, p. A26.
29. Jonah Goldberg, "Goldberg File," *The National Review,* http://www.nationalreview.com/goldberg/goldberg052103.asp.
30. Stephen Schwartz, "Trostkycons?," *The National Review,* http://www.nationalreview.com/comment/comment-schwartz061103.asp.
31. Heer, "Trotsky's Ghost Wandering the White House."
32. Blumenthal, *Our Long National Daydream,* p. 236.
33. Blumenthal, "Richard Perle's Nuclear Legacy; An Acolyte's Education and the Passing of the Torch," *Washington Post,* November 24, 1987, p. D1.
34. http://www.hollywoodhighalumni.com/about/history.html.
    http://books.google.com/books?id=hK6_LUZkB5oC&dq=famous+alumni+of+Hollywood+High&pg=PA513&ots=DzIs77BzYM&sig=jw9OU5JNBR0Q0-wjAwA2vtAbaXs&prev=http://www.google.com/search%3Fsourceid%3Dnavclient-ff%26ie%3DUTF-8%26rlz%3D1B2GGGL_enUS176%26q%3Dfamous%2Balumni%2Bof%2BHollywood%2BHigh&sa=X&oi=print&ct=result&cd=1.
    http://www.experiencela.com/destinations/fullprofile.asp?key=221.
    http://www.seeing-stars.com/Schools/HollywoodHigh.shtml.
    "Hollywood High School," http://en.wikipedia.org/wiki/Hollywood_High_School.
35. Blumenthal, *Our Long National Daydream,* p. 234.
36. *PBS Think Tank:* Transcript for "Richard Perle: The Making of a Neoconservative," Ben Wattenberg interview, http://www.pbs.org/thinktank/transcript1017.html.
37. It has been widely, and erroneously, reported that Richard Perle married Joan Wohlstetter. His wife is Leslie Barr. Joan Wohlstetter married Lorenzo Hall. Corrections, *New York Times,* May 7, 2003, http://query.nytimes.com/gst/fullpage.html?res=9503E6DF1F3CF934A35756C0A9659C8B63&n=Top%2FReference%2FTimes%20Topics%2FPeople%2FP%2FPerle%2C%20Richard%20N.
38. Wil S. Hylton, "The Big Bad Wolfowitz," http://men.style.com/gq/features/landing?id=content_5151.
39. Robert Zarate, "The First Lady of Intelligence," *Weekly Standard* 12, no. 18 (Janu-

ary 22, 2002), http://www.weeklystandard.com/Content/Protected/Articles/000/000/013/171kcloi.asp.

40. Author interview with Fred Kaplan, author of *The Wizards of Armageddon* (New York: Touchstone, 1983), June 2007.
41. Anne Hessing Cahn, *Killing Detente: The Right Attacks the CIA* (University Park: Pennsylvania State University, 1998), p. 13.
42. Ibid., p.14
43. Blumenthal, *Our Long National Daydream,* p. 234.
44. Anthony David, "The Apprentice," *American Prospect,* June 5, 2007, http://www.prospect.org/cs/articles?article=the_apprentice, and W. Patrick Lang, "Drinking the Kool-Aid," *Middle East Policy* 11, no. 2 (June 22, 2004), http://www.mepc.org/journal_vol11/0406_lang.asp.
45. Bruce Kuklick, "Wise Guys," *Los Angeles Times,* May 14, 2006, p. 1.
46. Interview with Fred Kaplan.
47. Ibid.
48. Leila Hudson, "The New Ivory Towers: Think Tanks, Strategic Studies, and 'Counterterrorism,'" *Middle East Policy* 12, no. 4 (December 22, 2005): p. 118.
49. Helene Cooper, "On to a New Trouble Spot," *New York Times,* January 6, 2007, p. A8.
50. Elizabeth Drew, "The Neocons in Power," *New York Review of Books* 50, no. 10 (June 12, 2003).
51. Hudson, "The New Ivory Towers," p. 118.
52. Ibid.
53. Lang, "Drinking the Kool-Aid," p. 39.
54. Ibid.
55. Sam Tanenhaus, "Deputy Secretary Wolfowitz Interview with Sam Tanenhaus," *Vanity Fair,* http://www.defenselink.mil/transcripts/2003/tr20030509-depsecdef0223.html.
56. Jason Vest, "Bush's War Hawk," *American Prospect,* http://www.prospect.org/print/V12/19/vest-j.html.
57. Author interview with Patrick Lang, March 2007. Lang's meeting with the Wohlstetters took place around 1992.
58. Lang, "Drinking the Kool-Aid," p. 39.
59. Khurram Husain, "Neocons: The Men Behind the Curtain: Undeterred by Their Encounters with Reality, the Strategists Who Pushed for War in Iraq Believed Then, and Still Believe, That Their Moment Has Come," *Bulletin of the Atomic Scientists* 59, no. 6 (November 1, 2003): 62.
60. Hudson, "The New Ivory Towers," p. 118.

Chapter Four: The Foreshadowing

1. Anne Hessing Cahn, *Killing Detente: The Right Attacks the CIA* (University Park: Pennsylvania State University, 1998), pp. 8, 9.
2. Stefan Halper and Jonathan Clarke, *America Alone: The Neoconservatives and the Global Order* (New York: Cambridge University Press, 2004), p. 55.
3. A more complete list of the principals in the Coalition for a Democratic Majority can be found at http://rightweb.irc-online.org/groupwatch/cdm/php.
4. Ben Wattenberg, "Passing of a Patriot," *Washington Times,* October 23, 2005, p. B4.
5. Joe Holley, "Political Activist Penn Kemble Dies at 64," *Washington Post,* October 18, 2005, http://www.washingtonpost.com/wp-dyn/content/article/2005/10/18/AR2005101801743.html
6. Halper and Clarke, *America Alone,* p. 56.
7. Cahn, *Killing Detente,* p. 9.
8. Ibid., p. 12.
9. Ibid., p. 14.
10. Ibid., p. 122.

11. Daniel Patrick Moynihan, *A Dangerous Place* (Boston: Little, Brown, 1978), p. 45.
12. Ibid., pp. 45, 46.
13. Joan Didion, "Cheney: The Fatal Touch," *New York Review of Books,* October 5, 2006, p. 51.
14. Andrew Cockburn, *Rumsfeld: His Rise, Fall, and Catastrophic Legacy* (New York: Scribner, 2007), p. 20.
    Also see Victor Gold, *Invasion of the Party Snatchers: How the Holy-Rollers and the Neo-Cons Destroyed the GOP* (Chicago: Sourcebooks Trade, 2007), pp. 19–20.
15. James Mann, *Rise of the Vulcans: The History of Bush's War Cabinet* (New York: Viking, 2004), p. 60.
16. Ibid., p. 61.
17. Cockburn, *Rumsfeld: His Rise, Fall, and Catastrophic Legacy,* p. 29.
18. Ibid., p. 27.
19. Ibid., p. 28.
20. Gold, *Invasion of the Party Snatchers,* pp. 86–87.
21. Mann, *Rise of the Vulcans,* p. 66.
22. Ibid.
23. Gold, *Invasion of the Party Snatchers,* p. 19.
24. Cockburn, *Rumsfeld,* p. 31.
25. Gold, *Invasion of the Party Snatchers,* pp. 19, 92.
26. Author interview with Pete Teeley, June 2007.
27. Cockburn, *Rumsfeld,* p. 31.
28. Sidney Blumenthal, "No Time to Heal," *Salon,* January 3, 2007, http://www.salon.com/opinion/blumenthal/2007/01/03/gerald_ford/index1.html.
29. Cahn, *Killing Detente,* pp. 46–47.
30. The BBC documentary, *The Power of Nightmares,* by Adam Curtis, can be seen online at http://news.bbc.co.uk/1/hi/programmes/3755686.stm. An article about it by Thomas Hartmann can be read at http://www.commondreams.org/views04/1207-26.htm.
31. Cahn, *Killing Detente,* p. 138.
32. Ibid., p. 139.
33. Paul C. Warnke, "Killing Detente: The Right Attacks the CIA," *Bulletin of the Atomic Scientists* 55, no. 1 (January 1999): p. 70.
34. Sam Tanenhaus, "The Hardliner," *Boston Globe,* November 2, 2003, http://www.boston.com/news/globe/ideas/articles/2003/11/02/the_hard_liner/.
35. Cahn, *Killing Detente,* p. 150.
36. Peter Boyer, "The Believer," *The New Yorker,* http://www.newyorker.com/archive/2004/11/01/041101fa_fact?currentPage=1.
37. Cahn, *Killing Detente,* p. 151.
38. Ibid., p. 152.
39. Anne H. Cahn, "Perspective on Arms Control; How We Got Oversold on Overkill," *Los Angeles Times,* July 23, 1993, p. B7.
40. Cahn, *Killing Detente,* p. 158.
41. Ibid.
42. Ibid., p. 88.
43. Fareed Zakaria, "Exaggerating the Threats," *Newsweek,* June 16, 2003, http://www.fareedzakaria.com/articles/newsweek/061603.html.
44. Leila Hudson, "The New Ivory Towers: Think Tanks, Strategic Studies and 'Counterrealism,'" *Middle East Policy* 12, no. 4 (December 22, 2005): p. 118.
45. Cahn, *Killing Detente,* p. 167.
46. Ibid., p. 165.
47. Ibid., p. 145.
48. Tom Barry, "US: Danger, Danger Everywhere," *Asia Times,* June 23, 2006, http://www.atimes.com/atimes/Front_Page/HF23Aa01.html.
49. Fareed Zakaria, "Exaggerating the Threats," *Newsweek,* June 16, 2003, http://www.fareedzakaria.com/articles/newsweek/061603.html.
50. Thom Hartmann, "Hyping Terror for Fun, Profit—And Power," Common-

Dreams.org, December 7, 2004, http://www.commondream.org/views04/1207-26.htm. (Retrieved on April 23, 2006.)

51. Sidney Blumenthal, "A Time to Heal."
52. Melvin Goodman, "Righting the CIA," *Baltimore Sun,* November 19, 2004, http://www.commondreams.org/views04/1119–20.htm.
53. Murrey Marder, "Carter to Inherit Intense Dispute on Soviet Intentions," *Washington Post,* January 2, 1977, p. A1.
54. Hudson, "The New Ivory Towers," p. 118.
55. Author interview with Richard Pipes, March 2007.
56. Eric Alterman, "Think Again, Team B," Center for American Progress, http://www.americanprogress.org/issues/2003/10/b11003.html.
57. Hartmann, "Hyping Terror for Fun, Profit—And Power."
58. Cahn, *Killing Detente,* p. 179.
59. Ibid., p. 177.
60. Thom Hartmann, "Rumsfeld and Cheney Revive Their 70's Terror Playbook," CommonDreams.org, February 13, 2006, www.commondreams.org/views06/0213-28.htm.
61. Halper and Clarke, *America Alone,* p. 14.
62. Sidney Blumenthal, *The Rise of the Counter-Establishment: From Conservative Ideology to Political Power* (New York: HarperCollins, 1988), p. 128.
63. Sidney Blumenthal, *Our Long National Daydream: A Political Pageant of the Reagan Era* (New York: HarperCollins, 1990), p. 226.
64. http://rightweb.irc-online.org/gw/1589. Rightweb cites as original sources John S. Saloma III, *Ominous Politics* (New York: Hill and Wang, 1984), and Thomas Bodenheimer and Robert Gould, *Rollback! Right-wing Power in U.S. Foreign Policy* (Boston: South End Press, 1989).
65. Melvin Goodman, "Right the CIA," *Baltimore Sun,* http://bailey83221.livejournal.com/51190.html.
66. Cahn, "Perspective on Arms Control," *Los Angeles Times,* July 23, 1993, p. B7.
67. Robert Kuttner, "Philanthropy and Movements," *American Prospect,* July 15, 2002, p. 2.
68. http://www.hudson.org/learn/index.cfm?fuseaction=staff_type#link3.
69. Halper and Clarke, *America Alone,* p. 48.
70. http://www.aei.org/scholars/filter.all/scholar_byname.asp.
71. Halper and Clarke, *America Alone,* p. 49.
72. http://rightweb.irc-online.org/profile/1431.
73. Blumenthal, *The Rise of the Counter-Establishment,* p. 37.
74. Kuttner, "Philanthropy and Movements," *American Prospect,* July 14, 2002, p. 2.
75. Ibid.

Chapter Five: Into the Fray

1. Liberty University, "From the Founder and President," http://64.233.161.104/search?q=cache:fK9LwcQoF30J:www.liberty.edu/media/1109/%255B791%255DLU_Graduate_Catalog_2004–2005.pdf+%22Jerry+Falwell+Museum%22+and+Jonathan+Edwards&hl=en.
2. Ibid.
3. Author interview with Jerry Falwell, May 2005.
4. Ibid.
5. Michael S. Lief and Mitchell Calwell, *And the Walls Came Tumbling Down: Greatest Closing Arguments Protecting Civil Liberties* (New York: Scribner, 2006), p. 250.
6. Interview with Jerry Falwell.
7. Ibid.
8. Thomas Road Baptist Church website, http://home.trbc.org/index.cfm?PID= 9059.

9. "Billy Graham Voices Shock Over Decision," *New York Times* (UPI), June 18, 1963, p. 17.
10. Interview wth Richard Land, "The Jesus Factor," *Frontline*, PBS, http://www.pbs.org/wgbh/pages/frontline/shows/jesus/interviews/land.html.
11. Interview with Jerry Falwell.
12. Michael Hamilton, "The Dissatisfaction of Francis Schaeffer," *Christianity Today*, 1997.
13. Louis Gifford Parkhurst Jr., *Francis Schaeffer: The Man and His Messsage* (Wheaton, IL: Tyndale House, 1985), p. 13.
14. Hamilton, "The Dissatisfaction of Francis Schaeffer."
15. Ibid.
16. Francis Schaeffer, *How Should We Then Live? The Rise and Decline of Western Thought and Culture* (Wheaton, Ill.: Crossway Books, 2005), pp. 170–71.
17. Hamilton, "The Dissatisfaction of Francis Schaeffer."
18. Author interview with John Berger, senior editor at Cambridge University Press, August 2006.
19. Hamilton, "The Dissatisfaction of Francis Schaeffer."
20. E-mail exchange with Wayne White, February 2006.
21. John Fischer, "Learning to Cry for the Culture," *Christianity Today*, http://www.christianitytoday.com/ct/2007/april/13.40.html.
22. Schaeffer, *How Should We Then Live?*, p. 258.
23. Francis Schaeffer and C. Everett Koop, *Whatever Happened to the Human Race?* (Wheaton, Ill.: Crossway Books, 1983), p. 1.
24. Frederick Clarkson, "Theocratic Dominioniam Gains Influence," Part 2, Public-Eye.org, March/June 2004, http://publiceye.org/magazine/v08n1/chrisre2 .html.
25. Peter and Rochelle Schweizer, *The Bushes: Portrait of a Dynasty* (New York: Random House, 2004), p. 335.
26. Fischer, "Learning to Cry for the Culture."
27. Interview with Jerry Falwell.
28. Ibid.
29. Ibid.
30. Doug Willis, "Church Groups Promise Anti-Homosexual Drive," Associated Press, January 13, 1981.
31. Nancy Shepherdson, "Writing for Godot," *Los Angeles Times Magazine*, April 25, 2004, p. 16.
32. "Evangelical Power Couple," *Atlanta Journal-Constitution*, July 7, 2001.
33. "Moral Majority Leader Quits Church," Associated Press, February 12, 1981.
34. Author interview with Tim LaHaye in Israel, May 2005.
35. Tim LaHaye, *Battle for the Mind* (Grand Rapids, Mich.: Baker Book House, 1980).
36. Glenn Shuck, *Marks of the Beast: The Left Behind Novels and the Struggle for Evangelical Identity* (New York: New York University Press, 2004), p. 66.
37. LaHaye, *Battle for the Mind*, pp. 37, 57.
38. Gayle White, "LaHayes Make Mark on America," *Washington Times*, July 26, 2001.
39. LaHaye, *Battle for the Mind*, rear flap.
40. Author interview with Chip Berlet, May 2005.
41. Kenneth E. Woodward and Eloise Salhoz, "The Right's New Bogyman," *Newsweek*, July 6, 1981, p. 48.
42. Calvin Skaggs and David Van Taylor, *With God on Our Side: George W. Bush and the Rise of the Religious Right*, documentary film, 2004.
43. Jason DeParle, "Sheila Burke Is the Militant Feminist Commie Peacenik Who's Telling Bob Dole What to Think," *New York Times*, November 12, 1995, sec. 6, p. 32.
44. William J. Lanouette, "The New Right—'Revolutionaries' Out After the 'Lunch-Pail' Vote," *National Journal* 10, no. 3 (January 21, 1978): 88.
45. DeParle, "Sheila Burke Is the Militant Feminist Commie Peacenik Who's Telling Bob Dole What to Think."

46. John W. Dean, *Conservatives Without Conscience* (New York: Viking, 2006), p. 91.
47. David Grann, "Robespierre of the Right," *New Republic,* October 27, 1997, p. 20.
48. Ibid.
49. Skaggs and Taylor, *With God on Our Side*.
50. Ibid.
51. Karen Armstrong, *The Battle for God* (New York: Ballantine, 2001), pp. 266–67.
52. Interview with Jerry Falwell.
53. Skaggs and Taylor, *With God on Our Side*.
54. Ibid.
55. Ibid.
56. Dudley Clendinen, *New York Times,* August 18, 1980, p. B7.
57. Author interview with Pete Teeley, April 2007.
58. Armstrong, *The Battle for God,* p. 26.
59. The Family Research Council was founded in 1981, but it was not incorporated until 1983. Theocracy Watch, http://www.theocracywatch.org/taking_over.htm.
60. E-mail exchange with Max Blumenthal, February 2007.
61. Rob Boston, "The Top Ten Powerbrokers of the Religious Right," http://www.alternet.org/story/38467/.
62. William Martin, *With God on Our Side: The Rise of the Religious Right in America* (New York: Broadway, 2005), p. 169.
63. Gayle White, "Evangelical Power Couples," *Atlanta Journal-Constitution,* July 7, 2001, p. B1.
64. http://www.truthout.org/docs_2005/050905A.shtml.
65. Megan Rosenfeld, "Reining in the Right," *Washington Post,* May 19, 1981, p. C1.
66. DeParle, "Sheila Burke Is the Militant Feminist Commie Peacenik Who's Telling Bob Dole What to Think."
67. "Apocalypse!" *Frontline,* PBS, http://www.pbs.org/wgbh/pages/frontline/shows/apocalypse/etc/cron2.html.
68. Ronald Reagan, Remarks to the National Association of Evangelicals, http://www.americanrhetoric.com/speeches/ronaldreaganevilempire.htm.
69. Martin, *With God on Our Side,* p. 249.
70. Ibid., p. 250.
71. Ibid., p. 251.
72. Skaggs and Taylor, *With God on Our Side*.

Chapter Six: The Prodigal Son

1. George W. Bush and Mickey Herskowitz, *A Charge to Keep: My Journey to the White House* (New York: Harper Paperbacks, 2001), p. 136.
2. Peter Schweizer and Rochelle Schweizer, *The Bushes: Portrait of a Dynasty* (New York: Random House, 2004), p. 333.
3. http://www.msnbc.msn.com/id/6560118/.
4. Author interview with Mickey Herskowitz, February 2007.
5. Ibid.
6. Russ Baker, "Two Years Before 9/11, Candidate Bush Was Already Talking Privately About Attacking Iraq, According to His Former Ghost Writer," CommonDreams.org, http://www.commondreams.org/headlines04/1028-01.htm.
7. Interview with Mickey Herskowitz.
8. http://www.blessitt.com/hisplace.html.
9. http://www.blessitt.com/books/turned.html.
10. Stephen Prothro, *American Jesus: How the Son of God Became a National Icon* (New York: Farrar, Straus and Giroux, 2004), pp. 128, 129.
11. http://www.blessitt.com/hisplace.html.
12. http://www.blessitt.com/bush.html.
13. Interview with Mickey Herskowitz.

14. Mary Jacoby, "George Bush's Missing Year," *Salon,* http://dir.salon.com/story/news/feature/2004/09/02/allison/index.html.
15. Ibid.
16. Ibid.
17. Robert Parry, *Secrecy & Privilege: Rise of the Bush Dynasty from Watergate to Iraq* (Arlington, Va.: The Media Consortium, 2004), p. 291. First reported in Bill Minutaglio's *First Son: George W. Bush and the Bush Family Dynasty* (New York: Crown, 1999).
18. George Lardner Jr. and Lois Romano, "At Height of Vietnam, Bush Picks Guard," *Washington Post,* July 28, 1999, http://www.washingtonpost.com/wp-srv/politics/campaigns/wh2000/stories/bush072899.htm.
19. Calvin Skaggs and David Van Taylor, *With God on Our Side: George W. Bush and the Rise of the Religious Right,* documentary film, 2004.
20. Interview with Mickey Herskowitz.
21. http://www.tsha.utexas.edu/handbook/online/articles/MM/hdm3.html.
22. "The Jesus Factor: Midland's Community Bible Study," *Frontline,* PBS, http://www.pbs.org/wgbh/pages/frontline/shows/jesus/president/cbs.html.
23. Ibid.
24. David Aikman, *A Man of Faith: The Spiritual Journey of George W. Bush* (Nashville, Tenn.: Thomas Nelson, 2004), p. 69.
25. Ibid., p. 72
26. Skaggs and Taylor, *With God on Our Side.*
27. Stephen Mansfield, *The Faith of George W. Bush* (New York: Penguin, 2004), p. 60.
28. Author interview with Jim Sale, November 2006.
29. http://www.blessitt.com/bush.html.
30. Author interview with Arthur Blessitt, November 2006.
31. http://www.blessitt.com/bush.html.
32. Ibid.
33. Ibid.
34. Interview with Jim Sale.
35. Skaggs and Taylor, *With God on Our Side.*
36. http://www.blessitt.com/bush.html.
37. http://www.wheaton.edu/isae/Gallup-Bar-graph.html.
38. Robert D. Putnam, *Bowling Alone: The Collapse and Revival of American Community* (New York: Simon & Schuster, 2001).
39. "Generational Differences," http://www.barna.org/FlexPage.aspx?Page=Topic&TopicID=22.
40. Putnam, *Bowling Alone,* p. 438 ff.
41. Ron Fournier, Douglas B. Sosnik, and Matthew J. Dowd, *Applebee's America: How Successful Political, Business, and Religious Leaders Connect with the New American Community* (New York: Simon & Schuster, 2006), p. 118.
42. Alexandra Pelosi, *Friends of God,* a documentary for HBO, 2007.
43. Prothro, *American Jesus,* p. 145.
44. http://sbno.illicitohio.com/heritage/thestory.html.
45. In 2005, the author interviewed a group of five young evangelicals and former evangelicals who had moved to New York and who agreed to be interviewed on a not-for-attribution basis. Quotations from "former evangelicals" and "young evangelicals" refer to members of this group.
46. Tim LaHaye and Beverly LaHaye, *The Act of Marriage* (Grand Rapids, Mich.: Zondervan, 1998), p. 245.
47. http://www.book22.com/merchant2/merchant.mvc?Screen=PROD&Store_Code=Book22&Product_Code=RRFCAF&Category_Code=CO.
48. http://www.book22.com/merchant2/merchant.mvc?Screen=CTGY&Store_Code=Book22&Category_Code=ED.
49. Joan Didion, "Mr. Bush & the Divine," *New York Review of Books,* November 6, 2003.
50. Interview with Mickey Herskowitz.

51. William Yardley, "Jeb Bush: His Early Values Shape His Policies," *Miami Herald*, September 22, 2002, p. A1.
52. George Lardner Jr. and Lois Romano, "George Walker Bush," *Washington Post*, July 30, 1999, p. A1.
53. Interview with Mickey Herskowitz.
54. "Absalom Willis Robertson," http://en.wikipedia.org/wiki/Absalom_Willis_Robertson.
55. William Martin, *With God on Our Side: The Rise of the Religious Right in America* (New York: Broadway, 2005), p. 258.
56. Michael Lind, "Rev. Robertson's Grand International Conspiracy Theory," *New York Review of Books* 42, no. 2 (February 2, 1995).
57. Interview with Molly Ivins, http://www.geocities.com/CapitolHill/7027/quotes.html.
58. Pat Robertson, *The 700 Club*, Dec. 30, 1981.
59. Pat Robertson, *The New World Order* (Dallas, Tex.: Word Publishing, 1991), p. 218.
60. Ron Hutcheson, "With Call for Assassination, Robertson Again Raises Eyebrows," Knight Ridder Washington Bureau, August 24, 2005.
61. Lind, "Rev. Robertson's Grand International Conspiracy Theory."
62. Robertson, *The New World Order*, p. 37.
63. Jeanne Pugh, "Robertson Has Been Recruiting His 'Invisible Army' for Years," *St. Petersburg Times* (Florida), February 13, 1988, p. E2.
64. Ibid.
65. Ibid.
66. *Time*, February 8, 1988, p. 30.
67. Michael Kramer, "The Tortoise and the Hare," *U.S. News & World Report*, February 8, 1988, p. 14.
68. Dick Kirschten, "The Back of the Pack," *National Journal* 20, no. 4 (January 23, 1988): p. 18.
69. "The GOP's Wild Card," *National Journal* 20, no. 9 (February 27, 1988): 519.
70. Douglas A. Harbrecht and Richard Fly, "For Bush, a Moment of Truth in New Hampshire," *BusinessWeek*, February 22, 1988, p. 28.
71. "Robertson's Grand Design," *U.S. News & World Report*, February 22, 1988, p. 14.
72. Ibid.
73. "Quotes of the Week," *U.S. News & World Report*, December 7, 1987, p. 13.
74. Hanna Rosin, "Applying Personal Faith to Public Policy," *Washington Post*, July 24, 2000, p. A1.
75. http://www.dailyhowler.com/h051401_1.shtml.
76. Maureen Dowd, "Amnesia in the Garden," *New York Times*, September 5, 2004, http://www.nytimes.com/2004/09/05/opinion/05dowd.html?ex=1252123200&en=2996960f258fca85&ei=5090&partner=rssuserland
77. Rosin, "Applying Personal Faith to Public Policy," *Washington Post*, July 24, 2000, p. A1.
78. Robert Shogan, *Constant Conflict: Politics, Culture and the Struggle for America's Future* (Bolder, Colo.: Basic Books, 2004), p. 233.
79. Didion, "Mr. Bush & the Divine."
80. Rosin, "Applying Personal Faith to Public Policy."
81. Skaggs and Taylor, *With God on Our Side*.
82. Ibid.
83. "The Jesus Factor: Interview: Doug Wead," *Frontline*, PBS, http://www.pbs.org/wgbh/pages/frontline/shows/jesus/interviews/wead.html.
84. James A. Albert, *Jim Bakker: Miscarriage of Justice?* (Chicago: Open Court, 1999).
85. Skaggs and Taylor, *With God on Our Side*.
86. Susan Schindehette, "A First Family That Just Won't Quit," *People*, January 30, 1989, p. 52.
87. R. W. Apple Jr. and Colin MacKenzie, "The 41st President: Washington Memo," *New York Times*, January 20, 1989, p. A1.
88. Marjorie Hyer, "Week of Parties, Pomp Closes With a Prayer," *Washington Post*, January 23, 1989, p. B1.

89. Robert Reno, "Common Thread of All the New President's Men," *Newsday,* January 18, 1989, Nassau and Suffolk Edition, p. 39.
90. John Rockwell, "The 41st President: Music; Pop (as in Populism) Go the Festivities," *New York Times,* January 20, 1989.
91. Stephanie Mansfield, Kara Swisher, Carla Hall, Lois Romano, Ann Devroy, Sarah Booth Conroy, "Fetes and Lines in Texas Time," *Washington Post,* January 20, 1989, p. B1.
92. Skaggs and Taylor, *With God on Our Side.*
93. Sidney Blumenthal, "Fall of the Rovean Empire?" *Salon,* October 6, 2005, http://dir.salon.com/story/opinion/blumenthal/2005/10/06/rovean_empire/index2.html?pn=1.
94. Sidney Blumenthal, "Republican Tremors," http://www.opendemocracy.net/democracy/republican_2899.jsp.
95. Skaggs and Taylor, *With God on Our Side.*
96. Ibid.
97. Michelle Goldberg, *Kingdom Coming: The Rise of Christian Nationalism* (New York: W. W. Norton and Co., 2007), p. 41.

Chapter Seven: The Age of Unreason

1. Author interview with Mickey Herskowitz, February 2007.
2. "The Jesus Factor: Interview: Doug Wead," *Frontline,* PBS, http://www.pbs.org/wgbh/pages/frontline/shows/jesus/interviews/wead.html.
3. Andrew Rosenthal, "Cheney, a Conservative, Is Also a Compromiser," *New York Times,* March 12, 1989, p. A24.
4. John W. Mashek, "Cheney Tapped for Defense Post; Congressman Seen as Conciliatory Pick," *Boston Globe,* March 11, 1989, p. 1.
5. Ibid.
6. T. D. Allman, "The Curse of Dick Cheney," *Rolling Stone,* http://www.rollingstone.com/politics/story/6450422/the_curse_of_dic_cheney/print.
7. Sidney Blumenthal, "The Long March of Dick Cheney," *Salon,* November 25, 2005.
8. *The Fifty Years War: Israel and the Arabs,* directed by David Ash and Dai Richards, PBS Home Video, 2001.
9. Abraham Foxman, "Evangelical Support for Israeli Is a Good Thing," http://www.adl.org/Israel/evangelical.asp.
10. Dan Cohn-Sherbok, *The Politics of the Apocalypse: The History and Influence of Christian Zionism* (Oxford, England: Oneworld Publications, 2006), pp. 144, 145.
11. Sunday, a Ku Klux Klan sympathizer, asserted that "Jew blood means the capacity for making money." William Martin, *With God on Our Side: The Rise of the Religious Right in America* (New York: Broadway, 2005), p. 9.
12. Nancy K. MacLean, *Behind the Mask of Chivalry: The Making of the Second Ku Klux Klan* (New York: Oxford University Press, 1995), p. 94.
13. Martin, *With God on Our Side,* pp. 19, 20.
14. "Tuning Out," *Time,* September 29, 1980, http://www.time.com/time/magazine/article/0,9171,952810,00.html.
15. Michael Lind, "Rev. Robertson's Grand International Conspiracy Theory," *New York Review of Books,* February 2, 1995. And Colin Shindler, "Likud and the Christian Dispensationalists: A Symbiotic Relationship," *Israel Studies* 5, no. 1 (March 31, 2000): 153.
16. Donald Wagner, "Christians and Zion," *Daily Star,* http://www.informationclearinghouse.info/article4959.htm.
17. David Parson, *Swords into Ploughshares: Christian Zionism and the Battle of Armageddon,* a publication of the International Christian Embassy in Jerusalem, p. 6.
18. Barbara Tuchman, *Bible and Sword: England and Palestine from the Bronze Age to Balfour* (New York: Ballantine, 1984), p. 122. And Donald Wagner, "Evangelicals

and Israel: Theological Roots of a Political Alliance," *Christian Century,* November 4, 1998; and "Short Fuse to Apocalypse?," *Sojourner Magazine,* http://www.sojo.net/index.cfm?action=magazine.article&issue=soj0307&article=030710.
19. Wagner, "Christians and Zion."
20. Wagner, "Short Fuse to Apocalypse?"
21. Christopher Sykes, *Two Studies in Virtue* (New York: Knopf, 1953), pp. 150–51.
22. Donald Wagner, "Reagan and Begin, Bibi and Jerry: The Theopolitical Alliance of the Likud Party with the American Christian 'Right,'" *Arab Studies Quarterly* (ASQ) 20, no. 4 (September 22, 1998): p. 33.
23. Ibid.
24. Sykes, *Two Studies in Virtue,* p. 170.
25. David Fromkin, *A Peace to End All Peace: The Fall of the Ottoman Empire and the Creation of the Modern Middle East* (New York: Owl Books, 2001), p. 271.
26. Wagner, "Reagan and Begin, Bibi and Jerry."
27. Fromkin, *A Peace to End All Peace,* pp. 271, 273.
28. Ibid., p. 270.
29. "Dead Aristocrats Hidden Flu Clue," http://news.bbc.co.uk/2/hi/uk_news/england/humber/6402539.stm.
30. Fromkin, *A Peace to End All Peace,* p. 27.
31. Ibid., pp. 269–70. And Sykes, *Two Studies in Virtue,* p. 189.
32. Ristov Stefov, "Paris Peace Talks of 1919, Part II, End of the Ottoman Empire," http://www.maknews.com/html/articles/stefov/stefov49.html.
33. Wagner, "Reagan and Begin, Bibi and Jerry."
34. http://www.mideastweb.org/palpop.htm.
35. Timothy P. Weber, *On the Road to Armageddon: How Evangelicals Became Israel's Best Friend* (Grand Rapids, Mich.: Baker Academic, 2005), p. 173.
36. Ibid.
37. Author interview with Yechiel Eckstein, July 2005.
38. Author interview with Yechiel Kadishai, Jerusalem, June 2005.
39. Author interview with Jerry Falwell, May 2005.
40. Ibid.
41. Grace Halsell, *Prophecy and Politics* (Chicago: Lawrence Hill, 1986), pp. 75, 76.
42. Interview with Jerry Falwell.
43. Craig Unger, *House of Bush, House of Saud: The Secret Relationship Between the World's Two Most Powerful Dynasties* (New York: Scribner, 2004), p. 67.
44. Weber, *On the Road to Armageddon,* p. 219.
45. Interview with Jerry Falwell.
46. Weber, *On the Road to Armageddon,* p. 219.
47. Ibid., p. 214.
48. Ibid., pp. 216–17.
49. Ibid., p. 215.
50. International Fellowship of Christians and Jews, http://www.ifcj.org/site/PageServer.
51. Ibid.
52. Interview with Yechiel Eckstein.
53. Wagner, "Reagan and Begin, Bibi and Jerry."
54. Ibid.
55. Author interview with Benny Elon, June 2005.
56. Author interview with Uzi Arad, June 2005.
57. Edward Tivnan, *The Lobby: Jewish Political Power and American Foreign Policy* (New York: Simon & Schuster, 1987), p. 182.
58. Allan C. Brownfeld, "Fundamentalists and the Millennium: A Potential Threat to Middle Eastern Peace," *Washington Report on Middle East Affairs* 18, no. 4 (June 30, 1999): 82.
59. Author interview with Gershom Gorenberg, June 2005.
60. Ray Moseley, "Arens rejects U.S. idea on PLO," *Chicago Tribune,* March 15, 1989, p. 1.

61. "U.S. Plan Urges Israel to Pull Out of Occupied Cities," *The Toronto Star,* March 30, 1989, p. A34.
62. Richard C. Hottelet, "Pushing US Mideast Policy Out of Its Rut," *Christian Science Monitor,* August 10, 1989, p. 18.
63. Ray Moseley, "Shamir: Israel Can't Yield Lands," *Chicago Tribune,* April 7, 1989, p. 1.
64. Stephen Pizzo, "Bush Family Values," *Mother Jones,* http://www.motherjones.com/news/feature/1992/09/bushboys.html.
65. "The War Behind Closed Doors," *Frontline,* PBS, http://www.pbs.org/wgbh/pages/frontline/shows/iraq/etc/synopsis.html.
66. Jim Lobe, "Chickenhawk Groupthink?" Inter Press Service, CommonDreams.org, May 12, 2004, http://www.commondreams.org/headlines04/0512–02.htm.
67. Kenneth T. Walsh, "A Rough Road for 'Scooter,'" *U.S. News & World Report,* October 23, 2005, http://www.usnews.com/usnews/news/articles/051031/31libby.htm.
68. "The War Behind Closed Doors."
69. "Excerpts From Pentagon's Plan: 'Prevent the Re-Emergence of a New Rival,'" *New York Times,* March 8, 1992, sec. 1, p. 14.
70. Patrick Tyler, "U.S. Strategy Plan Calls for Insuring No Rivals Develop a One-Superpower World," *New York Times,* March 8, 1992, http://work.colum.edu/~amiller/wolfowitz1992.htm.
71. Frances FitzGerald, "George Bush & the World," *New York Review of Books,* September 26, 2002.
72. Stefan Halper and Jonathan Clarke, *America Alone: The Neoconservatives and the Global Order* (New York: Cambridge University Press, 2004), p. 145.
73. Barton Gellman, "Keeping the U.S. First, Pentagon Would Preclude a Rival Superpower," *Washington Post,* March 11, 1992.
74. Ibid.
75. "Excerpts from 1992 Draft 'Defense Planning Guidance," *Frontline,* PBS, http://www.pbs.org/wgbh/pages/frontline/shows/iraq/etc/wolf.html.
76. Tyler, "U.S. Strategy Plan Calls for Insuring No Rivals Develop a One-Superpower World."
77. Anthony David, "The Apprentice," *American Prospect,* http://www.prospect.org/cs/articles?article=the_apprentice.
78. Ibid.
79. Tyler, "U.S. Strategy Plan Calls for Insuring No Rivals Develop a One-Superpower World."
80. R. W. Apple Jr., "The 1992 Campaign; Suddenly, the Choices Are Clearer," *New York Times,* March 20, 1992, p. A1.

Chapter Eight: First Son

1. Calvin Skaggs and David Van Taylor, *With God on Our Side: George W. Bush and the Rise of the Religious Right,* documentary film, 2004.
2. "October: Mayday for Marriage," http://www.pfaw.org/pfaw/general/default.aspx?oid=17324&print=yes.
3. David Jackson, "George W. Bush Considering '94 Race for Governor," *Dallas Morning News,* February 12, 1993, p. A29.
4. David Hackworth, "Saddam Fiddles, We Dance," *Newsweek,* January 25, 1993 , p. 44.
5. "2 U.S. Jets Attack Radar Site in Iraq; Clinton Aides Cite Provocation," *Buffalo News,* January 21, 1993, p. 1.
6. Scott Sherman, "The Avenger; Sy Hersh, Then and Now,"*Columbia Journalism Review,* July-August 2003, p. 34. And Chaim Kaufmann, "Threat Inflation and the Failure of the Marketplace of Ideas; The Selling of the Iraq War," *International Security,* Summer 2004, p. 5.
7. Jim Lobe, " So, Did Saddam Hussein Try to Kill Bush's Dad?," Inter Press Service,

October 19, 2004, http://www.commondreams.org/headlines04/ 1019–05.htm.

8. Thomas L. Friedman, "The World; Writing New Rules For Criticizing Israel from Afar," *New York Times,* July 11, 1993,sec. 4, p. 3.
9. Nahum Barnea, "Painful Acupuncture," *Yedioth Ahronoth,* November 7, 2005, http://www.ynetnews.com/articles/0,7340,L-3107115,00.html. And author interview with Rabinovitch.
10. Bill Clinton, *My Life* (New York: Random House, 2004), p. 543.
11. Ibid., p. 544.
12. Ibid.
13. "Zion's Christian Soldiers," *60 Minutes,* CBS, http://www.cbsnews.com/stories/2002/10/03/60minutes/main524268.shtml.
14. Michael Karpin and Ina Friedman, *Murder in the Name of God: The Plot to Kill Yitzhak Rabin* (New York: Metropolitan Books, 1998), p. 22.
15. Ibid.
16. W. Patrick Lang, "Drinking the Kool-Aid," *Middle East Policy* 11, no. 2 (June 22, 2004): 39, http://www.mepc.org/journal_vol11/0406_lang.asp.
17. John Dizard, "How Chalabi Conned the Neocons," *Salon,* http://dir.salon.com/story/news/feature/2004/05/04/chalabi/index.html?pn=3.
18. Jane Mayer, "A Reporter at Large: The Manipulator," *The New Yorker,* http://www.newyorker .com/archive/2004/06/07/040607fa_fact1?currentPage=3.
19. Ibid.
20. Dizard, "How Chalabi Conned the Neocons."
21. Mayer, "A Reporter at Large: The Manipulator."
22. Ibid.
23. Stefan Halper and Jonathan Clarke, *America Alone: The Neoconservatives and the Global Order* (New York: Cambridge University Press, 2004), p. 220.
24. David Ignatius, "The War of Choice and the One Who Chose It," *Washington Post,* November 2, 2003, p. B1.
25. Lang, "Drinking the Kool-Aid."
26. Halper and Clarke, *America Alone,* p. 220.
27. Dizard, "How Chalabi Conned the Neocons."
28. Mayer, "A Reporter at Large: The Manipulator."
29. Dizard, "How Chalabi Conned the Neocons."
30. Robert Dreyfuss, *American Prospect,* http://www.prospect.org/print/V13/21/dreyfuss-r.html.
31. Ibid.
32. Mayer, "A Reporter at Large: The Manipulator."
33. Ibid.
34. James Bamford, "The Man Who Sold the War," *Rolling Stone,* http:// www.rolling stone.com/politics/story/8798997/the_man_who_sold_the_war/3.
35. Author interview with Carlo Bonini, April 2007.
36. Dizard, "How Chalabi Conned the Neocons."
37. Lang, "Drinking the Kool-Aid."
38. Ibid.
39. Kim A. Lawton, "Whatever Happened to the Religious Right?" *Christianity Today* 33 (December 15, 1989).
40. Tom Strode, "SBC Resolution Rebukes Clinton for 'Gay Pride' Proclamation," http://www.sbcannualmeeting.org/sbc99/news55.htm.
41. Sidney Blumenthal, *The Clinton Wars* (New York: Penguin, 2004), p. 67.
42. *Reliable Sources,* CNN, July 3, 1994. Transcript #123.
43. Joe Conason and Gene Lyons, *The Hunting of the President: The Ten-Year Campaign to Destroy Bill and Hillary Clinton* (New York: St. Martin's Press. 2000), p. 137.
44. Ibid., p. 140.
45. Ibid., p. 142.
46. "Taking Over the Republican Party," http://www.theocracywatch.org/taking _over.htm.

47. Ibid.
48. Bob Woodward, *State of Denial: Bush at War, Part III* (New York: Simon & Schuster, 2006), p. 3.
49. Elsa Walsh, "The Prince: How the Saudi Ambassador Became Washington's indispensable operator," *The New Yorker,* March 24, 2003, p. 48.
50. David Brooks, "Texas Ranger; Did Running a Baseball Team Help Prepare George W. Bush to Run America?," *Weekly Standard,* December 13, 1999, p. 20.
51. Ibid.
52. Author interview with Mickey Herskowitz, February 2007.
53. Carl Cannon, Lou Dubose, and Jan Reid, *Boy Genius: Karl Rove, the Architect of George W. Bush's Remarkable Political Triumphs* (New York: Public Affairs, 2005), p. 68.
54. David S. Broder, "GOP Poised to Grab Control of Senate; Survey Shows Party with Edge in Close Races," *Washington Post,* November 6, 1994, p. A1. And R. W. Apple, "The 1994 Campaign: Key Contests; Watching the Results," *New York Times,* November 8, 1994, p. A21.
55. Cannon, Dubose, and Reid, *Boy Genius,* p. 72.
56. Sam Howe Verhovek, "The Bushes; Texas Elects George W. While Florida Rejects Jeb," *New York Times,* November 9, 1994, p. B4.

Chapter Nine: The Righteous Assassin

1. Author interview with Gershom Solomon, June 2005.
2. Michael Oren, *Six Days of War: June 1967 and the Making of the Modern Middle East* (New York: Presidio Press, 2003), p. 246.
3. Interview with Gershom Solomon.
4. Robert I. Friedman, *Zealots for Zion: Inside Israel's West Bank Settlement Movement* (Piscataway, N.J.: Rutgers University Press, 1994), p. 19.
5. Robert I. Friedman, *The False Prophet: Rabbi Meir Kahane—From FBI Informant to Knesset Member* (London: Faber and Faber, 1990), jacket blurb.
6. Donald Wagner, "Reagan and Begin, Bibi and Jerry: The Theopolitical Alliance of the Likud Party with the American Christian 'Right,'" *Arab Studies Quarterly* (ASQ) 20, no. 4 (September 22, 1998): 33.
7. Author interview with Yitzhak Fhantich, June 2005.
8. Ibid. Also, Allyn Fisher, "Shamir Calls for International Recognition of Jerusalem," Associated Press, May 30, 1984.
9. Interview with Yitzhak Fhantich.
10. Oren, *Six Days of War,* p. 309.
11. Ibid., p. 246.
12. Michael Karpin and Ina Friedman, *Murder in the Name of God: The Plot to Kill Yitzhak Rabin* (New York: Metropolitan Books, 1998), p. 22.
13. Ibid., p. 59.
14. Interview with Yitzhak Fhantich.
15. Karpin and Friedman, *Murder in the Name of God,* pp. 114–15.
16. Author interview with Itamar Rabinovitch, June 2005.
17. Karpin and Friedman, *Murder in the Name of God,* p. 60.
18. Ibid., pp. 44–45.
19. Jonathan Mark, "Hebron and the Bias Factor," *Jewish Week* 206, no. 47 (March 31, 1994): 19.
20. Ibid.
21. The *New York Times* reported only 10,000 demonstrators, but the Israeli press cited police officials saying over 100,000 people attended the rally. Herb Keinon and Bill Hutman, "100,000 Protest in Jerusalem Against Arafat," *Jerusalem Post,* July 3, 1994, p. 1.
22. Karpin and Friedman, *Murder in the Name of God,* p. 77.
23. Ibid., p. 78.

24. Joel Greenberg, "Angry Crowds in Jerusalem Protest Arafat's Visit to Gaza," *New York Times,* July 3, 1994, p. A8.
25. Karpin and Friedman, *Murder in the Name of God,* pp. 114–15.
26. Ibid., p. 105.
27. Interview with Yitzhak Fhantich.
28. Karpin and Friedman, *Murder in the Name of God,* p. 25.
29. Ibid., p. 27.
30. Interview with Yitzhak Fhantich.
31. Karpin and Friedman, *Murder in the Name of God,* pp. 147–48.
32. Cynthia Mann, "Behind the Headlines: Rabin's Death Leads U.S. Jews to Reflect on Impact of Rhetoric," *Jewish Telegraphic Agency,* November 5, 1995, p. 4.
33. Karpin and Friedman, *Murder in the Name of God,* p. 119.
34. Ibid., p. 130.
35. Ibid., p. 191.
36. "Knesset Debates Oslo B. Agreement," *BBC Summary of World Broadcasts,* October 7, 1995.
37. Karpin and Friedman, *Murder in the Name of God,* p. 93.
38. Hillell Halkin, "Israel & the Assassination: A Reckoning," *Commentary* 101, no. 1 (January 1996): 23.
39. Karpin and Friedman, *Murder in the Name of God,* p. 94.
40. Ibid., p. 97.
41. Ibid., p. 27.
42. Guilain Denoeuz, "Murder in the Name of God: The Plot to Kill Yitzhak Rabin," *Middle East Policy* 6, no. 4 (June 1, 1999): 203.
43. Karpin and Friedman, *Murder in the Name of God,* pp. 4, 5.
44. Ibid., pp. 90–91.
45. Barton Gellman, "Israeli Prime Minister Rabin Is Killed; Jewish Gunman Says He Acted Alone," *Washington Post,* November 5, 1995, p. A1.
46. Karpin and Friedman, *Murder in the Name of God,* p. 177.
47. Gellman, "Israeli Prime Minister Rabin Is Killed; Jewish Gunman Says He Acted Alone."
48. Ibid.
49. Derek Brown and David Hudson, "Political Theatre Ends in Tragedy; Tragic Ending to a Spectacular Political Display," *The Guardian* (London), November 6, 1995, p. 1.
50. Karpin and Friedman, *Murder in the Name of God,* p. 177.
51. Ibid.
52. Gellman, "Israeli Prime Minister Rabin Is Killed; Jewish Gunman Says He Acted Alone."
53. "Yitzak Rabbin Assasinated: Leaders Across the World Praise His Peace Efforts," *Providence Journal-Bulletin* (Rhode Island), November 5, 1995, p. A1.
54. Mark Matthews, "Clinton Mourns a Friend and Ally," *The Sun* (Baltimore), November 5, 1995, p. A1.
55. Jim Mann, "Assasination Aftermath; Policy; U.S. Seeks Hope Out of Israeli Tragedy," *Los Angeles Times,* November 6, 1995, p. A13.
56. "Zion's Christian Soldiers," *60 Minutes,* CBS, http://www.cbsnews.com/stories/2002/10/03/60minutes/main524268.shtml.
57. Karpin and Friedman, *Murder in the Name of God,* p. 177–93.
58. Barton Gellman, "Israel's Great Divide; Longing for Unity Vies with the Urge to Blame Divergent Views of Future Dampen Hopes in Israel," *Washington Post,* December 3, 1995, p. A29.
59. Leslie Susser, "Netanyahu's Israel," *The Jerusalem Report,* June 27, 1996, p. 14.
60. Yoram Hazony, "Arlosoroff's Assassination: A Cautionary Tale," *Weekly Standard* 1, no. 10 (November 20, 1995): 31.
61. Dennis Ross, *The Missing Peace: The Inside Story of the Fight for Middle East Peace* (New York: Farrar, Strauss and Giroux, 2005), p. 267.
62. Stephen Rodrick, "Puppetmaster: The Secret Life of Arthur J. Finkelstein," *Boston Magazine,* October 1996, p. 57ff.

63. Susser, "Netanyahu's Israel."
64. David Aikman, "Brethren in the Holy Land: Settlers and Christians," *Weekly Standard* 1, no. 8 (November 6, 1995): 34.
65. Donald Wagner, The Evangelical-Jewish Alliance, Religion-Online, http://www.religion-online.org/showarticle.asp?title=2717.
66. Susser, "Netanyahu's Israel."
67. Johanna McGreary, Lara Marloew, and J. F. O. McAllister, "The Right Way to Peace? Pledging to Both Support and Thwart the Process, Netanyahu Squeaked into Power," *Time,* June 1996.
68. Ross, *The Missing Peace,* p. 258.

Chapter Ten: Ripe for the Plucking

1. Richard Perle, James Colbert, Charles Fairbanks Jr., Douglas Feith, Robert Loewenberg, David Wurmser, and Meyrav Wurmser, "A Clean Break: A New Strategy for Securing the Realm," Institute for Advanced Strategic and Political Studies, http://www.iasps.org/strat1.htm.
2. Author interview with Meyrav Wurmser, September 2006.
3. Perle, Colbert, et al., "A Clean Break: A New Strategy for Securing the Realm."
4. Interview with Meyrav Wurmser.
5. Perle, Colbert, et al., "A Clean Break: A New Strategy for Securing the Realm."
6. Ibid.
7. Genesis 15:18.
8. Perle, Colbert, et al., "A Clean Break: A New Strategy for Securing the Realm."
9. Juan Cole, "Shiite Religious Parties Fill Vacuum in Southern Iraq," Middle East Report Online, http://www.merip.org/mero/mero042203.html.
10. Interview with Meyrav Wurmser.
11. William Kristol and Robery Kagan, "Toward a Neo-Reaganite Foreign Policy," *Foreign Affairs,* July-August 1996, p. 18.
12. William Kristol and Robert Kagan,"Saddam Must Go," *The Weekly Standard—A Reader: 1995–2005* (November 17, 1997), p. 218.
13. Charles Krauthammer, "A Weak Response to Saddam's Threat; Once Again, Neglect and Vacillation Characterize Foreign Policy in the Clinton Administration," *Pittsburgh Post-Gazette,* September 9, 1996, p. A11.
14. A. M. Rosenthal, "On My Mind; On Clinton's Watch," *New York Times,* September 10, 1996, p. A27.
15. Zalmay Khalilzad and Paul Wolfowitz, "We Must Lead the Way in Deposing Saddam," *Washington Post,* November 9, 1997, p. C9.
16. Robert Dreyfuss and Jason Vest, "The Lie Factory," *Mother Jones,* January 26, January-February 2004.
17. David Wurmser, *Tyranny's Ally: America's Failure to Defeat Saddam Hussein* (La Vergne, Tenn.: AEI Press, 1999).
18. Ibid., p. 74.
19. Ibid.
20. Robert Dreyfuss, "The Shia Fellas," *American Prospect,* http://www.prospect.org/cs/articles?article=the_shia_fellas.
21. William Kristol and David Brooks, "What Ails Conservatism," *Wall Street Journal,* September 15, 1997, referred to by Claes Ryn in *The Ideology of American Empire, Neo-conned!* (Vienna, Virg.: Again, Light in the Darkness Publications, 2005), p. 67.
22. Jonah Goldberg, "Baghdad Delenda Est," Part Two, *National Review,* http://www.nationalreview.com/goldberg/goldberg042302.asp.
23. Irving Kristol, *Neoconservatism: The Autobiography of an Idea* (New York: Free Press, 1995), p. 368, as cited in *Journal of Ecumenical Studies* 42, no. 1 (January 1, 2007): 76.
24. Norman Podhoretz, "In the Matter of Pat Robertson; Antisemitism Allegations," *Commentary* 100, no. 2 (August 1995): 27.

25. Hugh B. Urban, Machiavelli Meets the Religious Right: Michael Ledeen, the Neoconservatives, and the Political Uses of Fundamentalism," *Journal of Ecumenical Studies* 42, no. 1 (January 1, 2007): 76.
26. Michael Ledeen, *The First Duce: D'Annunzio at Fiume* (Baltimore, Md.: Johns Hopkins University Press, 1977), p. 202. See also Michael A. Ledeen, *Universal Fascism: The Theory and Practice of the Fascist International, 1928–1936* (New York: H. Fertig, 1972). Cited in "Machiavelli Meets the Religious Right: Michael Ledeen, the Neoconservatives, and the Political Uses of Fundamentalism," *Journal of Ecumenical Studies* 42, no. 1 (January 1, 2007): 76.
27. Jeff Sharlet, "Soldiers of Christ," *Harper's Magazine*, http://www.harpers.org/archive/2005/05/0080539.
28. Author interview with Jerry Falwell, May 2005. The $200 million in revenues was for 2005.
29. William C. Symonds with Brian Grow and John Cady, "Earthly Empires," http://www.businessweek.com/magazine/content/05_21/b3934001_mz001.htm?chan=search.
30. Ibid.
31. Charles Gibson, "NASCAR Takes Religion to the Raceway," http://abcnews.go.com/WNT/Beliefs/story?id=727941&page=1.
32. Heralds of the Cross Motorcycle Ministry, http://www.heraldsofthecross.org/.
33. "Christian rock," http://en.wikipedia.org/wiki/Christian_rock.
34. Alexandra Pelosi, *Friends of God*, a documentary for HBO, 2007.
35. Website for Christian furniture, http://www.bettybowers.com/marge.html.
36. Alex Fryer, "Scoop Jackson's Protégés Shaping Bush's Foreign Policy," http://seattletimes.nwsource.com/html/localnews/2001834779_jackson12m.html.
37. Karen R. Long, "'Left Behind' and the Rupture Over the Rapture," *Washington Post,* May 5, 2001, p. B9.
38. Joe Conason and Gene Lyons, *The Hunting of the President: The Ten-Year Campaign to Destroy Bill and Hillary Clinton* (New York: St. Martin's Griffin, 2000), p. 140.
39. Robert Dreyfuss, "Reverend Doomsday: According to Tim LaHaye, the Apocalypse Is Now," *http://www.rollingstone.com/politics/story/5939999/reverend_doomsday/*.
40. Conason and Lyons, *The Hunting of the President,* p. 138. And Frederick Clarkson, "The Clinton Contra's Smoke & Mirrors," *In These Times,* May 3, 1998, p. 11.
41. Author interview with Bob Strauss, February 2007.
42. Bob Woodward, *State of Denial: Bush at War, Part III* (New York: Simon & Schuster, 2006), p. 1.
43. Ibid., p. 3.
44. Ibid.
45. Ibid., p. 4.
46. Ibid.
47. Ibid., p. 5.
48. Ibid.
49. Ibid.
50. Yossi Beilin, "Only Clinton Can Break the Mideast Stalemate," *Washington Post,* January 19, 1998, p. A25.
51. Anthony Lewis, "Netanyahu Not Interested in Deal," *St. Louis Post-Dispatch* (Missouri), January 20, 1998, p. B7.
52. Interview with Jerry Falwell.
53. Sidney Blumenthal, *The Clinton Wars* (New York: Penguin, 2004), p. 321.
54. Ibid., p. 322.
55. Donald Wagner, "Reagan and Begin, Bibi and Jerry: The Theopolitical Alliance of the Likud Party with the American Christian 'Right,'" *Arab Studies Quarterly* (ASQ) 20, no. 4 (September 22, 1998), p. 33.
56. Barton Gellman, "Strong Ties of '48 Have Yielded to Today's Ambiguity," *Washington Post,* April 29, 1998, p. A1.

57. Hillel Kuttler, "Playing Politics with the President," *Jerusalem Post,* January 23, 1998, p. 13.
58. Ibid.
59. Interview with Jerry Falwell.
60. *Imus in the Morning,* January 20, 1998, Video Monitoring Services of America.
61. Interview with Jerry Falwell
62. "Letter to President Clinton on Iraq," The Project for a New American Century, http://www.newamericancentury.org/iraqclintonletter.htm.

Chapter Eleven: Dog Whistle Politics

1. Ted Cohen and Jack Beaudoin, "Kevin Costner Stops in to Visit Former President," *Portland Press Herald* (Maine), July 2, 1998, p. E1.
2. "Fantastic Four," *Texas Monthly,* August 1999, p. 111.
3. Stephen Mansfield, *The Faith of George W. Bush* (New York: Penguin, 2004), pp. 107–108
4. Fred Barnes, "The Gospel According to George W. Bush," *Weekly Standard,* March 22, 1999, p. 20.
5. John Dart, "Questions Linger on Robertson's Dropping Status as Minister," *Los Angeles Times,* October 17, 1987, p. B6.
6. Mansfield, *The Faith of George W. Bush,* p. 108.
7. Tom Hundley and Stephen J. Hodges, "Removal of Hussein Possible but Difficult," *Chicago Tribune,* November 22, 1998, p. 1.
8. Jane Mayer, "A Reporter at Large: The Manipulator," http://www.newyorker.com/archive/2004/06/07/040607fa_fact1?currentPage=7.
9. Charlie Rose interviews Zbigniew Brzezinksi, Henry Kissinger, and Brent Scowcroft, http://www.iht.com/articles/2007/06/18/america/web-rose.php?page=4.
10. Elizabeth Shogren, "U.S. Is Stuck with 'Grim' Iraq Options," *Los Angeles Times,* November 12, 1998, p. A1.
11. *CNN Early Edition,* CNN, November 16, 1998. Transcript # 98111609V08.
12. Alexander Moens, *The Foreign Policy of George W. Bush: Values, Strategy, and Loyalty* (Burlington, Vt.: Ashgate Publishing, 2004), p. 23.
13. Michael Kranish, "On the Road to the White House, Kosovo May Be a Speed Bump," *The Boston Globe,* April 6, 1999, p. 12.
14. Maria L. La Ganga, "Pop Quizzes Aside, Bush Continues Serious Study," *Los Angeles Times,* November 19, 1999, p. A1.
15. James Mann, *Rise of the Vulcans: The History of Bush's War Cabinet* (New York: Viking, 2004), p. 252.
16. Stefan Halper and Jonathan Clarke, *America Alone: The Neoconservatives and the Global Order* (New York: Cambridge University Press, 2004), pp. 112–13.
17. William Plummer, "Elliot Abrams Must Now Face the Music on Nicaragua and Congress Is Calling the Tune," *People,* June 8, 1987, p. 44.
18. Jim Lobe, "All in the Neocon Family," Inter Press Service, March 7, 2003, http://www.alternet.org/story/15481/?page=1.
19. Ibid.
20. http://www.commentarymagazine.com/cm/main/gotoArchive.html.
21. Edited by William Kristol, *The Weekly Standard: A Reader: 1995–2005* (New York: HarperCollins, 2005), p. vii-x.
22. Sidney Blumenthal, *Our Long National Daydream: A Political Pageant of the Reagan Era* (New York: HarperCollins, 1990), p. 226.
23. Author interview with Meyrav Wurmser, September 2005.
24. Blumenthal, *Our Long National Daydream,* p. 226.
25. Thomas Omestad, "An Alpha Hawk Spreads His Wings; Portrait: Richard Perle," November 17, 2002, http://www.usnews.com/usnews/news/articles/021125/archive_023377.htm. And http://www.amazon.com/Hard-Line-Richard-

Perle/dp/051710590X/ref=sr_1_7/102-6564264-6456152?ie=UTF8&s=books&qid=1184114075&sr=8-7.
26. Nicholas Lemann, "The Iraq Factor," *The New Yorker,* http://web.archive.org/web/20050211030910/newyorker.com/printable/?fact/010122fa_fact.
27. Interview with Meyrav Wurmser.
28. Dana Milbank, "What 'W' Stands For," *New Republic,* April 26, 1999, p. 66.
29. Jacob Weisberg, "The Misunderestimated Man," *Slate,* http://www.slate.com/id/2100064/.
30. Audrey Hudson, "Untraveled Bush Draws Global Mirth; Aide Hits 'Silly Standard,'" *Washington Times,* December 26, 2000, p. A4.
31. Lawrence F. Kaplan and William Kristol, *The War Over Iraq,* p. 63.
32. Halper and Clarke, *America Alone,* p. 4.
33. Robert Novak, "Inside the Bush 2000 Non-Campaign," *Austin American-Statesman,* Feburary 25, 1999, p. A19.
34. Bob Woodward, *State of Denial: Bush at War, Part III* (New York: Simon & Schuster, 2006), p. 13.
35. Peter Schweizer and Rochelle Schweizer, *The Bushes, Portrait of a Dynasty* (New York: Random House, 2004), p. 461.
36. George Bush and Brent Scowcroft, *A World Transformed,* p. 489.
37. Author interview with Mickey Herskowitz, May 2007.
38. Russ Baker interview with Mickey Herskowitz, http://www.gnn.tv/articles/article.php?id=761.
39. Interview with Mickey Herskowitz.
40. Milbank, "What 'W' Stands For."
41. Norman Podhoretz, "Strange Bedfellows: A Guide to the New Foreign—Policy Debates; Basic Issue of Whether the United States Is an Isolationist Country or Not," *Commentary* 108, no. 5 (December 1, 1999): 19.
42. Skipp Porteous, "Bush's Secret Religious Pandering," Public Eye, originally printed in *Penthouse,* http://www.publiceye.org/ifas/fw/0009/bush.html.
43. Ibid.
44. Joseph L. Conn, "Religious Right Interview Team Approves of GOP Candidate Bush," *Church & State* 52, no. 10 (November 1, 1999): 12.
45. Porteous, "Bush's Secret Religious Pandering."
46. Ibid.
47. Conn, "Religious Right Interview Team Approves of GOP Candidate Bush."
48. Author interview with Jerry Falwell, May 2005.
49. Jim Yardley, "The 2000 Campaign: The Governor's Speech; Bush's Words to Conservative Group Remain a Mystery," *New York Times,* May 19, 2000, p. A 22.
50. Ibid.
51. Stephen Bates, "The Christcoalition Nobody Knows," *Weekly Standard* 1, no. 2 (September 25, 1995): 36.
52. Joan Bokaer, Theocracy Watch, http://www.theocracywatch.org/taking_ over.htm.
53. Michelle Goldberg, *Kingdom Coming: The Rise of Christian Nationalism* (New York: W. W. Norton and Co., 2007), p. 110–11.
54. Ibid.
55. Ira Chernus, "'Compassionate Conservatism' Goes to War," http://www.commondreams.org/views01/1110-05.htm
56. Ibid.
57. Ibid.
58. E. J. Dionne Jr., "A Real Election," *Washington Post,* June 18, 1999, p. A41.
59. Dana Milbank, *Smashmouth: Two Years in the Gutter with Al Gore and George W. Bush* (New York: Basic Books, 2001), p. 84.
60. Ibid., p. 84.
61. Ann McFeatters, "Bush Drafting Call for Saddam's Ouster," *Pittsburgh Post-Gazette,* November 24, 1999, p. A12. Also, interview with Ann McFeatters.
62. McFeatters, "Bush Drafting Call for Saddam's Ouster."
63. Author interview with Ann McFeatters, April 2007.

64. *This Week,* ABC News, December 5, 1999.
65. David Nyhan, "A Bush Slip-up at the End," *Boston Globe,* December 3, 1999, p. A31.
66. Frank Bruni, "Bush Has Tough Words and Rough Enunciation for Iraqi Chief," *New York Times,* December 4, 1999, p. A12.
67. W. Patrick Lang, "Drinking the Kool-Aid," *Middle East Journal* 9, no. 2 (Summer 2004), http://www.mepc.org/journal_ vol11/0406_lang.asp.
68. *Decision 2000: The Presidential Debates,* MSNBC Special, October 3, 2000. Transcript #100301cb.455.
69. MSNBC Special, October 11, 2000. Transcript #101101cb.455.
70. Interview with Katherine Harris, *Florida Baptist Witness,* http://www.floridabaptistwitness.com/6298.article.
71. "The Woman in Charge," CBS News, http://www.cbsnews.com/stories/2000/11/15/politics/main249787.shtml.
72. Libby Copeland, "Campaign Gone South: Florida's Katherine Harris Continues Her Senate Race, Shedding Staff Along the Way," *Washington Post,* October 31, 2006, http://www.washingtonpost.com/wp-dyn/content/article/2006/10/30/AR2006103001311_pf.html.
73. Jeffrey Toobin, *Too Close to Call: The Thirty-Six-Day Battle to Decide the 2000 Election* (New York: Random House, 2001), p. 58.
74. Ibid., p. 61.
75. Ibid., p.75.
76. Ibid., p. 65.
77. Ibid., p. 66.
78. Ibid., p. 101.
79. Interview with Katherine Harris, *Florida Baptist Witness.*
80. Calvin Skaggs and David Van Taylor, *With God on Our Side: George W. Bush and the Rise of the Religious Right,* documentary film, 2004.
81. Ibid.
82. Ibid.

Chapter Twelve: Grandmaster Cheney

1. E-mail memo from a Pentagon official who served under Cheney during the administration of George H.W. Bush, June 2007.
2. Ibid.
3. "Remarks of the Secretary of Defense Dick Cheney to the Discovery Institute, Seattle, Washington," Federal News Service, August 14, 1992.
4. Victor Gold, *Invasion of the Party Snatchers: How the Holy-Rollers and the Neo-Cons Destroyed the GOP* (Chicago: Sourcebooks Trade, 2007), p. 62.
5. James Mann, "The True Rationale? It's a Decade Old," *Washington Post,* March 7, 2004.
6. E-mail memo from a Pentagon official who served under Cheney during the administration of George H.W. Bush.
7. http://select.nytimes.com/gst/abstract.html?res=F00612FD3F5D0C7A8CDDA80994DD494D81&n=Top%2fOpinion%2fEditorials%20and%20Op%2dEd%2fOp%2dEd%2fColumnists%2fMaureen%20Dowd.
8. Karen DeYoung, *Soldier: The Life of Colin Powell* (New York: Knopf, 2006), p. 243.
9. E-mail memo from a Pentagon official who served under Cheney during the administration of George H.W. Bush.
10. Lawrence K. Altman, "Cheney Is Likely to Recover Quickly, Hospital Says," *New York Times,* November 23, 2000, p. A35.
11. "Inauguration Eve: Washington Welcomes the Bushes," *Larry King Live,* CNN, January 19, 2001, Transcript #01011900V22.
12. E-mail memo from a Pentagon official who served under Cheney during the administration of George H.W. Bush.

13. DeYoung, *Soldier,* p. 295.
14. Ibid., p. 296.
15. Ibid., p. 297.
16. Ibid., p. 296.
17. E-mail memo from a Pentagon official who served under Cheney during the administration of George H.W. Bush.
18. DeYoung, *Soldier,* p. 297.
19. Dana Milbank and Mike Allen, "Powell Is Named Secretary of State; Nominee 1st African American Tapped for Post," *Washington Post,* December 17, 2000, p. A1.
20. Thomas L. Friedman, "The Powell Perplex," *New York Times,* December 19, 2000, p. A35.
21. Andrew Cockburn, *Rumsfeld: His Rise, Fall, and Catastrophic Legacy* (New York: Scribner, 2007), p. 96.
22. Vicki Haddock, "Son of a Preacher Man," *San Francisco Chronicle,* August 4, 2002, http://www.sfgate.com/cgi-bin/article.cgi?f=/c/a/2002/08/04/IN136349.DTL.
23. Calvin Skaggs and David Van Taylor, *With God on Our Side: George W. Bush and the Rise of Religious Right,* documentary film, 2004.
24. Jay Tolson, "The New School Spirit," *U.S. News & World Report,* February 14, 2005.
25. "Quotes of the Week," *U.S. News & World Report,* December 7, 1987, p. 13.
26. http://www.dailyhowler.com/h051401_1.shtml. Originally, Gerald M. Boyd, "Summit Aftermath: Gorbachev Visit Called Boon to Bush," *New York Times,* December 12, 1987, p. A9.
27. E-mail memo from a Pentagon official who served under Cheney during the administration of George H.W. Bush.
28. Author interview with Michael Janeway, April 2007.
29. Author interview with Pete Teeley, April 2007.
30. Cockburn, *Rumsfeld,* p. 31.
31. Ibid., p. 97.
32. Ibid., p. 80.
33. Sidney Blumenthal, *The Rise of the Counter-Establishment: From Conservative Ideology to Political Power* (New York: HarperCollins, 1988), p. 144.
34. Sharon Churcher and Annette Witheridge, "Will a British Divorcee cost 'Wolfie' His Job?" *Mail on Sunday,* March 20. 2005, http://www.dailymail .co.uk/pages/live/articles/news/news.html?in_article_id=342048&in_page_id=1770.
35. "Profile: Paul Wolfowitz: Hawk with a Lot of Loot Needs a Bit of Lady Luck," *Sunday Times,* http://www.timesonline.co.uk/tol/comment/article432446.ece.
36. Linton Weeks and Richard Leiby, "In the Shadow of a Scandal: Shaha Riza Remains the Mystery Woman from the World Bank," *Washington Post,* May 10, 2007, http://www.washingtonpost.com/wp-dyn/content/article/2007/05/09/AR2007050902501_pf.html.
37. Sidney Blumenthal, "Wolfowitz's Tomb," *Salon,* May 24, 2007, http://www.salon.com/opinion/blumenthal/2007/05/24/wolfowitz_aftermath/index1.html.
38. *The Nelson Report,* April 16, 2007.
39. Ward Harkavy, "Libby and Wolfie: A Story of Reacharounds," *Village Voice,* June 14, 2007, http://www.villagevoice.com/blogs/bushbeat/archive/2007/06/libby_and_wolfi.php; Blumenthal, "Wolfowitz's Tomb"; "Wolfowitz," World Bank President, http://www.worldbankpresident.org/archives/000742.php; and "Comments: What Happened to Wolfowitz the Strategist?" Washington Note, http://www.thewashingtonnote.com/mt/mt-comments.cgi?entry_id=2136.
40. "Comments: What Happened to Wolfowitz the Strategist?" Washington Note.
41. Michael Isikoff and Evan Thomas, "Follow the Yellowcake Road," *Newsweek,* July 28, 2003.
42. Brian Blomquist, "Bushed W. Kicks Back After a Two-Week Rush," *New York Post,* December 24, 2000, p. 8.
43. Peter Nicholas, "Fortune 500 Gives Big for Inauguration; 250 Firms and People Gave $100,000 Each as Bush's Camp Raised $40 Million for the Bash, *Philadelphia Inquirer,* January 19, 2001, p. A23.

44. Thomas B. Edsall and Dana Milbank, "White House's Roving Eye for Politics; President's Most Powerful Adviser May Also Be the Most Connected," *Washington Post,* March 10, 2003, p. A1, http://www.washingtonpost.com/ac2/wp-dyn?pagename=article&contentId=A2674-2003Mar9¬Found=true.
45. "Comments: What Happened to Wolfowitz the Strategist."
46. Thomas E. Ricks, "Rumsfeld Impresses Armed Services Panel; Wolfowitz, a Cheney Protégé, Is Chosen Over Powell Ally for No. 2 Post at the Pentagon," *Washington Post,* January 12, 2001p. A16.
47. Richard Clarke, *Against All Enemies: Inside America's War on Terror* (New York: Free Press, 2004), p. 30.
48. *ABC News Special Report: The 2001 Inauguration,* January 20, 2001.
49. Peter Nicholas, "Clinton's Final 24 Hours? Frantic Amid a Flurry of Activity, He Actually Found Time to Pack," *Philadelphia Inquirer,* January 21, 2001, p. A12.
50. Olivia Barker, Maria Montoya, and Maria Puente, "Grand Old Party Parties Down," *USA Today,* January 22, 2001, p. D2.
51. William Glanz, "Lanham, Va.–Based Inaugural Float Creations Date Back to Truman," *Washington Times,* January 17, 2001.
52. Robert Salladay, "A Two-Step Program for D.C. Society," *San Francisco Chronicle,* December 24, 2000, p. A25.
53. Michael Kilian, "Inaugural Ball; Bush Goings-on Go Clinton One Better Except in Celebrity Dept.," *Chicago Tribune,* January 18, 2001, p. 5.
54. Barker, Montoya, and Puente, "Grand Old Party Parties Down."
55. Larry Witham, "Bush Speech Retains Religious Flourishes," *Washington Times,* January 21, 2001, p. A6.
56. http://www.whitehouse.gov/news/inaugural-address.html.
57. Ibid.
58. Sidney Blumenthal, "Happy Talk," *The Guardian* (London), http://www.guardian.co.uk/usa/story/0,12271,1390294,00.html.
59. Book of Nahum, 1:3.
60. Book of Nahum, 1:5.
61. http://www.whitehouse.gov/news/inaugural-address.html.

Chapter Thirteen: Cheney's Gambit

1. *George W. Bush: Faith in the White House,* DVD documentary, executive producers Ted Beckett and Audrey Beckett. And Dana Milbank, "One Committee's Three Hours of Inquiry, in Surreal Time," *Washington Post, June 23,* 2005, http://www.washingtonpost.com/wpdyn/content/article/2005/06/22/AR200506220 2100_pf.html
2. *George W. Bush: Faith in the White House,* DVD documentary.
3. Steve Erickson, "George Bush and the Treacherous Country," *LA Weekly,* http://www.laweekly.com/general/features/george-bush-and-the-treacherous-country/1987/.
4. Hanna Rosin, "The Annals of Education: God and Country," *The New Yorker,* June 26, 2005, http://www.newyorker.com/archive/2005/06/27/050627fa_fact?printable=true.
5. Max Blumenthal, "Monica Goodling, One of 150 Pat Robertson Cadres in the Bush Administration," *Huffington Post,* March 30, 2007, http://www.huffingtonpost.com/max-blumenthal/monica-goodling-one-of-1_b_44588.html. Plus Frederick Clarkson, "Theocratic Dominionism Gains Influence: No Longer Without Sheep," PublicEye.org, March/June 1994, http://www.publiceye.org/magazine/v08n1/chrisre3.html, and http://www.regent.edu/about_us/quick_ facts.cfm.
6. Calvin Skaggs and David Van Taylor, *With God on Our Side: George W. Bush and the Rise of the Religious Right,* documentary film, 2004.
7. Author interview with Jerry Falwell, May 2005.

8. Skaggs and Taylor, *With God on Our Side*.
9. Barton Gellman and Jo Becker, "A Different Understanding with the President," *Washington Post,* June 24, 2007, p. A1.
10. Sebastian Mallaby, "The Character Question," *Washington Post,* August 30, 2004, http://www.washingtonpost.com/wp-dyn/articles/A45418-2004Aug29.html.
11. Bob Woodward, *State of Denial: Bush at War, Part III* (New York, Simon & Schuster, 2006), p. 12.
12. Walter Pincus, "Under Bush, the Briefing Gets Briefer: Key Intelligence Report by CIA and FBI Is Shorter, 'More Targeted,' Limited to Smaller Circle of Top Officials and Advisers, May 24, 2002, http://www.washingtonpost.com/ ac2/wp-dyn?page name=article&node=&contentId=A2130–2002May23¬Found=true.
13. Gellman and Becker, "A Different Understanding with the President."
14. Paul Kiel, "Cheney: You Can't Touch This," originally appeared in *Washington Post,* May 17, 2007, http://www.tpmmuckraker.com/archives/003245.php.
15. Gellman and Becker, "A Different Understanding with the President."
16. Ibid.
17. Daniel Benjamin, "President Cheney: His Office Really Does Run National Security," *Slate,* November 5, 2005, http://www.slate.com/id/2129686/#return,
18. Eric Schmitt, "Cheney Assembles Formidable Team," *New York Times,* February 3, 2001.
19. Thomas Ricks, "Iraq War Planner Downplays Role," *Washington Post,* October 22, 2003.
20. Jane Mayer, "The Hidden Power; The Legal Mind Behind the White House's War on Terror, *The New Yorker,* July 3, 2006, p. 44.
21. Ibid.
22. Nicholas Lemann, "The Quiet Man: Dick Cheney's Discreet Rise to Unprecedented Power," *The New Yorker,* May 7, 2001, p. 56.
23. Ron Suskind, *The Price of Loyalty: George W. Bush, the White House and the Education of Paul O'Neill* (New York: Pocket Books, 2004), p. 72–73.
24. Ibid.
25. "Reversal of Course," ABC's *Nightline,* March 30, 2004.
26. James Bamford, *A Pretext for War: 9/11, Iraq, and the Abuse of America's Intelligence Agencies* (New York: Doubleday, 2004), p. 267.
27. Energy Information Administration, http://www.eia.doe.gov/emeu/cabs/Iraq/ Oil.html.
28. "Bush Sought 'Way' to Invade Iraq? O'Neill Tells '60 Minutes' Iraq Was 'Topic A' 8 Months Before 9-11," http://www.cbsnews.com/stories/2004/01/09/ 60minutes/ main592330.shtml?source=search_story.
29. "Iraq's Oil," Online NewsHour, http://www.pbs.org/newshour/bb/middle_ east/iraq/oil_4–24–03.html.
30. Antonia Juhasz, "Whose Oil Is It, Anyway?" *New York Times,* March 13, 2007, http://www.nytimes.com/2007/03/13/opinion/13juhasz.html?ex=1331438400&en= 8289271df648123e&ei=5088&partner=rssnyt&emc=rss.
31. The full text of Dick Cheney's speech at the Institute of Petroleum autumn lunch, 1999, is available at http://www.energybulletin.net/559.html.
32. Michael Abramowitz and Steven Mufson, "Papers Detail Industry's Role in Cheney's Energy Report," *Washington Post,* July 18, 2007, http://www.washington post.com/wp-dyn/content/article/2007/07/17/AR2007071701987_3.html?nav=rss_ politics.
33. Mark LeVine, "War for Oil? George Clooney's New Movie Hits Uncomfortably Close to Home," *Mother Jones,* November 30, 2005, http://www.motherjones .com/arts/feature/2005/11/syriana.html.
34. Jane Mayer, "Contract Sport; What Did the Vice-President Do for Halliburton?" *The New Yorker,* February 16, 2004, p. 80.
35. Ibid.
36. Jane Mayer, "Reporter at Large: The Manipulator," *The New Yorker,* http:// www.newyorker.com/archive/2004/06/07/040607fa_fact1?printable=true.

37. http://www.whitehouse.gov/news/releases/20010206-7.html.
38. Lawrence F. Kaplan, "Drill Sergeant," *New Republic,* March 26, 2001, p. 17.
39. Al Kamen, "A Bent for Brent," *Washington Post,* May 11, 2001, p. A43.
40. Walsh, Iran-Contra Report, Chapter 25, http://www.fas.org/irp/offdocs/walsh/chap_25.htm.
41. Robert Kagan and William Kristol, "Foreign Policy Challenges of the New Administration; While Bush May Delegate, He Must Chart the Course," *San Diego Union-Tribune,* January 21, 2001, p. G4.
42. Jane Perlez, "Capitol Hawks Seek Tougher Line on Iraq," *New York Times,* March 7, 2001, p. 10.
43. "Military Strikes Against Iraq: A Show of Power or an Exercise in Futility?" CNN's *Crossfire,* February 16, 2001, Transcript #01021600V20.
44. Jim Hoagland, "Policy Wars Over Iraq," *Washington Post,* April 8, 2001, p. B7.
45. William Kristol, ed., *The Weekly Standard, A Reader: 1995–2005,* p. 251
46. Author interview with Patrick Lang, March 2007.
47. Ibid.
48. Joshua Micah Marshall, "The Italian Connection, Part III," Talking Points Memo, November 10, 2005, http://www.talkingpointsmemo.com/archives/006975.php.
49. Craig Unger, "The War They Wanted, the Lies They Needed," *Vanity Fair,* July 2006, http://www.vanityfair.com/politics/features/2006/07/yellowcake200607?printable=true&CurrentPage=all.
50. Author interview with Ray McGovern, September 2006.
51. Author interview with Larry Wilkerson, October 2006, and Unger, "The War They Wanted, the Lies They Needed."
52. Author interview with Karen Kwiatkowski, December 2005.
53. Author interview with Joseph Wilson, September 2006.
54. Tom Hamburger, Peter Wallsten, and Bob Drogin, "French Told CIA of Bogus Intelligence," *Los Angeles Times,* December 11, 2005, p. A1.
55. Elsa Walsh, "The Prince," *The New Yorker,* March 24, 2003.
56. Jane Perlez, "Bush and Sharon Differ on Ending Violence," *New York Times,* June 27, 2001, p. A9.
57. Ibid.
58. Jane Perlez, "Bush Senior, on His Son's Behalf, Reassures Saudi Leader," *New York Times,* July 15, 2001, p. A6.
59. Ibid.
60. Richard L. Berke, "G.O.P. Defends Bush in Face of Dip in Poll Ratings," *New York Times,* June 29, 2001, p. A19.
61. http://www.drpeppermuseum.com/.
62. "Ted Nugent to Fellow NRAers: Get Hardcore," Associated Press article, April 17, 2005. And "Ted Nugent: Off His Rocker?" *The Independent,* May 28, 2006.
63. David Jackson, "After Initial Success, Bush Agenda Falters," *Dallas Morning News,* August 6, 2001
64. http://www.whitehouse.gov/news/releases/2001/08/20010809–2.html.
65. "National Roundup," Herald Wire Services, *Miami Herald,* August 22, 2001, p. A20.
66. Michael Elliott, "How the U.S. Missed the Clues," *Time,* May 27, 2002.
67. "Bin Laden Determined to Strike in US," http://www.cnn.com/2004/ALLPOLITICS/04/10/august6.memo/index.html.
68. Johanna McGeary et al., "Odd Man Out; Colin Powell Is a Global Eminence," *Time,* September 10, 2001, p. 24.
69. Ibid.
70. Project for the New American Century, accessed May 30, 2007, www.newamericancentury.org/RebuildingAmericasDefenses.pdf.

Chapter Fourteen: In the Shadows

1. Michael Ledeen, "Code Alpha: And Other Useless Things That Got Us Here," *National Review Online,* September 11, 2001, http://www.nationalreview.com/contributors/ledeen091101.shtml.
2. "Plans For Iraq Attack Began On 9/11: Exclusive: Rumsfeld Sought Plan For Iraq Strike Hours After 9/11 Attack," http://www.cbsnews.com/stories/2002/09/04/september11/main520830.shtml.
3. Peter Bergen, "Armchair Provocateur: Laurie Mylroie: The Neocon's Favorite Conspiracy Theorist," *Washington Monthly,* December 2003, http://www .washingtonmonthly.com/features/2003/0312.bergen.html. And Daniel Benjamin and Steven Simon, *The Next Attack: The Failure of the War on Terror and a Strategy for Getting It Right* (New York: Owl Books, 2006), p. 145.
4. Michael Elliott, "'We're at War'; Washington Builds a Global Coalition and Prepares for Military Action in Afghanistan," *Time,* September 24, 2001, p. 40.
5. Jim Hoagland, "Hidden Hand of Horror," *Washington Post,* September 12, 2001, p. A31.
6. *CBS News Special Report,* "Continuing Coverage of Terrorist Attacs on America," September 12, 2001.
7. *ABC News Special Report,* "America Under Attack," September 12, 2001.
8. *The O'Reilly Factor,* Fox News Network, September 14, 2001. Transcript #091404cb.256.
9. Michael Barone, "War We Must Win," *U.S. News & World Report,* September 14, 2001, p. 55.
10. *Crossfire,* CNN, September 28, 2001. Transcript #092800CN.V20.
11. Richard Beeston, Saddam Should Not Be Ruled Out, Says US Intelligence," *The Times* (London), September 15, 2001.
12. Jim Landers, "Iraq Could Be Siding Bin Laden; Experts Believe Complex Operation Would Require Hussein's Knowledge," *Dallas Morning News,* September 15, 2001, p. A26.
13. *CBS News Special Report,* "Continuing Coverage of Terrorist Attacks on America," September 12, 2001.
14. Woodward, *Bush at War* (New York: Simon & Schuster, 2002), pp. 60–61.
15. Ibid., p. 61.
16. "Perle After Five Days: 'We Do Know Saddam Hussein Has Ties to Osama Bin Laden,'" *Think Progress,* http://thinkprogress.org/2007/04/30/perle-five-days-after/.
17. *Crossfire,* CNN, September 17, 2001. Transcript #091700CN.V20.
18. *Wolf Blitzer Reports,* "Investigation Continues into Terrorist Attacks," CNN, September 18, 2001. Transcript #091800CN.V67.
19. Geoffrey Wawro, United States Navy's *Naval War College Review* 55, no. 4 (Autumn 2002).
20. William Kristol, "Bush vs. Powell," *Washington Post,* September 25, 2001, p. A23.
21. "So Many Targets, So Little Time," *Washington Times,* September 26, 2001, p. A20.
22. "Closing the Gateway; Racial Profiling; Chicago Shudders; How Wide a War," *The NewsHour with Jim Lehrer,* PBS, September 26, 2001, Transcript #7165.
23. "Falwell Apologizes to Gays, Feminists, Lesbians," Cnn.com, September 14, 2001, http://archives.cnn.com/2001/US/09/14/Falwell.apology/.
24. http://www.raptureready.com/rap13a.html.
25. Calvin Skaggs and David Van Taylor, *With God on Our Side: George W. Bush and the Rise of the Religious Right,* documentary film, 2004.
26. Jerry Falwell in Skaggs and Taylor, *With God on Our Side.*
27. http://www.sourcewatch.org/index.php?title=Instruments_of_evil. And John Darby, "John Darby's Synopsis of the New Testament: Commentary on Reveleation 13," http://www.searchgodsword.org/com/dsn/view.cgi?book=re& chapter=013.
28. "Pearl's Kidnappers Extend Execution Deadline; NRC Warns Nuclear Power

Plants to Be on Alert: Florida Vot rs Concerned About Reno's Health," CNN.com, http://transcripts.cnn.com/TRANSCRIPTS/0201/31/ip.00.html.

29. "Remarks by the President in Photo Opportunity with the National Security Teams," http://www.whitehouse.gov/news/releases/2001/09/20010912–4.html.
30. Max Blumenthal, "The Christian Right's Humble Servant," November 15, 2004, http://www.alternet.org/election04/20499/.
31. "President Bush's Approval Ratings," http://www.washingtonpost.com/wp-dyn/content/custom/2006/02/02/CU2006020201345.html. Plus "President Bush: Job Ratings," http://www.pollingreport.com/BushJob1.htm; and "Historical Bush Approval Ratings," http://www.hist.umn.edu/~ruggles/ Approval.htm.
32. Barton Gellman and Jo Becker, "A Different Understanding with the President," *Washington Post,* June 24, 2007, p. A1.
33. Ibid.
34. R. Jeffrey Smith, "Tyco Exec: Abramoff Claimed Ties to Administration," Washingtonpost.com, http://www.washingtonpost.com/wp-dyn/content/article/2005/09/22/AR2005092202204.html.
35. Joint Resolution: Authorization for Use of Military Force, Public Law 107-40 [S. J. RES. 23], 107th CONGRESS, September 18, 2001, http://news.findlaw.com/wp/docs/terrorism/sjres23.es.html.
36. Gellman and Becker, "A Different Understanding with the President."
37. "The Vice President Appears on Meet the Press with Tim Russert," http://www.whitehouse.gov/vicepresident/news-speeches/speeches/vp20010916.html.
38. Gellman and Becker, "A Different Understanding with the President."
39. Ibid.
40. Ibid.
41. Bruce Fein, "Restrain This White House," *Washington Monthly,* October 2006, http://www.washingtonmonthly.com/features/2006/0610.fein.html.
42. Gellman and Becker, "A Different Understanding with the President."
43. Ibid.
44. Ibid.
45. Ibid.
46. "Remarks by Vice President Dick Cheney to the U.S. Chamber of Commerce," http://www.whitehouse.gov/vicepresident/news-speeches/speeches/vp20011114–1.html.
47. Bruce Fein, "Impeach Cheney," *Slate,* http://slate.com/id/2169292/.
48. Ibid.
49. *Bill Moyers Journal,* July 13, 2007, http://www.pbs.org/moyers/journal/07132007/transcript4.html.
50. Jeffrey Goldberg, "Breaking Ranks: What Turned Brent Scowcroft Against the Bush Administration?," *The New Yorker,* August 4, 2007, http://www.newyorker.com/archive/2005/10/31/051031fa_fact2.
51. Ibid.
52. Ibid.
53. Ron Suskind, *The One Percent Doctrine: Deep Inside America's Pursuit of Its Enemies Since 9/11* (New York: Simon & Schuster, 2006), p. 29.
54. Ibid., p. 33.
55. Ibid., p. 32–34.
56. James Risen, "How Pair's Finding on Terror Led to Clash on Shaping Intelligence," *New York Times,* April 28, 2004, http://select.nytimes.com/search/ restricted/article?res=FB0E16FF345E0C7B8EDDAD0894DC404482
57. "Who's Who?" A monthly column on Washington personalities by Paul Glastris, http://www.washingtonmonthly.com/features/2001/0111.whoswho.html.
58. "The Dark Side Interview: F. Michael Maloof," *Frontline,* PBS, http://www.pbs.org/wgbh/pages/frontline/darkside/interviews/maloof.html.
59. Risen, "How Pair's Finding on Terror Led to Clash on Shaping Intelligence."
60. "The Dark Side Interview: F. Michael Maloof," *Frontline.*
61. Risen, "How Pair's Finding on Terror Led to Clash on Shaping Intelligence."

62. Robert Kagan, "The Powell Papers," *Washington Post,* October 3, 2001, p. A31.
63. Charles Krauthammer, "Clear Thinking on Coalitions," *Washington Post,* October 19, 2001, p. A29.
64. Rich Lowry, "The Limits of Patience," *National Review,* October 2, 2001.
65. William Safire, "Essay: Advance the Story," *New York Times,* October 22, 2001, p. A19.
66. Lowry, "The Limits of Patience."
67. Brent Scowcroft, "Build a Coalition," *Washington Post,* October 16, 2001, http://www.ffip.com/opeds101601.htm.
68. *La Repubblica.* For an English translation of the series in *La Repubblica,* see the website of Nur al-Cubicle at http://nuralcubicle.blogspot.com/2005/10/berlusconi-behind-fake-yellowcake.html.
69. Andrew Buncombe, John Phillips, and Raymond Whitaker, "The Niger Connection; The British Believed It, President Bush Quoted It, and It Helped," *The Independent,* November 6, 2005, p. 43.
70. *La Repubblica,* http://nuralcubicle.blogspot.com/2005/10/berlusconi-behind-fake-yellowcake.html.
71. Philip Giraldi, "Forging the Case for War," *American Conservative,* November 21, 2005 http://www.amconmag.com/2005/2005_11_07/feature.html.
72. "The Original Niger Reporting," http://www.globalsecurity.org/intell/library/congress/2004_rpt/iraq-wmd-intell_chapter2.htm.
73. Ibid.
74. James Bamford, *A Pretext for War: 9/11, Iraq, and the Abuse of America's Intelligence Agencies* (New York: Doubleday, 2004), p. 305.
75. Michael Ledeen, "My Family Friend, Walt Disney," *Jewish World Review,* http://www.jewishworldreview.com/michael/ledeen120601.asp.
76. Author interview with Larry Johnson, September 2006.
77. Author interview with Michael Ledeen, January 2006.
78. E-mail from Michael Ledeen to the author, February 11, 2006.
79. Jewish Institute for National Security Affairs, http://www.jinsa.org/about/adboard/adboard.html?documentid=742.
80. Thomas B. Edsall and Dana Milbank, "White House's Roving Eye for Politics; President's Most Powerful Adviser May Also Be the Most Connected," *Washington Post,* March 10, 2003, p. A1.
81. Author interview with Meyrav Wurmser, September 2006.
82. Interview with Michael Ledeen.
83. Michael Ledeen, "Scowcroft Strikes Out," *National Review,* August 6, 2002.
84. Michael Ledeen, "Creative Destruction," *National Review,* September 20, 2001.
85. Michael Ledeen, "How to Lose It; We Must Be Imperious, Ruthless, and Relentless. No Compromise with Evil; We Want Total Surrender," *National Review Online,* December 7, 2001.
86. Author interview with Guido Moltedo, January 2006.
87. Jonathan Kwitny, "How an Italian Ex-Spy Who Also Helped the U.S. Landed in Prison Here," *Wall Street Journal,* August 8, 1985.
88. Ibid.
89. Ibid.
90. Ibid.
91. Ibid.
92. Interview with Michael Ledeen.
93. Ibid.
94. The Billygate scandal began after Billy Carter mortified President Carter in 1979 by going to Tripoli at a time when Libya's leader, Muammar Qaddafi, was reviled as a radical Arab dictator who supported terrorism. Coupled with Billy's later admission that he had received a $220,000 loan from Qaddafi's regime, the ensuing scandal made headlines across America and led to a Senate investigation. But it had died down as the November 1980 elections approached.

    Then, in the last week of October 1980, just two weeks before the election, *New*

Republic in Washington and *Now* magazine in Great Britain published a story coauthored by Michael Ledeen and Arnaud de Borchgrave, now an editor-at-large at the *Washington Times* and United Press International. According to the story, headlined "Qaddafi, Arafat and Billy Carter," the president's brother had been given an additional $50,000 by Qaddafi, on top of the loan, and had met secretly with Palestine Liberation Organization leader Yasser Arafat. The story had come dramatically back to life. The new charges were disputed by Billy Carter and many others, and were never corroborated.

In 1981, Ledeen played a role in what has been widely characterized as another disinformation operation. Once again his alleged ties to SISMI were front and center. The episode began after Mehmet Ali Agca, the right-wing terrorist who shot Pope John Paul II that May, told authorities that he had been taking orders from the Soviet Union's KGB and Bulgaria's secret service. With Ronald Reagan newly installed in the White House, the so-called Bulgarian connection made perfect Cold War propaganda. Michael Ledeen was one of its most vocal proponents, promoting it on TV and in newspapers all over the world. In light of the ascendancy of the Solidarity Movement in Poland, the Pope's homeland, the Bulgarian connection played a role in the demise of communism in 1989.

There was just one problem: it probably wasn't true. "It just doesn't pass the giggle test," said Frank Brodhead, coauthor of *The Rise and Fall of the Bulgarian Connection.* "Agca, the shooter, had been deeply embedded in a Turkish youth group of the Fascist National Action Party known as the Gray Wolves. It seemed illogical that a Turkish Fascist would work with Bulgarian Communists."

The only real source for the Bulgarian connection theory was Agca himself, a pathological liar given to delusional proclamations such as his insistence that he was Jesus Christ. When eight men were later tried in Italian courts as part of the Bulgarian connection case, all were acquitted for lack of evidence. One reason was that Agca had changed his story repeatedly. On the witness stand, he said he had put forth the Bulgarian connection theory after Francesco Pazienza offered him freedom in exchange for the testimony. He subsequently changed that story as well.

Years later, *Washington Post* reporter Michael Dobbs, who had initially believed the theory, wrote that "I became convinced . . . that the Bulgarian connection was invented by Agca with the hope of winning his release from prison. . . . He was aided and abetted in this scheme by right-wing conspiracy theorists in the United States and William Casey's Central Intelligence Agency, which became a victim of its own disinformation campaign."

Exactly which Americans might have been behind such a campaign? According to a 1987 article in *The Nation,* Francesco Pazienza said Ledeen "was the person responsible for dreaming up the 'Bulgarian connection' behind the plot to kill the Pope." Similarly, according to *The Rise and Fall of the Bulgarian Connection,* Pazienza claimed that Ledeen had worked closely with the SISMI team that coached Agca on his testimony.

But Ledeen angrily denied the charges. "It's all a lie," he said. He added that he protested to the *Wall Street Journal* when it first reported on his alleged relationship with Pazienza: "If one-tenth of it were true, I would not have security clearances, but I do."

Not long before his death, in 2005, Pope John Paul II announced that he did not believe the Bulgarian connection theory. But that wasn't the end of it. In March 2006 an Italian commission run by Paolo Guzzanti, a senator in the right-wing Forza Italia Party, reopened the case and concluded that the Bulgarian connection was real. According to Frank Brodhead, however, the new conclusions are based on the same old information, which is "bogus at best and at worst deliberately misleading."

In the wake of Billygate and the Bulgarian connection, Ledeen allegedly began to play a role as a behind-the-scenes operative with the ascendant Reagan-Bush team. According to *Mission Italy,* by former ambassador to Italy Richard Gardner, after Reagan's victory but while Jimmy Carter was still president, "Ledeen and Pazienza set themselves up as the preferred channel between Italian political leaders and members of the new administration." Ledeen responded, "Gardner was wrong. And, by the way, he had every opportunity to raise it with me and never did."

See also Michael J. Sniffen, Associated Press, July 14, 1980; interview with Frank Brodhead; Wolfgang Achtner and Tony Barber, "Search for a Plot to kill the Pope; Who Put the Gun in Agca's Hand?" *The Independent* (London), May 19, 1991, p. 18; Joel Kovel, "The Rise and Fall of the Bulgarian Connection," book review, Monthly Review 38 (April 1987): 52; and Edward S. Herman and Frank Brodhead, *The Rise and Fall of the Bulgarian Connection* (New York: Sheridan Square Press, 1986). Alexander Cockburn, "The Gospel According to Ali Agca," *The Nation* 241 (July 6, 1985): 1. Also Michael Dobbs, "Conspiracy Theorist; The Truth About the Pope's Would-Be Assassin Remains Unknown," *Washington Post*, June 15, 2000, p. A33; Stefano Delle Chiaie, "Delle Chiaie: From Bologna to Bolivia; a Terrorist Odyssey," *The Nation* 244 (April 25, 1987): 525; and interview with Frank Brodhead.

95. John Phillips, "Berlusconi's Hour of Reckoning," *The Times*, December 22, 1994.
96. Author interview with Philip Giraldi, October 2006.
97. Interview with Michael Ledeen.
98. Joshua Micah Marshall, Laura Rozen, and Paul Gastris, "Iran-Contra II? Fresh scrutiny on a rogue Pentagon operation," http://www.washingtonmonthly.com/features/2004/0410.marshallrozen.html.
99. Gareth Porter, "Burnt Offering: How a 2003 Secret Overture from Tehran Might Have Led to a Deal on Iran's Nuclear Capacity—If the Bush Administration Hadn't Rebuffed It," American Prospect, May 21, 2006, http://www.prospect.org/web/page.ww?section=root&name=ViewPrint&articleID=11539.
100. Author interview with a State Department official, July 2007.
101. http://www.repubblica.it/2005/j/sezioni/esteri/iraq69/bodv/bodv.html.
102. *La Repubblica*, http://nuralcubicle.blogspot.com/2005/10/berlusconi-behind-fake-yellowcake.html.
103. Author interview with Carlo Bonini, January 2006. For an English translation of the Bonino-D'Avanzio articles, http://nuralcubicle.blogspot.com/2005/10/yellowcake-dossier-not-work-of-cia.html.
104. Tracy Wilkinson, "Italians Debate Role in Operative's Saga," *New York Times*, October 28, 2005, p. 24.
105. Author interview with Patrick Lang, October 2006.
106. Ibid., p. 301.
107. Julian Borger, "Bush Aide Misled FBI, Say Reports," *The Guardian* (London), July 23, 2005, p. 17.
108. Author interview with Frank Brodhead, March 2006.
109. Jbalazs, "Niger Forgery and Michael Ledeen," http://www.dailykos.com/story/2005/7/4/10362/61305.
110. Michael Yglesias, "Ledeen, Cannistraro, Niger," http://warrenreports.tpmcafe.com/story/2005/10/30/111254/16.
111. Eriposte, "Treasongate: The *Real* Significance of the Niger Uranium Forgery Stories in *La Repubblica*," http://www.theleftcoaster.com/archives/005844.php.
112. Larisa Alexandrovna, "American Who Advised Pentagon Says He Wrote for Magazine That Forged Niger Documents," http://rawstory.com/news/2005/American_who_consulted_for_Pentagon_says_0117.html.
113. Tyler Drumheller, *On the Brink: An Insider's Account of How the White House Compromised American Intelligence* (New York: Carroll and Graf Publishers, 2006), p. 123.
114. Author interview with Milt Bearden, April 2006.
115. Ibid.
116. Interview wth Patrick Lang.
117. SSCI, http://www.globalsecurity.org/intell/library/congress/2004_rpt/iraq-wmd-intell_chapter2.htm.
118. Ibid.

119. Author interview with Ray McGovern, March 2006.
120. http://www.globalsecurity.org/intell/library/congress/2004_rpt/iraq-wmd-intell_toc.htm.
121. Author interview with Joseph Wilson, September 2006.
122. Ibid.
123. Ibid.
124. Ibid.
125. Declassified State Dept memo, http://www.judicialwatch.org/archive/niger-uranium.pdf.
126. Tom Hamburger, Peter Wallsten, and Bob Drogin, "French Told CIA of Bogus Intelligence," originally published in *New York Times,* October 17, 2004, http://www.truthout.org/docs_2005/121105B.shtml.
127. John Daniszewsi, "Indignation Grows in U.S. Over British Prewar Documents," originally published in *Los Angeles Times,* May 12, 2005, http:// www.commondreams.org/headlines05/0512–01.htm.
128. Ron Suskind, "Without a Doubt," originally published in *New York Times,* October 17, 2004, http://www.truthout.org/docs_04/101704A.shtml.
129. Interview with Colin Powell on *ABC's This Week with Sam Donaldson and Cokie Roberts,* http://www.state.gov/secretary/former/powell/remarks/2002/9941.htm.
130. Peter Schweizer and Rochelle Schweizer, *The Bushes: Portrait of a Dynasty* (New York: Random House, 2004), p.535

Chapter Fifteen: Fear: The Marketing Campaign

1. Sidney Blumenthal, *How Bush Rules: Chronicles of a Radical Regime* (Princeton, N.J.: Princeton University Press, 2006), p. 143.
2. Bob Graham, "What I Knew Before the Invasion," *Washington Post,* November 20, 2005, http://www.washingtonpost.com/wp-dyn/content/article/2005/11/18/AR2005111802397.html.
3. "The Dark Side: The October '02 National Intelligence Estimate (NIE)," *Frontline,* PBS, http://www.flworld.org/wgbh/pages/frontline/darkside/themes/nie.html.
4. Ibid.
5. Ibid.
6. Ibid.
7. "A Spy Speaks Out," *60 Minutes,* CBS, April 23, 2006, http://www.cbsnews.com/stories/2006/04/21/60minutes/main1527749_page2.shtml.
8. Sidney Blumenthal, "Bush Knew Saddam Had No Weapons of Mass Destruction," *Salon,* September 6, 2007.
9. "The Dark Side: Paul Pillar," *Frontline,* PBS, http://www.flworld.org/wgbh/pages/frontline/darkside/interviews/pillar.html.
10. "The Dark Side: The October '02 National Intelligence Estimate (NIE)."
11. Commission on the Intelligence Capability of the United States Regarding Weapons of Mass Destruction, Robb Silberman, http://www.wmd.gov/report/report.html.
12. Bob Drogin and John Goetz, "The Curveball Saga; How U.S. Fell Under the Spell of 'Curveball,'" *Los Angeles Times,* November 20, 2005, p. A1.
13. Commission on the Intelligence Capability of the United States Regarding Weapons of Mass Destruction.
14. Joby Warrick, "Warnings on WMD 'Fabricator' Were Ignored, Ex-CIA Aid Says," *Washington Post,* June 25, 2006, http://www.washingtonpost.com/wp-dyn/content/article/2006/06/24/AR2006062401081.html. And Commission on the Intelligence Capability of the United States Regarding Weapons of Mass Destruction.
15. Commission on the Intelligence Capability of the United States Regarding Weapons of Mass Destruction.
16. Drogin and Goetz, "The Curveball Saga; How U.S. Fell Under the Spell of 'Curveball.'"
17. Warrick, "Warnings on WMD 'Fabricator' Were Ignored, Ex-CIA Aid Says."

18. Drogin and Goetz, "The Curveball Saga; How U.S. Fell Under the Spell of 'Curveball.'"
19. Ibid.
20. Ibid.
21. Bryan Burrough, Evgenia Peretz, David Rose, and David Wise, "The Path to War," *Vanity Fair,* May 2004, p. 228.
22. Warrick, "Warnings on WMD 'Fabricator' Were Ignored, Ex-CIA Aid Says."
23. "The Dark Side: Interview: Tyler Drumheller," *Frontline,* PBS, http://www.flworld.org/wgbh/pages/frontline/darkside/interviews/drumheller.html.
24. "The Dark Side: The October '02 National Intelligence Estimate (NIE)."
25. "The Dark Side: Interview: Tyler Drumheller."
26. Ibid.
27. "The Dark Side: Paul Pillar."
28. Blumenthal, *How Bush Rules,* p. 143.
29. http://www.pbs.org/wgbh/pages/frontline/darkside/etc/network.html.
30. Author interview with Lawrence Wilkerson, October 2006.
31. Memo from an official who served in the Pentagon while Cheney was secretary of defense, July 2007. The memo points out that the observation that Bush's approach to the presidency resembled his tenure at the Texas Rangers was first made in the *National Review.*
32. "Marketing Iraq: Why Now?" September 12, 2002, www.cnn.com/2002/ALLPOLITICS/09/12/schneider.iraq/index.html.
33. "Top Bush Officials Push Case Against Saddam," September 8, 2002, http://archives.cnn.com/2002/ALLPOLITICS/09/08/iraq.debate/.
34. "Conservative Media Silent on Prior Publications of Leaks Favorable to White House," June 30, 2006, http://mediamatters.org/items/200607010007.
35. Michael Massing, *Now They Tell Us,* February 26, 2004, http://www.nybooks.com/authors/71.
36. "Buying the War," *Bill Moyers Journal,* http://www.pbs.org/moyers/journal/btw/watch.html.
37. "Trying Times," *New York Post,* Page Six, January 26, 2007, http:// www.nypost.com/seven/01262007/gossip/pagesix/trying_times_pagesix_.htm.
38. Massing, *Now They Tell Us.*
39. "Buying the War."
40. Jimmy Burns, Guy Dinmore, et al., "Did Intelligence Agencies Rely Too Much on Unreliable Data from Iraqi Exiles," *Financial Times* (London), June 4, 2003, p. 19.
41. *Hardball with Chris Matthews,* MSNBC, November 1, 2005, http://www.msnbc.msn.com/id/9896575/.
42. Joby Warrick, "Evidence on Iraq Challenged," *Washington Post,* September 19, 2002, http://www.washingtonpost.com/ac2/wp-dyn/A36348-2002Sep18?language=printer.
43. "Buying the War."
44. Ibid.
45. Ibid.
46. Ibid.
47. Ibid.
48. Ibid.
49. "Poll: No Rush to War," CBS News, August 14, 2007, http://www.cbsnews.com/stories/2002/09/24/opinion/polls/main523130.shtml.
50. "Condoleeza Rice on Fox News Sunday," September 16, 2002, http://www.foxnews.com/story/0,2933,63125,00.html.
51. "Buying the War."
52. "Republican Pollster David Winston Discusses Antiwar Sentiment in U.S.," *The News with Brian Williams,* CNBC, September 24, 2002, CNBC News Transcripts.
53. Ron Hutcheson and Diego Ibarguen, "Bush: 'Can't Distinguish' Between Iraq and al-Qaida," Knight Ridder Newspapers, September 25, 2002, http://web.archive.org/web/20030621092642/http://www.kansas.com/mld/kansas/4150487.htm.

54. E. J. Dionne Jr., " . . . Not to Talk," *Washington Post,* October 4, 2002, p. A29.
55. Alex Tresniowski, Diane Herbst, and Melody Simmons, "Very Contrary Ex-Arms Inspector Scott Ritter Draws Fire by Flip-Flopping About Iraq," *People,* September 30, 2002, p. 91.
56. *Donahue,* MSNBC, September 27, 2002. Transcript #092700cb.466.
57. *Late Edition with Wolf Blitzer,* CNN, March 9, 2003, http://transcripts.cnn.com/TRANSCRIPTS/0303/09/le.00.html.
58. *Hannity & Colmes,* interview with Ann Coulter and Susan Estrich, Fox News Network, September 27, 2002. Transcript #092701cb.253.
59. *Morning Edition,* "Price of Keeping Government Secrets Secret," National Public Radio (NPR), September 23, 2002.
60. Maureen Dowd, "Rapture and Rupture," *New York Times,* October 6, 2002, p. D13.
61. James Ridgeway, "I Hear America Sinking," *Village Voice,* September 17, 2002, p. 29.
62. Ibid.
63. Samuel Muwakkil, "The Forgotten History of Islam in America, *In These Times,* September 16, 2002, p. 14.
64. Jeffrey Weiss, "Does Any Religion Speak with One Voice?" *Dallas Morning News,* September 21, 2002, p. G1.
65. Ann Coulter, "This Is War," *National Review,* September 13, 2001, http://www.nationalreview.com/coulter/coulter.shtml.
66. "Al Gore Speaks Out Against Mr. Bush's Pre-Emptive Policy on Iraq," *All Things Considered,* NPR, September 23, 2002, http://www.npr.org/programs/atc/transcripts/2002/sep/020923.marinucci.html.
68. Susan Milligan, "Confronting Iraq/Congress; Despite Reservations, OK Is Expected," *Boston Globe,* October 8, 2002, p. A19.
69. *La Repubblica.* For an English translation of the series in *La Repubblica,* see the website of Nur al-Cubicle at http://nuralcubicle.blogspot.com/2005/10/berlusconi-behind-fake-yellowcake.html.
70. Joshua Micah Marshall, blog, Talking Points Memo, http://www.talkingpointsmemo.com/archives/week_2005_10_30.php#006908.
71. Author interview with Philip Giraldi, October 2006.
72. Author interview with Melvin Goodman, May 2006.
73. Author interview with Michael Ledeen, January 2006.
74. Author interview with former senior intelligence official.
75. "Continuing Analysis," http://www.globalsecurity.org/intell/library/congress/2004_rpt/iraq-wmd-intell_chapter2-c.htm.
76. Bob Drogin and Tom Hamburger, "Niger Uranium Rumors Wouldn't Die," *Los Angeles Times,* February 17, 2006 p. A1.
77. Author interview with Milt Bearden, September 2006.
78. Drogin and Hamburger, "Niger Uranium Rumors Wouldn't Die."
79. Author interview with Carlo Rosella, January 2006.
80. Bryan Bender and Michael Kranish, "FBI: Iraq-Niger Papers Part of Scheme, Says Forgeries Committed for Profit," *Boston Globe,* November 5, 2006.
81. "The Cincinnati Speech," http://www.globalsecurity.org/intell/library/congress/2004_rpt/iraq-wmd-intell_chapter2-f.htm.
82. Ibid.
83. Ibid.
84. Ibid.
85. Neil McKay, "You're Damned if You Do George, and You're Damned if You Don't," *Sunday Tribune* (Ireland), October 13, 2002, p. 21.
86. Interview with Larry Wilkerson.
87. Sidney Blumenthal, "Weapons of Mass Dissembling," *Salon,* http://dir.salon.com/story/opinion/blumenthal/2004/02/05/wmd/index.html.
88. Ibid.
89. "The Dark Side," *Frontline,* PBS, http://www.pbs.org/wgbh/pages/frontline/darkside/etc/script.html.

90. Ibid.
91. Seymour Hersh, "The Stovepipe," *The New Yorker*, http://www.newyorker.com/fact/content/?031027fa_fact.
92. "Connie Chung: Skeptical of Skepticism," *Fair & Accuracy in Reporting*, October 10, 2002, http://www.fair.org/index.php?page=1642.
93. Peter Beaumont and Faisal Islam, "Carve-up of Oil Riches Begins," *The Observer*, November 3, 2003, http://observer.guardian.co.uk/international/story/0,6903,825103,00.html.
94. Jim Wolf, "U.S. Carrier Deployments Could Hold Iraq Key, A 'Concerted Effort' to Discredit Bush Critic," Reuters, October 3, 2002.
95. Jim Garamone, "Rumsfeld Says Link Between Iraq, al Qaeda 'Not Debatable,'" American Forces Press Service, http://www.defenselink.mil/news/newsarticle.aspx?id=43413.
96. Radio Address by the President to the Nation, September 28, 2002, http://www.whitehouse.gov/news/releases/2002/09/20020928.html.
97. Remarks by the President at John Cornyn for Senate Reception, September 26, 2002, http://www.whitehouse.gov/news/releases/2002/09/20020926-17.html.
98. Andrew Gumbel, "Case for War Confected, Say Top US Officials," *Independent/UK*, November 9, 2003, http://www.commondreams.org/headlines03/1109-12.htm.
99. A declassified but heavily redacted version of the NIE can be found at the Federation of American Scientists website, http://www.fas.org/irp/cia/product/iraq wmd.html.
100. Graham, "What I Knew Before the Invasion."
101. Dana Priest, "Congressional Oversight of Intelligence Criticized; Committee Members, Others Cite Lack of Attention to Reports on Iraqi Arms, Al Qaeda Threat," April 27, 2004, p. A1, http://www.washingtonpost.com/ac2/wp-dyn/A44837–2004Apr26?language=printer.
102. Ibid.
103. "Iraq's Continuing Programs for Weapons of Mass Destruction," http://www.fas.org/irp/cia/product/iraq-wmd.html.
104. Excerpts from an October 9 statement by U.S. senator Russ Feingold (D-Wis.), opposing Senate authorization to use force against Iraq, (Madison) *Wisconsin State Journal*, October 13, 2002. House vote: 296–133.
105. Wendy S. Ross, "Bush Says House Vote Sends 'Clear Message' to Iraq: Disarm," *Washington File*, October 10, 2002, http://www.globalsecurity.org/wmd/library/news/iraq/2002/iraq-021010-usia02.htm.
106. "President: Terrorism Insurance Agreement Needed by Friday," October 1, 2002, http://www.whitehouse.gov/news/releases/2002/10/20021001-1.html.
107. Ibid.
108. Jonathan Marcus, "US Intensifies Iraq Build-up," *BBC News*, October 29, 2002, http://news.bbc.co.uk/2/hi/americas/2371379.stm.
109. Stefano Ambrogi, "U.S. Orders Large Volume of Ammunition to Gulf," Reuters, November 1, 2002, http://vredessite.nl/andernieuws/2002/week45/11-01_ammunition.html.
110. "Rumsfeld: It Would Be a Short War," November 15, 2002, http://www.cbsnews.com/stories/2002/11/15/world/main529569.shtml.
111. Knut Royce, "Plan: Tap Iraq's Oil," *Newsday*, January 10, 2003, http://www.commondreams.org/headlines03/0110-01.htm.
112. CNN Special Report, http://www.cnn.com/SPECIALS/2002/iraq/documents/page1.html.
113. "US Secretary of State Colin Powell's Statement on Iraq's Weapons Declaration," *Guardian Unlimited*, December 20, 2002, http://www.guardian.co.uk/Iraq/Story/0,,863575,00.html.
114. "Illustrative Examples of Omissions from the Iraqi Declaration to United Nations Security Council," State Department, December 19, 2002, http://www.state.gov/r/pa/prs/ps/2002/16118.htm.
115. "President Discusses Iraq and North Korea with Reporters," December 31, 2002, http://www.whitehouse.gov/news/releases/2002/12/20021231-1.html.

116. Barton Gellman and Dafna Linzer, "Prosecutor Describes Cheney, Libby as Key Voices Pitching Iraq-Niger Story," *Washington Post,* April 9, 2006, p. A1, http://www.washingtonpost.com/wp-dyn/content/article/2006/04/08/AR2006040800916_2.html
117. "Report on the U.S. Intelligence Community's Prewar Intelligence Assessments on Iraq," http://www.globalsecurity.org/intell/library/congress/ 2004_rpt/iraq-wmd-intell_chapter2-h.htm.
118. Interview with Milt Bearden.
119. President Delivers "State of the Union," January 28, 2003, http://www.whitehouse.gov/news/releases/2003/01/20030128-19.html.
120. Ibid.
121. Ibid.
122. Ibid.
123. Ibid.
124. Ibid.
125. Ibid.
126. Ibid.
127. Michael R. Gordon, "State of the Union; The Iraq Issue; Bush Enlarges Case for War by Linking Iraq with Terrorists," *New York Times,* January 29, 2003, p. A1.

Chapter Sixteen: The Good Soldier

1. President Delivers "State of the Union," January 28, 2003, http://www.whitehouse.gov/news/releases/2003/01/20030128-19.html.
2. Peter Eisner and Knut Royce, *The Italian Letter: How the Bush Administration Used a Fake Letter to Build the Case for War in Iraq* (New York: Rodale, 2007), p. 118–19.
3. Author interview with Lawrence Wilkerson, October 2006.
4. Ibid.
5. "The Powell Doctrine Revisited," *The Economist,* February 1, 2003.
6. Richard Leiby, "Breaking Ranks Larry Wilkerson Attacked the Iraq War. In the Process, He Lost the Friendship of Colin Powell." *Washington Post,* January 19, 2006, p. C1, C01http://www.washingtonpost.com/wp-dyn/content/article/2006/01/18/AR2006011802607_pf.html.
7. Interview with Lawrence Wilkerson.
8. Seymour Hersh, "The Stovepipe," *The New Yorker,* August 14, 2007, http://www.newyorker.com/archive/2003/10/27/031027fa_fact.
9. Karen DeYoung, *Soldier: The Life of Colin Powell* (New York: Knopf, 2006), p. 441.
10. "Iraq Nuke Claim Was Echoed," CBS News, August 14, 2007, http://www.cbsnews.com/stories/2003/08/09/iraq/main567481.shtml.
11. Interview with Lawrence Wilkerson.
12. "An Oversight Hearing on Pre-War Intelligence Relating to Iraq," Senate Democratic Policy Hearing, June 26, 2006, http://pdf2html.spawncamp.net/pdf2html.php?url=http://democrats.senate.gov/dpc/hearings/hearing33/wilkerson.pdf.
13. James Mann, *Rise of the Vulcans: The History of Bush's War Cabinet* (New York: Viking, 2004), p. 353.
14. DeYoung, *Soldier,* p. 442.
15. Michael Isikoff and David Corn, *Hubris: The Inside Story of Spin, Scandal, and the Selling of the Iraq War* (New York: Crown, 2006), p. 176.
16. Bryan Burrough, Evgenia Peretz, David Rose, and David Wise, "The Path to War," *Vanity Fair,* May 2004, p. 228.
17. DeYoung, *Soldier,* p. 541.
18. Although the CIA's budget is once again classified, FAS's Steven Aftergood estimates that the agency will receive $5 billion out of the $35 billion to $40 billion for all intelligence agencies this fiscal year (http://www.govexec.com/dailyfed/0303/031703cdam1.htm).

19. "The Dark Side," *Frontline,* PBS, http://www.pbs.org/wgbh/pages/frontline/darkside/etc/script.html.
20. "Former Aide: Powell WMD Speech 'Lowest Point in My life'" CNN.com, August 19, 2005.
21. Interview with Lawrence Wilkerson.
22. "Key Players in the Plame Affair," *Washington Post,* October 20, 2005.
23. Interview with Lawrence Wilkerson.
24. Douglas Jehl, "Through Indictment, a Glimpse into a Secretive and Influential White House Office," *New York Times,* October 30, 2005, p. 28.
25. Bryan Burrough, Evgenia Peretz, David Rose, and David Wise, "The Path to War," *Vanity Fair,* May 2004, p. 228.
26. "Former Aide: Powell WMD Speech 'Lowest Point in My Life,'" CNN.com, August 19, 2005.
27. Interview with Lawrence Wilkerson.
28. "National Intelligence Estimate," http://www.sourcewatch.org/index.php?title=National_Intelligence_Estimate.
29. DeYoung, *Soldier,* p. 442.
30. Isikoff and Corn, *Hubris,* p. 179.
31. "The Dark Side."
32. Ibid.
33. Burrough, Peretz, Rose, and Wise, "The Path to War."
34. *Meet the Press,* NBC, December 9, 2001.
35. Evan Thomas, Richard Wolffe, and Michael Isikoff, "Where Are Iraq's WMDs?" *Truth Out,* June 1, 2003, http://www.truthout.org/docs_03/060203A .shtml.
36. "The Dark Side."
37. "Iran Offers Talks with U.S. About Iraq Situation; a Look at Operation Swarmer," CNN, March 17, 2006. Transcript #031703CN.V16.
38. "The Dark Side: Interview with Lawrence Wilkerson," *Frontline,* PBS, http://www.pbs.org/wgbh/pages/frontline/darkside/interviews/wilkerson.html.
39. Interview with Lawrence Wilkerson.
40. Ibid.
41. Burrough, Peretz, Rose, and Wise, "The Path to War."
42. Interview with Lawrence Wilkerson.
43. Ibid.
44. Ibid.
45. Ibid.
46. Ibid.
47. Joby Warrick, "Warnings on WMD 'Fabricator' Were Ignored, Ex-CIA Aide Says," *Washington Post,* June 25, 2006, http://www.washingtonpost.com/wp-dyn/content/article/2006/06/24/AR2006062401081.html.
48. "Remarks to the United Nations Security Council, Secretary Colin L. Powell," February 5, 2003, http://www.globalsecurity.org/wmd/library/news/iraq/2003/iraq-030205-powell-un-17300pf.htm.
49. Warrick, "Warnings on WMD 'Fabricator' Were Ignored, Ex-CIA Aide Says."
50. David Ensor, "Encore Presentation: Dead Wrong," *CNN Presents,* June 18, 2006. Transcript # 061802CN.V79.
51. "The Dark Side."
52. Isikoff and Corn, *Hubris,* p. 179.
53. Michael R. Gordon and James Risen, "Threats and Responses: Nuclear Report; Findings of U.N. Group Undercut U.S. Assertion," *New York Times,* January 28, 2003.
54. "More Work Remains on Iraqi Disarmament, Blix Says," Washington File, January 27, 2003, http://usinfo.org/wf-archive/2003/030127/epf112.htm.
55. Warrick, "Warnings on WMD 'Fabricator' Were Ignored, Ex-CIA Aide Says."
56. Ibid.
57. "The Dark Side."
58. Interview with Lawrence Wilkerson.
59. Warrick, "Warnings on WMD 'Fabricator' Were Ignored, Ex-CIA Aide Says."

60. George Tenet, *At the Center of the Storm: My Years at the CIA* (New York: HarperCollins, 2007), p. 381.
61. Burrough, Peretz, Rose, and Wise, "The Path to War."
62. Ibid.
63. Ibid.
64. Ibid.
65. Interview with Lawrence Wilkerson.
66. "CNN/USA Today/Gallup: Powell, Blair Fav Ratings Higher Than Bush," The National Journal Group, Inc., The Hotline, February 4, 2003.
67. Interview with Larry Wilkerson.
68. "Remarks to the United Nations Security Council, Secretary Colin L. Powell."
69. Michael R. Gordon, "Piling Up Evidence on Baghdad," *New York Times,* February 8, 2003, p. A1.
70. "Powell Presents U.S. Case to Security Council of Iraq's Failure to Disarm," February 5, 2003, http://www.un.org/apps/news/storyAr.asp?NewsID=6079&Cr=iraq&Cr1 =inspect.
71. Based on a search of LexisNexis database between February 6, 2003 and February 8, 2003.
72. "Washington Post—ABC News Poll: Iraq," February 11, 2003, http://www.washingtonpost.com/wp-srv/politics/polls/vault/stories/data021103.htm.
73. Greg Mitchell, "Shoptalk: Why We Are in Iraq," *Editor & Publisher Magazine,* September 8, 2003.
74. Ibid.
75. Ibid.
76. William Safire, "Irrefutable and Undeniable," *New York Times,* February 6, 2003, p. A39.
77. Mitchell, "Shoptalk: Why We Are in Iraq."
78. Bob Drogin and John Goetz, "How US Fell Under the Spell of 'Curveball,'" *Los Angeles Times,* November 20, 2005 (http://www.commondreams.org/headlines05/1120–01.htm).
79. E-mail from Lawrence Wilkerson, June 2006.
80. Sidney Blumenthal, "There Was No Failure of Intelligence," *The Guardian* (London), February 5, 2004 http://www.guardian.co.uk/Iraq/Story/ 0,2763,1141401,00.html.
81. Sidney Blumenthal, *How Bush Rules: Chronicles of a Radical Regime* (Princeton, N.J.: Princeton University Press, 2006), p. 36.
82. Ibid.
83. "IAEA Says It Has No Evidence of Prohibited Iraqi Nuclear Activities," December 19, 2002, http://www.usembassy.it/file2002_12/alia/a2121908.htm.
84. Drogin and Goetz, "How US Fell Under the Spell of 'Curveball.'"
85. Report on the U.S. Intelligence Community's Prewar Intelligence Assessments on Iraq," http://www.globalsecurity.org/intell/library/congress/2004_rpt/iraq-wmd-intell_chapter2-j.htm.
86. Bob Drogin and Tom Hamburger, "Niger Uranium Rumors Wouldn't Die," *Los Angeles Times,* February 17, 2006, p. A1.
87. Ron Laurenzo, "A Gathering Storm, by the Numbers," *Defense Week,* February 10, 2003.
88. John Hendren, "39,000 More Reservists Called to Active Duty," *Los Angeles Times,* February 13, 2003.
89. Carola Hoyos, "Big Players Rub Hands in Anticipation of Iraq's Return to Fold," *Financial Times* (London), February 25, 2003, p. 6.
90. Press Briefing by Ari Fleischer, February 18, 2003, http://www.whitehouse.gov/news/releases/2003/02/20030218-4.html.
91. http://atrios.blogspot.com/2002_08_04_atrios_archive.html.
92. http://www.un.org/Depts/unmovic/blix14Febasdel.htm.
93. Richard Perle, "Take Out Saddam—It's the Only Way," *AEI,* posted February 25, 2003, http://www.aei.org/publications/pubID.16100,filter.all/pub_ detail.asp.

94. Abbas J. Ali, "From Economy to Militarization: The end of the Nation State," *Competitiveness Review* 13, no. 1 (2003): 1–6.
95. Robert Dreyfuss, "Just the Beginning; Is Iraq the Opening Salvo in a War to Remake the World?," *The American Prospect,* April 2003, p. 26.
96. Bourke Kennedy, "Drive for 'Total War' Puts Nation in Danger," *Post-Standard* (Syracuse, N.Y.), February 13, 2003, p. A13.
97. Dreyfuss, "Just the Beginning; Is Iraq the Opening Salvo in a War to Remake the World?"
98. Jonah Micah Marshall, "Practice to Deceive," *Washington Monthly,* April 2003, http://www.washingtonmonthly.com/features/2003/0304.marshall.html.
99. William Houston, "Canadians Not Bombastic, and That's Good," *Globe and Mail* (Canada), February 15, 2003, p. A10.
100. "Unsolved Problem—Interview with Richard Hanley and Steve Adubato," *The O'Reilly Factor,* Fox News Network, April 1, 2003. Transcript #040104cb.256.
101. Joe Garofoli and Edward Epstein, "Lifting the Veil on Anti-War Groups; Opposition to War Is Only Shared Theme," *San Francisco Chronicle,* February 13, 2003, p. A1.
102. Eric Alterman, *What Liberal Media? The Truth About Bias and News* (New York: Basic Books, 2003), p. 271.
103. Tumbler, "Washington Watch: All Change at the State Department as the Neocons Take Over," *Prospect,* March 20, 2003.
104. Nick Turse, "Casualties of the Bush Administration," *Asia Times Online,* October 18, 2005, http://www.atimes.com/atimes/Front_Page/GJ18Aa01.html.
105. Tumbler, "Washington Watch: All Change at the State Department as the Neocons Take Over."
106. Turse, "Casualties of the Bush Administration."
107. Eric Schmitt, "Pentagon Contradicts General on Iraq Occupation Force's Size," *Global Policy Reform,* originally published in the *New York Times,* February 28, 2003, http://www.globalpolicy.org/security/issues/iraq/attack/consequences/2003/0228pentagoncontra.htm.
108. John C. Henry, "Army Secretary White Resigns; Pentagon Gives No Reason or Date," *Houston Chronicle,* April 26, 2003.
109. Author interview with Maura Moynihan, October 2006.
110. Ibid.
111. "Full Text of Joint Declaration," posted March 6, 2003, http://www.guardian.co.uk/Iraq/Story/0,2763,908441,00.html.
112. James Kitfield, "Foreign Affairs: Fractured Alliances," *National Journal,* March 8, 2003.
113. *Meet the Press,* NBC, March 16, 2003.
114. Andrew Bushell and Brent Cunningham, "Being There; Suddenly the Pentagon Grants Access to the Action, but the Devil's in the Details," *Columbia Journalism Review,* March 2003/April 2003, p. 18.
115. Ibid.
116. Press Briefing by Ari Fleischer, March 11, 2003, http://www.whitehouse.gov/news/releases/2003/03/20030311-5.html.
117. "United Nations Withdraws Staff from Iraq," Washington File, March 17, 2003, http://usinfo.org/wf-archive/2003/030317/epf114.htm.
118. "President Says Saddam Hussein Must Leave Iraq Within 48 Hours," March 17, 2003, http://www.whitehouse.gov/news/releases/2003/03/20030317-7.html.
119. "US Forces Order of Battle—17 March," http://www.globalsecurity.org/military/ops/iraq_orbat_030317.htm.
120. Presidential Letter, March 19, 2003, http://www.whitehouse.gov/news/releases/2003/03/20030319-1.html.
121. "Brink of War; Speaking with Former President Bush and Barbara Bush," *Good Morning America,* March 18, 2003.
122. Ibid.
123. Ibid.

124. Bob Woodward, *Plan of Attack* (New York: Simon & Schuster, 2004), p. 376.
125. Ibid., p. 379.
126. President Delivers "State of the Union," January 28, 2003, http://www.whitehouse.gov/news/releases/2003/01/20030128-19.html.
127. Woodward, *Plan of Attack,* p. 379.
128. "Dora Farms," http://www.globalsecurity.org/military/world/iraq/dora.htm.

Chapter Seventeen: Season of Mirth

1. "Lynch: Military Played Up Rescue Too Much," November 7, 2003, http://www.cnn.com/2003/US/11/07/lynch.interview/; Susan Schmidt and Vernon Loeb, "She Was Fighting to the Death; Details Emerging of W. Va. Soldier's Capture and Rescue," *Washington Post,* April 3, 2003, p. A1; "Lie by Lie Timeline," *Mother Jones,* http://www.motherjones.com/bush_war_timeline/; David D Kirkpatrick, "Jessica Lynch Criticizes U.S. Accounts of Her Ordeal," *New York Times,* November 7, 2003, p. 25.
2. Tamara Jones, "Hope in a Hollow for a Girl Who Dreamed; West Virginia Town Prays for the Return of Missing U.S. Soldier," *Washington Post,* March 26, 2003, p. A1.
3. Byron York, "The Little-Noticed Order That Gave Dick Cheney New Power," *National Review,* February 16, 2006.
4. Ibid.
5. W. Patrick Lang, "Drinking the Kool-Aid," *Middle East Policy* 11, no. 2 (June 22, 2004), p. 39.
6. Ibid.
7. Robert Dreyfuss, "Vice Squad," *American Prospect,* May 2006, p. 30.
8. Author interview with Larry Wilkerson, October 2006.
9. David Ignatius, "Cheney's Cheney," *Washington Post,* January 6, 2006, p. A19, http://www.washingtonpost.com/wp-dyn/content/article/2006/01/05/AR2006010501902_pf.html.
10. Sidney Blumenthal, "George Bush's Rough Justice; the Career of the Latest Supreme Court Nominee Has Been Marked by His Hatred of Liberalism," *The Guardian* (London), January 12, 2006, http://www.guardian.co.uk/usa/story/0,12271,1684464,00.html.
11. Lang, "Drinking the Kool-Aid."
12. Ibid.
13. Dana Milbank and Claudia Deane, "Hussein Link in 9/11 Lingers in Many Minds," *Washington Post,* September 6, 2003, p. A1, http://www.washingtonpost.com/ac2/wp-dyn/A32862-2003Sep5?language=printer.
14. "Poll: Americans Less Positive on Iraq," CNN.com, July 1, 2003, http://www.cnn.com/2003/US/06/30/sprj.irq.iraq.poll/.
15. "Fighting Reported Across Baghdad," CNN.com, April 8, 2003, http://www.cnn.com/2003/WORLD/meast/04/08/sprj.irq.baghdad.fighting/index.html.
16. Joel Campagna and Rhonda Roumani, "Permission to Fire: CPJ Investigates the Attack on the Palestine Hotel," http://www.cpj.org/Briefings/2003/palestine_hotel/palestine_hotel.html.
17. "Lie By Lie Timeline," *Mother Jones,* http://www.motherjones.com/bush_war_timeline/.
18. Valentinas Mite, "Iraq: Culture Minister Asks That Art, Music, Theater Not Be Forgotten," RadioFreeEurope, February 24, 2004, http://www.rferl.org/featuresarticle/2004/2/6D8C1A2D-FCF1-441F-9D65-EA64FDC445D0.html.
19. "Baghdad Falls to US Forces," BBC News, April 9, 2003, http://news.bbc.co.uk/2/hi/middle_east/2930913.stm. And Frank Rich, *The Greatest Story Ever Sold: The Decline and Fall of Truth* (New York: Penguin, 2006), p. 84.

20. "Focus of Battle Shifts North," *Baltimore Sun,* April 11, 2003, p. A1.
21. "President Bush: Job Ratings," http://www.pollingreport.com/BushJob1.htm.
22. Maureen Dowd, "Perle's Plunder Blunder," *New York Times,* March 23, 2003, p. 13.
23. Stephen Schwartz, "Fear Not the Shias; Their Tradition Recognizes the Rights of Minorities, Because They Have Always Been a Minority," *Weekly Standard* 8, no. 27 (March 24, 2003).
24. David Brooks, "Today's Progressive Spirit; The scenes in Baghdad Flow from Understandings Realized at the American Founding," *Daily Standard,* April 9, 2003.
25. Ken Adelman, "Cakewalk In Iraq," *Washington Post,* February 13, 2002, p. A27.
26. Ken Adelman, "'Cakewalk' Revisited," *Washington Post,* April 10, 2003. p. A29.
27. Bob Woodward, *Plan of Attack* (New York: Simon & Schuster, 2004), p. 409.
28. Ibid., pp. 409–10.
29. Charles Marsh, "Wayward Christian Soldiers," *International Herald Tribune,* January 22, 2006.
30. Elisabeth Bumiller, "With Echoes of Presidents Past; Bush Uses Striking Images to Show He Is Shifting His Focus," *New York Times,* May 3, 2003, p. 1.
31. Rich, *The Greatest Story Ever Sold,* p. 89.
32. Scott Lindlaw, "Tailhook Landing Wasn't Needed," *Cincinnati Post* (Ohio), May 3, 2003, p. A2.
33. Roland Watson and David Charter, "Bush the Pilot Performs Ace Publicity Stunt," *The Times* (London), May 3, 2003, p. 20.
34. John Ritter, "Jet Co-pilot Bush: 'Yes, I Flew It!,'" *USA Today,* May 2, 2003, http://www.globalsecurity.org/org/news/2003/030502-bush-fly01.htm.
35. Brian Dickerson, "Bush Plays Dress-Up for Troops," *Detroit Free Press,* May 5, 2003, p. B1.
36. "David Broder, Doris Kearns Goodwin, and Robert Novak Discuss the Politics of War, Tax Cuts, and the Presidential Countenders," *Meet the Press,* NBC, May 4, 2003
37. Rich, *The Greatest Story Ever Sold,* p. 89.
38. Maureen Dowd, "Disney On Parade," *New York Times,* September 17, 2005, p. A15.
39. Rich, *The Greatest Story Ever Sold,* p. 90.
40. "Text of Bush Speech: President Declares End to Major Combat in Iraq," http://www.cbsnews.com/stories/2003/05/01/iraq/main551946.shtml. And Woodward, *State of Denial: Bush at War, Part III* (New York: Simon & Schuster, 2006), p. 186.
41. Richard Perle, "Relax, Celebrate Victory," *USA Today,* May 1, 2003, http://www.usatoday.com/news/opinion/editorials/2003-05-01-oppose_x.htm.
42. "Endgame: Timeline," *Frontline,* PBS, http://www.pbs.org/wgbh/pages/frontline/endgame/cron/#1.
43. "Endgame: Interview: Lt. Col. Andrew Krepinevich," *Frontline,* PBS, http://www.pbs.org/wgbh/pages/frontline/endgame/interviews/krepinevich.html.
44. "Endgame: Timeline."
45. "CNN Presents: Dead Wrong," *CNN Presents: Encore Presentation,* aired August 27, 2005, http://transcripts.cnn.com/TRANSCRIPTS/0508/27/ cp.01.html.
46. Byron York, "The 'Bush Lied' Case Falls Apart: The President's Critics Have to Ignore Too Much Evidence," *National Review,* June 26, 2003, http://www.national review.com/york/york062603.asp.
47. "The Dark Side: Interview: John Brennan," *Frontline,* PBS, http://www.pbs.org/wgbh/pages/frontline/darkside/interviews/brennan.html.
48. "Former Aide: Powell WMD Speech 'Lowest Point in My Life,'" CNN.com, August 23, 2005, http://www.cnn.com/2005/WORLD/meast/08/19/powell.un/.
49. Author interview with Uzi Arad, June 2005.
50. "Iran Poses Greater Threat Than Iraq, Israelis Warn," *Houston Chronicle,* February 07, 2002, p. A18.

51. Michael Ledeen, "One Battle in a Wider, Longer War," *New York Sun,* March 19, 2003, p. 6.
52. Gareth Porter, "Burnt Offering: How a 2003 Secret Overture from Tehran Might Have Led to a Deal on Iran's Nuclear Capacity—If the Bush Administration Hadn't Rebuffed It," *American Prospect,* May 21, 2006, http:// www.prospect.org/web/page.ww?section=root&name=ViewPrint&articleId= 11539.
53. Ibid.
54. Ibid.
55. Ibid.
56. Barton Gellman and Dafna Linzer, "Unprecedented Perils Forces Tough Calls," *Washington Post,* October 26, 2004, p. A1, http://www.washingtonpost.com/ac2/wp-dyn/A62727-2004Oct25?language=printer.
57. Nicholas D. Kristof, "Missing in Action: Truth," *New York Times,* May 6, 2003, p. A31.
58. Bob Woodward, "Testifying in the CIA Leak Case," *Washington Post,* November 16, 2005, http://www.washingtonpost.com/wp-dyn/content/article/2005/11/15/AR2005111501829.html.
59. Michael Isikoff and David Corn, *Hubris: The Inside Story of Spin, Scandal, and the Selling of the Iraq War* (New York: Crown, 2006), p. 249.
60. Ibid., pp. 251–52.
61. Joseph C. Wilson, "What I Didn't Find in Africa," *New York Times,* July 7, 2003.
62. Vicky Ward, "Double Exposure," *Vanity Fair,* January 2004, http://www.vanityfair.com/politics/features/2004/01/plame200401?printable=true¤tPage=all.
63. Joseph C. Wilson, *The Politics of Truth: Inside the Lies That Led to War and Betrayed My Wife's CIA Identity* (New York: Carroll and Graf, 2004), p. 161.
64. Michael Isikoff, "Matt Cooper's Source: What Karl Rove Told Time Magazine's Reporter," *Newsweek,* July 18, 2005, http://www.msnbc.msn.com/id/8525978/site/newsweek/.
65. "Senator Roberts' Statement on the Niger Documents," http://www.fas.org/irp/news/2003/07/ssci071103.html.
66. Dan Fromkin, "The Second Source," *Washington Post,* July 15, 2005, http://www.washingtonpost.com/wp-dyn/content/blog/2005/07/15/BL2005071500978_pf.html
67. David E. Rosenbaum, "Debating a Leak: Political Memo; First Leak, Then a Predictable Pattern," *New York Times,* October 3, 2003, p. a16.
68. George Packer, "Letter from Baghdad: War After the War," *The New Yorker,* November 24, 2003, http://www.newyorker.com/archive/2003/11/24/ 031124fa_fact1?printable=true.
69. James Risen and David Johnston, "The Reach of War: The Offense; Chalabi Reportedly Told Iran That U.S. Had Code," *New York Times,* June 2, 2004, http://select.nytimes.com/search/restricted/article?res=F00815FF3F550C718CDDAF0894DC404482.
70. "Halliburton's $400 Million Payday," CBS News, http://www.cbsnews.com/stories/2003/05/07/politics/main552688.shtml.
71. *Face the Nation,* CBS, October 19, 2003, http://www.state.gov/secretary/former/powell/remarks/2003/25350.htm.
72. Donald Rumsfeld interview, *Meet the Press,* NBC, November 2, 2003, http://www.defenselink.mil/transcripts/transcript.aspx?transcriptid=2859.
73. Roundtable Interview of the President by British Print Journalists in the Oval Office, November 12, 2003, http://www.whitehouse.gov/news/releases/ 2003/11/20031114-2.html.
74. Judy Keen, "Cheney Says It's Too Soon to Tell on Iraq Arms," *USA Today,* January 19, 2004, http://www.usatoday.com/news/washington/executive/president/2004-01-19-cheney-weapons_x.htm.
75. Jeffrey Gettleman, "Enraged Mob in Falluja Kills 4 American Contractors," *New York Times,* March 31, 2004, http://www.nytimes.com/2004/03/31/international/

worldspecial/31CND-IRAQ.html?ex=1396155600&en=d2943196 cca8fc85&ei=5007&partner=USERLAND.

76. "Endgame: Timeline: Struggling to Find Success in Iraq," *Frontline*, PBS, http://www.pbs.org/wgbh/pages/frontline/endgame/cron/.
77. Thomas L. Friedman, "The Chant Not Heard," *New York Times*, November 30, 2003, p. 9.
78. "Media Advisory: Tom Friedman's Flexible Deadlines: Iraq's 'Decisive' Six Months Have Lasted Two and a Half Years," May 6, 2006, FAIR, http:// www.fair.org/index.php?page=2884.
79. "Thomas Friedman Discusses Outsourcing and His Views on Iraq," *Fresh Air*, NPR, June 3, 2004; "Thomas Friedman Discusses the Presidential Race and Iraq," *Face the Nation*, October 3, 2004, CBS News Transcripts; Thomas L. Friedman, "The Last Mile," *New York Times*, November 28, 2004, p. D11; David Brooks, Maureen Dowd, and Tom Friedman discuss how the hurricane season has affected the political landscape, *Meet the Press*, NBC, September 25, 2005.

Friedman's other predictions, as compiled by FAIR:

> *New York Times*, September 28, 2005: "Maybe the cynical Europeans were right. Maybe this neighborhood is just beyond transformation. That will become clear in the next few months as we see just what kind of minority the Sunnis in Iraq intend to be. If they come around, a decent outcome in Iraq is still possible, and we should stay to help build it. If they won't, then we are wasting our time."
>
> *Face the Nation*, CBS, December 18, 2005: "We've teed up this situation for Iraqis, and I think the next six months really are going to determine whether this country is going to collapse into three parts or more or whether it's going to come together."
>
> *Charlie Rose*, PBS, December 20, 2005: "We're at the beginning of I think the decisive I would say six months in Iraq, OK, because I feel like this election—you know, I felt from the beginning Iraq was going to be ultimately, Charlie, what Iraqis make of it."
>
> *New York Times*, December 21, 2005: "The only thing I am certain of is that in the wake of this election, Iraq will be what Iraqis make of it—and the next six months will tell us a lot. I remain guardedly hopeful."
>
> *Oprah Winfrey Show*, January 23, 2006: "I think that we're going to know after six to nine months whether this project has any chance of succeeding. In which case, I think the American people as a whole will want to play it out or whether it really is a fool's errand."
>
> CBS, January 31, 2006: "I think we're in the end game there, in the next three to six months, Bob. We've got for the first time an Iraqi government elected on the basis of an Iraqi constitution. Either they're going to produce the kind of inclusive consensual government that we aspire to in the near term, in which case America will stick with it, or they're not, in which case I think the bottom's going to fall out."
>
> *Today*, NBC, March 2, 2006: "I think we are in the end game. The next six to nine months are going to tell whether we can produce a decent outcome in Iraq."
>
> CNN, April 23, 2006: "Can Iraqis get this government together? If they do, I think the American public will continue to want to support the effort there to try to produce a decent, stable Iraq. But if they don't, then I think the bottom is going to fall out of public support here for the whole Iraq endeavor. So one way or another, I think we're in the end game in the sense it's going to be decided in the next weeks or months whether there's an Iraq there worth investing in. And that is something only Iraqis can tell us."
>
> *Hardball*, MSNBC, May 11, 2006: "Well, I think that we're going to find out, Chris, in the next year to six months—probably sooner—whether a

decent outcome is possible there, and I think we're going to have to just let this play out."

80. Romenesko, "WSJ Reporter Fassihi's E-mail to Friends," Poynter Online, September 29, 2004, http://www.poynter.org/column.asp?id=45&aid=72659.
81. Sidney Blumenthal, "Far Graver Than Vietnam: Most Senior US Military Officers Now Believe the War on Iraq Has Turned into a Disaster on an Unprecedented Scale," *The Guardian* (London), September 16, 2004, http://www.guardian.co.uk/comment/story/0,3604,1305360,00.html. And John Judis, "Death Grip," *The New Republic,* August 27, 2007, http://www.tnr.com/doc.mhtml?i=20070827&s=judis082707.
82. "President Bush: Job Ratings," http://www.pollingreport.com/BushJob1.htm.
83. Isikoff and Corn, *Hubris,* p. 310-11.
84. Realclear Politics.com, http://www.realclearpolitics.com/Presidential_04/bush_vs_kerry_historical.html.
85. Bill Schneider, "Political Plays of the Year," CNN Political Unit, December 31, 2004, http://www.cnn.com/2004/ALLPOLITICS/12/31/plays.years/ index.html.
86. "Remarks by Secretary of Homeland Security Tom Ridge Regarding Recent Threat Reports," http://www.whitehouse.gov/news/releases/2004/08/20040801.html.
87. "Limbaugh Back to Labeling Abu Ghraib Prisoner Abuse 'Hazing' and 'a Fraternity Prank,'" Media Matters for America, August 4, 2004, http://mediamatters.org/items/200408050011.
88. David Broder, "Karl Rove's Trajectory; the Presidential Adviser, Famously Smart, Is Getting Too Much Attention," *Pittsburgh Post-Gazette,* May 15, 2003, p. A13.
89. "Sizeable Minorities Still Believe Saddam Hussein Had Strong Links to Al Qaeda, Helped Plan 9/11 and Had Weapons of Mass Destruction," http://www.prnewswire.com/cgi-bin/stories.pl?ACCT=104&STORY=/www/story/12-29-2005/0004240417&EDATE.
90. Ken Herman and Scrott Shepard, "Video Renews Terror Focus; Candidates Seek Edge from Airing," *Atlanta Journal-Constitution,* October 31, 2004, p. B1.
91. George W. Bush, "Remarks in Pontiac, Michigan," Weekly Compilation of Presidential Documents 40, no. 44 (November 1, 2004): 2641.
92. Author interview with Jerry Falwell, May 2005.
93. James Dao, "After Victory, Crusader Against Same-Sex Marriage Thinks Big," *New York Times,* November 26, 2004, http://www.nytimes.com/2004/11/26/national/26gay.html?ex=1187236800&en=cb9a5076f5f46ca5&ei=5070.
94. David D. Kirkpatrick, "The 2004 Campaign: Strategy; Bush Allies Till Fertile Soil, Among Baptists, for Votes," *New York Times,* June 18, 2004, http:// select.nytimes.com/search/restricted/article?res=FB0714FB3C5D0C7B8DDDAF0894DC404482.
95. Ibid.
96. Ibid.
97. Author interview with evangelical, not for attribution, June 2005.
98. Ewen MacAskill and Julian Borger, "Iraq War Was Illegal and Breached UN Charter, Says Annan," *The Guardian* (London), September 16, 2004, http://www.guardian.co.uk/Iraq/Story/0,2763,1305709,00.html.
99. Brian Faler, "Election Turnout in 2004 Was Highest Since 1968," *Washington Post,* January 15, 2005, p. A5, http://www.washingtonpost.com/wp-dyn/articles/A10492-2005Jan14.html.
100. Robert Parry, Sam Parry, and Nat Parry, "Neck Deep," The Media Consortium, Arlington, Virginia, 2007.
101. Collin Hansen, "Weblog: Moral Values Carry Bush to Victory," *Christianity Today,* November 1, 2004, http://www.ctlibrary.com/ct/2004/novemberweb-only/11-1-31.0.html.
102. Bob Jones III, "'A Reprieve from Paganism; Bob Jones III Urges President to 'Put Imprint of Righteousness on Nation,'" *Charlotte Observer,* November 22, 2004, p. A12.
103. President Holds Press Conference, White House Press Release, Office of the Press

Secretary, November 4, 2004, http://www.whitehouse.gov/news/releases/2004/11/20041104-5.html.

Chapter Eighteen: An Angel Directs the Storm

1. Dana Priest and Bradley Graham, "Missing Antiaircraft Missiles Alarm Aides," *Washington Post,* November 7, 2004, http://www.washingtonpost.com/wp-dyn/articles/A31050-2004Nov6.html.
2. "Endgame: Timeline," *Frontline,* PBS, http://www.pbs.org/wgbh/pages/frontline/endgame/cron/.
3. Robert H. Reid, "Over 30 Killed in Iraq Insurgent Attacks," Associated Press, November 6, 2004.
4. "Oil Well Fire Out of Control," Associated Press, November, 15, 2004.
5. Mark Matthews, "Experts Give Dim Prognosis on Iraqi Oil Industry; Sabotage, Poor Facilities Leave U.S. Goals Unlikely," *Baltimore Sun,* October 31, 2004.
6. Brian Whitaker, "Militants Call for Muslim Uprising," *The Guardian* (London), November 11, 2004, http://www.guardian.co.uk/Iraq/Story/0,2763,1348394,00.html.
7. Karen DeYoung, *Soldier: The Life of Colin Powell* (New York: Knopf, 2006), p. 6.
8. Ibid., p. 8.
9. Ibid., p. 509.
10. Douglas Jehl, "New C.I.A. Chief Tells Workers to Back Administration Policies," *New York Times,* November 17, 2004, http://www.nytimes.com/2004/11/17/politics/17intel.html?ex=1258434000&en=a6c2854b242c7139&ei=5088&partner=rssnyt.
11. "Poll Success Eclipses Past Blunders for U.S.," *Irish Times,* February 1, 2005, p. 9.
12. *American Morning,* CNN, February 4, 2005. Transcript #020403CN.V74.
13. George W. Bush, State of the Union Address, February 2, 2005, http:// www.whitehouse.gov/news/releases/2005/02/20050202-11.html.
14. "President Sworn-In to Second Term," January 20, 2005, http://www.whitehouse.gov/news/releases/2005/01/20050120-1.html.
15. Doyle McManus, "Bush Pulls 'Neo-Cons' Out of the Shadows," *Los Angeles Times,* January 22, 2005.
16. Ibid.
17. Maureen Dowd, "Defining Victory Down," originally published in *New York Times,* January 9, 2005, http://www.commondreams.org/views05/0109-23.htm.
18. "Endgame Interviews: General Jack Keane," *Frontline,* PBS, http:// www.pbs.org/wgbh/pages/frontline/endgame/interviews/keane.html.
19. Haider Abbas, "Suicide Bomb Kills 115 Near Iraq Marketplace," Reuters, February 28, 2005, http://www.truthout.org/cgi-bin/artman/exec/view.cgi/ 37/9274.
20. Juan Cole, "Dawa Party Background, Informed Comment," March 1, 2005, http://www.juancole.com/2005/03/dawa-party-background-aaron-glantz.html.
21. Mark Kukis, "Ethnic Cleansing in a Baghdad Neighborhood?," *Time,* October 25, 2006, http://www.time.com/time/world/article/0,8599,1550441,00.html.
22. "Gangs of Iraq," *Frontline,* PBS, http://www.pbs.org/wgbh/pages/frontline/gangsofiraq/etc/script.html.
23. Ibid.
24. "The World According to Coulter," Media Matters, October 4, 2004, http://mediamatters.org/items/200410040009.
25. A. J. Walzer, "Limbaugh: Democrats are 'PR spokespeople for Al Qaeda,'" Media Matters, August 1, 2007, http://mediamatters.org/items/200708010008.
26. "Content Analysis of O'Reilly's Rhetoric Finds Spin to Be a 'Factor,'" May 2, 2007, http://newsinfo.iu.edu/news/page/normal/5535.html. For the complete study on O'Reilly, go to "Villains, Victims, and the Virtuous" in Bill O'Reilly's No Spin Zone, http://journalism.indiana.edu/papers/oreilly.html.

27. "Grand Canyon: Monument to the World-wide Flood," Answeringingenesis .org, http://www.answersingenesis.org/creation/v18/i2/grand_canyon.asp.
28. "They Will Be Your Doctors When You're Old," *The Revealer,* January 26, 2005, http://www.therevealer.org/archives/today_001552.php.
29. Bill Keller, "God and George W. Bush," *New York Times,* May 17, 2003, p. A17.
30. Howard Dean, "Bush's War on Science," *Boulder Daily Camera,* July 5, 2004, http://www.commondreams.org/views04/0705-04.htm.
31. "Robertson: God 'Will Remove Judges from the Supreme Court Quickly," Media Matters, January 4, 2005, http://mediamatters.org/items/200501040010.
32. http://www.theocracywatch.org/index_before_nov_06.htm. For the Christian Coalition scorecard, go to http://cc.org.2004scorecard.pdf.
33. Lampman, "Mixing Prophecy and Politics."
34. "Inhofe 'Very Proud' . . ." Thinkprogress, June 6, 2006, http://thinkprogress .org/2006/06/06/inhofe-gay-marriage/.
35. Gary Leupp, "Nights in White House Satin: The Comings and Goings of Jeff Gannon," *Counterpunch,* May 21–22, 2005. For secret service reports of Gannon's White House visits go to http://www.counterpunch.org/leupp05212005 .html.
36. Greg Toppo, "Education Dept. Paid Commentator to Promote Law," *USA Today,* January 7, 2005, http://www.usatoday.com/news/washington/2005-01-06-williams-whitehouse_x.htm.
37. Ann Scott Tyson, "Two Years Later, Iraq War Drains Military," *Washington Post,* March 19, 2005, http://www.washingtonpost.com/wp-dyn/articles/A48306-2005Mar18.html?nav=rss_topnews.
38. "President Thanks DHS Secretary Chertoff at Swearing-In Ceremony," March 3, 2005, http://www.whitehouse.gov/news/releases/2005/03/20050303-1.html.
39. "Children 'Starving' in New Iraq," BBC, March 30, 2005, http://news.bbc.co.uk/2/hi/middle_east/4395525.stm.
40. "Iraq Insurgency in 'Last Throes,' Cheney Says," CNN.com, June 20, 2005, http://www.cnn.com/2005/US/05/30/cheney.iraq/.
41. Patrick Quinn, "In Iraq Baghdad's Mayor Complains About Crumbling Capital," Associated Press, June 30, 2005, http://www.signonsandiego.com/news/world/iraq/20050630-1454-iraq.html.
42. Sarah Lyall, "Rush-Hour Strike Wounds up to 1,000; Blair Sees G-8 Link," *New York Times,* July 7, 2005, http://select.nytimes.com/search/restricted/article?res=F30B17F63D590C748CDDAE0894DD404482.
43. Richard Norton-Taylor and Duncan Campbell, "Intelligence Officials Were Braced for an Offensive—but Lowered Threat Levels," *The Guardian* (London), July 8, 2005, http://www.guardian.co.uk/uk_news/story/0,3604,1523647 ,00.html.
44. Peter Bergen and Paul Cruickshank, "The Iraq Effect: War Has Increased Terrorism Seven Fold Worldwide," *Mother Jones,* March 1, 2007, http://www.motherjones .com/news/featurex/2007/03/iraq_effect_1.html.
45. Hala Jaber, "Allawi: This Is the Start of Civil War," *Sunday Times,* July 10, 2005 http://www.timesonline.co.uk/tol/news/world/article542335.ece.
46. Laura Rozen, "Curt Weldon's Deep Throat," *American Prospect,* web only, June 10, 2005, http://www.prospect.org/cs/articles?articleId=9836.
47. Statement by Raymond Tanter, Iran Policy Committee, http://www.nci.org/05nci/11/Tanter-Statement-Nov-21.htm.
48. Ibid.
49. Guy Dinmore and Gareth Smyth, "US and UK Develop Democracy Strategy for Iran," *Financial Times* (London), April 22, 2006, p. 9.
50. "President Bush: Job Ratings," http://www.pollingreport.com/BushJob1.htm.
51. Frank Rich, "Someone Tell the President the War Is Over," *New York Times,* August 14, 2005, http://select.nytimes.com/search/restricted/article?res=F10E17F83E5A0C778DDDA10894DD404482.
52. Robert H. Reid, "Rebel Offensive Kills 48 US Soldiers in 12 Days," Associated Press, August 4, 2005.
53. Author interview with Grover Norquist, October 2006.

54. David Wurmser, *Tyranny's Ally: America's Failure to Defeat Saddam Hussein* (La Vergne, Tenn.: AEI Press, 1999), p.72.
55. Ibid., p.74.
56. Author interview with Martin Indyk, October 2006.
57. Author interview with Larry Wilkerson, October 2006.
58. Ibid.
59. Author interview with Uzi Arad, June 2005.
60. Anonymous, "The State Department's Extreme Makeover," *Salon*, October 4, 2004, http://dir.salon.com/story/opinion/feature/2004/10/04/foggybottom/index.html?pn=2.
61. "Iran Wants China as Top Oil Importer," Agence France Presse (AFP), November 8, 2004.
62. Robin Wright "Iran's New Alliance with China Could Cost U.S. Leverage," *Washington Post*, November 17, 2004, p. A21.
63. "Recognizing Iran as a Strategic Threat," Staff Report of the House Permanent Select Committee on Intelligence, Subcommittee on Intelligence Policy, August 23, 2006, http://intelligence.house.gov/Media/PDFS/IranReport082206v2.pdf#search=%22House%20intelligence%20committee%20and%20Iran%22.
64. Author interview with Gary Sick, January 2007.
65. Dafna Linzer, *Washington Post*, September 14, 2006, p. A17.
66. Thomas L. Friedman, "Time for Plan B," *New York Times*, August 4, 2006, p. A17.
67. David R. Jones, "Why the Democrats Won," CBS News, November 8, 2006, http://www.cbsnews.com/stories/2006/11/08/politics/main2161309.shtml.
68. Sidney Blumenthal, "Shuttle Without Diplomacy," *Salon*, January 10, 2007.
69. Author interview with Frederick Kagan, January 2007.
70. Ibid.
71. Author interview with Philip Giraldi, October 2006.
72. Author interview with Meyrav Wurmser, September 2006.
73. Author interview with Sam Gardiner, October 2006.
74. Ibid.
75. Author interview with Patrick Lang, October 2006
76. David Ignatius, "An Iranian Missile Crisis?" *Washington Post*, April 12, 2006, p. A17.
77. Jim Rutenberg, "Ex-Aide Details a Loss of Faith in the President," *New York Times*, April 1, 2007, http://select.nytimes.com/search/restricted/article?res=F20911FF35540C728CDDAD0894DF404482.
78. "Was Tillman Murdered? AP Gets New Documents," Associated Press, July 26, 2007, http://www.editorandpublisher.com/eandp/news/article_display.jsp?vnu_content_id=1003617692.
79. Steven R. Hurst, "Iraq Body Count Running at Double Pace," Associated Press, August 25, 2007, http://news.yahoo.com/s/ap/20070825/ap_on_re_mi_ea/iraq_counting_the_dead.
80. Steven Lee Myers and Megan Thee, "Americans Feel Military Is Best at Ending War," *New York Times*, September 10, 2007, p. A1.
81. Sean Wilentz, "The Worst President in History?" *Rolling Stone*, April 21, 2006, http://www.rollingstone.com/news/profile/story/9961300/the_worst_president_in_history.
82. http://www.backwardsbush.com/.
83. Kelly Wallace, *CBS Evening News*, September 2, 2007.
84. Robert Draper, *Dead Certain: The Presidency of George W. Bush* (New York: Free Press, 2007), pp. 305–6.
85. Molly Hennessy-Fiske, "Bush Can't Recall Why Iraqi Army Disbanded," *Los Angeles Times*, September 3, 2007, http://www.latimes.com/news/printedition/asection/la-na-bush3sep03,1,2046900.story?coll=la-news-a_section.
86. Letter from George W. Bush to L. Paul Bremer, May 23, 2003, *New York Times*, http://www.nytimes.com/ref/washington/04bremer-text2.html.

87. Draper, *Dead Certain*, p. 406.
88. Sayed Salahuddin, "U.S. Show of Force in Gulf Alarming: Afghan Paper," Reuters, May 26, 2007.
89. "President Bush Addresses the 89th Annual National Convention of the American Legion," Reno-Sparks Convention Center, Reno, Nevada, August 28, 2007, http://www.whitehouse.gov/news/releases/2007/08/20070828-2.html.
90. Warren P. Strobel, John Walcott, and Nancy A. Youssef, "Cheney Urging Strikes on Iran," McClatchy Washington Bureau, August 9, 2007, http://www.mcclatchydc.com/227/story/18834.html.
91. James Rosen, "U.S. Officials Begin Crafting Iran Bombing Plan," Fox News, September 12, 2007, http://www.foxnews.com/story/0,2933,296450,0o.hmtl.
92. Uzi Mahnaimi and Sarah Baxter, "Israelis Seized Nuclear Material in Syrian Raid," *Sunday Times* (London), September 23, 2007, http://www.timesonline.co.u/tol/news/world/middle_east/article2512380.ece.

BIBLIOGRAPHY

Abrams, Elliot. *Faith or Fear: How Jews Can Survive in a Christian America.* New York: Free Press, 1997.

Achcar, Gilbert. *The Clash of Barbarisms: Sept. 11 and the Making of the New World Order.* New York: Monthly Review Press, 2002.

Aikman, David. *A Man of Faith: The Spiritual Journey of George W. Bush.* Nashville, Tenn.: Thomas Nelson, 2004.

Albert, James A. *Jim Bakker: Miscarriage of Justice?* Chicago: Open Court, 1999.

Ali, Tariq. *The Clash of Fundamentalisms: Crusades, Jihads, and Modernity.* London: Verso, 2003.

Alterman, Eric. *What Liberal Media? The Truth About Bias and News.* New York: Basic Books, 2003.

Armstrong, Karen. *The Battle for God.* New York: Ballantine, 2001.

———. *Jerusalem: One City, Three Faiths.* New York: Knopf, 1996.

Bamford, James. *A Pretext for War: 9/11, Iraq, and the Abuse of America's Intelligence Agencies.* New York: Doubleday, 2004.

Beeman, William O. *The "Great Satan" vs. the "Mad Mullahs:" How the United States and Iran Demonize Each Other.* Westport, Conn.: Praeger, 2005.

Bell, Daniel, ed. *The Radical Right.* New York: Anchor Books, 1964.

Benjamin, Daniel, and Steven Simon. *The Next Attack: The Failure of the War on Terror and a Strategy for Getting It Right.* New York: Owl Books, 2006.

Bercovitch, Sacvan. *The American Jeremiad.* Madison, Wis.: The University of Wisconsin Press, 1978.

Berlet, Chip, and Matthew N. Lyons. *Right-Wing Populism in America: Too Close for Comfort.* New York: Guilford Press, 2000.

Blaker, Kimberly, ed. *The Fundamentals of Extremism: The Christian Right in America.* New Boston, Mich.: New Boston Books, 2003.

Bloom, Allan, *The Closing of the American Mind.* New York: Touchstone, 1987.

Blumenthal, Sidney. *The Clinton Wars.* New York: Penguin, 2004.

. *How Bush Rules: Chronicles of a Radical Regime.* Princeton, N.J.: Princeton University Press, 2006.

———. *Our Long National Daydream: A Political Pageant of the Reagan Era.* New York: HarperCollins, 1990.

———. *The Rise of the Counter-Establishment: From Conservative Ideology to Political Power.* New York: HarperCollins, 1988.

Boehlert, Eric, *Lapdogs: How the Press Rolled Over for Bush.* New York: Free Press, 2006.

Bonini, Carlo, and Giuseppe D'Avvanzo. *Collusion: International Espionage and the War on Terror.* Hoboken, N.J.: Melville House, 2007.

Boston, Robert, *Close Encounters with the Religious Right: Journeys into the Twilight Zone of Religion and Politics.* New York: Prometheus Books, 2000.

———. *The Most Dangerous Man in America? Pat Robertson and the Rise of the Christian Coalition.* New York: Prometheus Books, 1996.

Boyer, Paul. *When Time Shall Be No More: Prophecy Belief in Modern American Culture.* Cambridge, Mass.: Belknap Press, 1994.
Breines, Paul. *Tough Jews: Political Fantasies and the Moral Dilemma of American Jewry.* New York: Basic Books, 1990.
Brock, David. *Blinded by the Right: The Conscience of an Ex-Conservative.* New York: Crown, 2002.
Brzezinksi, Zbigniew. *The Grand Chessboard: American Primacy and Its Geostrategic Imperatives.* New York: Basic Books, 1997.
———. *Second Chance: Three Presidents and the Crisis of American Superpower.* New York: Basic Books, 2007.
Bush, Barbara. *Reflections: Life After the White House.* New York: Scribner, 2003.
Bush, George [H.W.], and Brent Scowcroft. *A World Transformed.* New York: Vintage Books, 1998.
Bush, George W., with Victor Gold. *Looking Forward: The George Bush Story.* Garden City, N.Y.: Doubleday, 1987.
Bush, George W., and Mickey Herskowitz. *A Charge to Keep: My Journey to the White House.* New York: Harper Paperbacks, 2001.
Butler, Richard. *The Greatest Threat: Iraq, Weapons of Mass Destruction, and the Crisis of Global Security.* New York: Public Affairs, 2000.
Cahn, Anne Hessing. *Killing Détente: The Right Attacks the CIA.* University Park: Pennsylvania State University Press, 1998.
Cannon, Carl M., Lou Dubose, and Jan Reid. *Boy Genius: Karl Rove, the Architect of George W. Bush's Remarkable Political Triumph.* New York: Public Affairs, 2005.
Carter, Graydon. *What We've Lost: How the Bush Administration Has Curtailed Our Freedoms, Mortgaged Our Economy, Ravaged Our Environment, and Damaged Our Standing in the World.* New York: Farrar, Strauss and Giroux, 2004.
Carter, Jimmy. *Keeping Faith: Memoirs of a President.* New York: Bantam Books, 1982.
———. *Palestine: Peace Not Apartheid.* New York: Simon & Schuster, 2006.
Chamdrasekaran, Rajiv. *Imperial Life in the Emerald City: Inside Iraq's Green Zone.* New York: Knopf, 2006.
Chatterjee, Pratap. *Iraq, Inc.: A Profitable Occupation.* New York: Seven Stories Press, 2004.
Cherry, Conrad, ed. *God's New Israel: Religious Interpretations of American Destiny.* Englewood Cliffs, N.J.: Prentice-Hall, Inc., 1971.
Chubin, Shahram. *Iran's Nuclear Ambitions.* Washington, D.C.: Carnegie Endowment for National Peace, 2006.
Church, Forrest. *The Separation of Church and State: Writings on a Fundamental Freedom by America's Founders.* Boston: Beacon Press, 2004.
Clarke, Richard. *Against All Enemies: Inside America's War on Terror.* New York: Free Press, 2004.
Clarkson, Frederick. *Eternal Hostility: The Struggle Between Theocracy and Democracy.* Monroe, Maine: Common Courage Press, 1997.
Clarridge, Duane R. *A Spy for All Seasons: My Life in the CIA.* New York: Scribner, 1997.
Clinton, Bill. *My Life.* New York: Random House, 2004.
Cockburn, Andrew. *Rumsfeld: His Rise, Fall, and Catastrophic Legacy.* New York: Scribner, 2007.
Cockburn, Andrew, and Patrick Cockburn. *Out of the Ashes: The Resurrection of Saddam Hussein.* New York: HarperCollins, 1999.
Cockburn, Leslie. *Out of Control: The Story of the Reagan Administration's Secret War in Nicaragua, the Illegal Arms Pipeline, and the Contra Drug Connection.* New York: Atlantic Monthly Press, 1987.
Cockburn, Patrick. *The Occupation: War and Resistance in Iraq.* London: Verso, 2006.
Cohn, Norman. *Cosmos, Chaos, and the World to Come: The Ancient Roots of Apocalyptic Faith.* New Haven: Yale University Press, 1999.
———. *The Pursuit of the Millennium.* New York: Oxford University Press, 1974.
Cohn-Sherbok, Dan. *The Politics of the Apocalypse: The History and Influence of Christian Zionism.* Oxford, England: Oneworld Publications, 2006.

Conason, Joe. *Big Lies: The Right-Wing Propaganda Machine and How It Distorts the Truth.* New York: St. Martin's Press, 2003.

Conason, Joe, and Gene Lyons. *The Hunting of the President: The Ten-Year Campaign to Destroy Bill and Hillary Clinton.* New York: St. Martin's Griffin, 2000.

Cooley, John K. *An Alliance Against Babylon: The U.S., Israel, and Iraq.* London: Pluto Press, 2005.

———. *Unholy Wars: Afghanistan, America and International Terrorism.* Sterling, Va.: Pluto Press, 1999.

Corson, William R. *The Betrayal.* New York: W. W. Norton and Co., 1968.

Cox, Harvey. *The Secular City: Secularization and Urbanization in Theological Perspective.* New York: Collier Books, 1990.

Cramer, Richard Ben. *What It Takes: The Way to the White House.* New York: Random House, 1992.

Darwish, Adel, *Unholy Babylon: The Secret History of Saddam's War.* London: Gollancz, 1991.

de Marenches, Alexandre, and David Andelman, *The Fourth World War: Diplomacy and Espionage in the Age of Terrorism.* New York: William Morrow, 1992.

Dean, John W. *Conservatives Without Conscience.* New York: Viking, 2006.

Deffeyes, Kenneth S. *Beyond Oil: The View from Hubbert's Peak.* New York: Hill and Wang, 2005.

Dennett, Daniel C. *Breaking the Spell: Religion as a Natural Phenomenon.* New York: Viking, 2006.

DeYoung, Karen. *Soldier: The Life of Colin Powell.* New York: Knopf, 2006.

Diamond, Larry. *Squandered Victory: The American Occupation and the Bungled Effort to Bring Democracy to Iraq.* New York: Times Books, 2005.

Diamond, Sara. *Roads to Dominion: Right-Wing Movements and Political Power in the United States.* New York: The Guilford Press, 1995.

Draper, Robert. *Dead Certain: The Presidency of George W. Bush.* New York: Free Press, 2007.

Draper, Theodore. *A Very Thin Line—the Iran-Contra Affairs.* New York: Hill and Wang, 1991.

Drumheller, Tyler, and Elaine Monaghan. *On the Brink: An Insider's Account of How the White House Compromised American Intelligence.* New York: Carroll and Graf Publishers, 2006.

Drury, Shadia B. *Leo Strauss and the American Right.* New York: St. Martin's Press, 1999.

Duffy, Michael, and Dan Goodgame. *Marching in Place: The Status Quo Presidency of George Bush.* New York: Simon & Schuster, 1992.

Easton, Nina J. *Gang of Five: Leaders at the Center of the Conservative Crusade.* New York: Simon & Schuster, 2000.

Edsall, Thomas B. *Building Red America: The New Conservative Coalition and the Drive for Permanent Power.* New York: Basic Books, 2006.

Eisner, Peter, and Knut Royce. *The Italian Letter: How the Bush Administration Used a Fake Letter to Build the Case for War in Iraq.* New York: Rodale, 2007.

Feldman, Noah. *Divided by God: America's Church-State Problem—and What We Should Do About It.* New York: Farrar, Straus and Giroux, 2005.

Ferling, John. *Adams vs. Jefferson: The Tumultuous Election of 1800.* New York: Oxford University Press, 2004.

Fitzgerald, Frances. *Cities on a Hill: A Journey Through Contemporary American Cultures.* New York: Simon & Schuster, 1986.

Fournier, Ron, Douglas B. Sosnik, and Matthew J. Dowd. *Applebee's America: How Successful Political, Business, and Religious Leaders Connect with the New American Community.* New York: Simon & Schuster, 2006.

Frank, Justin A. *Bush on the Couch.* New York: ReganBooks, 2004.

Frank, Thomas. *What's the Matter with Kansas? How Conservatives Won the Heart of America.* New York: Metropolitan Books, 2004.

Freiling, Thomas M., ed. *George W. Bush: On God and Country.* Washington, D.C.: Allegiance Press, Inc. 2004.

Friedman, Robert I. *The False Prophet: Rabbi Meir Kahane—From FBI Informant to Knesset Member.* London: Faber and Faber, 1990.

———. *Zealots for Zion: Inside Israel's West Bank Settlement Movement.* Piscataway, N.J.: Rutgers University Press, 1994.

Friedman, Thomas L. *From Beirut to Jerusalem.* New York: Anchor Books, 1989.

———. *The World Is Flat: A Brief History of the Twenty-First Century.* New York: Farrar, Straus and Giroux, 2006.

Fromkin, David. *A Peace to End All Peace: The Fall of the Ottoman Empire and the Creation of the Modern Middle East.* New York: Owl Books, 2001.

Frum, David. *The Right Man: The Surprise Presidency of George W. Bush.* New York: Random House, 2003.

Fuller, Robert. *Religious Revolutionaries: The Rebels Who Reshaped American Religion.* New York: Palgrave Macmillan, 2004.

Germond, Jack, and Jules Witcover. *Whose Broad Stripes and Bright Stars? The Trivial Pursuit of the Presidency, 1988.* New York: Warner Books, 1989.

Gold, Victor. *Invasion of the Party Snatchers: How the Holy-Rollers and the Neo-Cons Destroyed the GOP.* Chicago: Sourcebooks Trade, 2007.

Goldberg, J. J. *Jewish Power: Inside the American Jewish Establishment.* Reading, Mass: Addison-Wesley, 1996.

Goldberg, Michelle. *Kingdom Coming: The Rise of Christian Nationalism.* New York: W. W. Norton and Co., 2007.

Goldhill, Simon, *The Temple of Jerusalem.* Cambridge: Harvard University Press, 2005.

Gonen, Jay Y. *A Psycho-History of Zionism.* New York: Meridian Books, 1975.

Gordon, Michael R., and Gen. Bernard E. Trainor. *Cobra II: The Inside Story of the Invasion and Occupation of Iraq.* New York: Pantheon, 2006.

Gore, Al. *An Inconvenient Truth: The Planetary Emergence of Global Warming and What We Can Do About It.* New York: Rodale, 2006.

Gorenberg, Gershom. *The Accidental Empire: Israel and the Birth of the Settlements, 1967–1977.* New York: Times Books, 2006.

———. *The End of Days: Fundamentalism and the Struggle for the Temple Mount.* New York: Oxford University Press, 2000.

Graubard, Stephen. *Mr. Bush's War: Adventures in the Politics of Illusion.* New York: Hill and Wang, 1992.

Green, Stephen. *Living by the Sword: America and Israel in the Middle East.* Brattleboro, Vt.: Amana Books, 1988.

Hagee, John. *Beginning of the End: The Assassination of Yitzhak Rabin and the Coming Antichrist.* Nashville, Tenn.: Thomas Nelson Publishers, 1996.

Halper, Stefan, and Jonathan Clarke. *America Alone: The Neoconservatives and the Global Order.* New York: Cambridge University Press, 2004.

Halpern, Ben. *The American Jew.* New York: Theodor Herzl Foundation, 1956.

Halsell, Grace. *Forcing God's Hand: Why Millions Pray for a Quick Rapture—and Destruction of Planet Earth.* Beltsville, Md.: International Graphics, 2003.

———. *Prophecy and Politics.* Chicago: Lawrence Hill, 1986.

Harris, Sam. *The End of Faith: Religion, Terror, and the Future of Reason.* New York: W. W. Norton and Co., 2005.

Hatch, Nathan O. *The Democratization of American Christianity.* New Haven: Yale University Press, 1989.

Hatch, Nathan O., and Harry S. Stout, eds. *Jonathan Edwards and the American Experience.* New York: Oxford University Press, 1988.

Hayes, Stephen F. *Cheney: The Untold Story of America's Most Powerful and Controversial Vice President.* New York: HarperCollins, 2007.

Henzel, Ronald M. *Darby, Dualism, and the Decline of Dispensationalism.* Tucson, Ariz: Fenestra Books, 2003.

Herman, Edward S., and Frank Brodhead. *The Rise and Fall of the Bulgarian Connection.* New York: Sheridan Square Publications, 1986.

Hersh, Seymour M. *Chain of Command: The Road From 9/11 to Abu Ghraib.* New York: HarperCollins, 2004.

Hiro, Dilip. *The Longest War: The Iran-Iraq Military Conflict.* New York: Routledge, 1991.
Hitchcock, Mark, and Thomas Ice. *The Truth Behind Left Behind: A Biblical View of the End Times.* Sisters, Oreg.: Multnomah, 2004.
Holmes, Stephen. *The Anatomy of Antiliberalism.* Cambridge: Harvard University Press, 1993.
———. *The Matador's Cape: America's Reckless Response to Terror.* New York: Cambridge University Press, 2007.
Huntington, Samuel. *The Clash of Civilizations and the Remaking of World Order.* New York: Touchstone, 1997.
Ide, Arthur Frederick. *The Father's Son: George W. Bush.* Las Colinas, Tex.: Sepore, 1998.
Isikoff, Michael, and David Corn. *Hubris: The Inside Story of Spin, Scandal, and the Selling of the Iraq War.* New York: Crown, 2006.
Ivins, Molly. *Molly Ivins Can't Say That, Can She?* New York: Random House, 1991.
Ivins, Molly, and Lou Dubose. *Shrub.* New York: Vintage Books, 2000.
Jacoby, Susan. *Freethinkers: A History of American Secularism.* New York: Metropolitan Books, 2004.
Judis, John B. *The Folly of Empire: What George W. Bush Could Learn from Theodore Roosevelt and Woodrow Wilson.* New York: Oxford University Press, 2004.
Kagan, Robert, and William Kristol, eds. *Present Dangers: Crisis and Opportunity in American Foreign and Defense Policy.* San Francisco: Encounter Books, 2000.
Kaplan, Esther. *With God on Their Side: How Christian Fundamentalists Trampled Science, Policy, and Democracy in George W. Bush's White House.* New York: The New Press, 2004.
Kaplan, Fred. *The Wizards of Armageddon.* New York: Touchstone, 1983.
Kaplan, Lawrence, and William Kristol. *The War Over Iraq: Saddam's Tyranny and America's Mission.* San Francisco: Encounter Books, 2003.
Karpin, Michael, and Ina Friedman. *Murder in the Name of God: The Plot to Kill Yitzhak Rabin.* New York: Metropolitan Books, 1998.
Kelly, Michael. *Martyr's Day: Chronicle of a Small War.* New York: Random House, 1993.
Klare, Michael T. *Blood and Oil: The Dangers and Consequences of America's Growing Dependency on Imported Petroleum.* New York: Metropolitan Books, 2004.
Kornbluh, Peter, and Malcolm Byrne. *The Iran-Contra Scandal: The Declassified History.* New York: New Press, 1993.
Kovel, Joel. *Red Hunting in the Promised Land: Anticommunism and the Making of America.* London, Cassell, 1997.
Kristol, William, ed. *The Weekly Standard: A Reader: 1995–2005.* New York: HarperCollins, 2005.
Kuo, David. *Tempting Faith: An Inside Story of Political Seduction.* New York: Free Press, 2006.
Kwitny, Jonathan. *The Crimes of Patriots: A True Tale of Dope, Dirty Money, and the CIA.* New York: W. W. Norton and Co., 1987.
LaHaye, Tim. *Battle for the Mind.* Grand Rapids, Mich.: Baker Book House, 1980.
———. *Understanding Bible Prophecy for Yourself.* Eugene, Oreg.: Harvest House Publishers, 1982.
LaHaye, Tim, and Jerry B. Jenkins. *Are We Living in the End Times? Current Events Foretold in Scripture . . . And What They Mean.* Wheaton, Ill.: Tyndale House Publishers, 1999.
———. *Glorious Appearing: The End of Days.* Wheaton, Ill.: Tyndale House Publishers, 2004.
———. *Left Behind: A Novel of the Earth's Last Days.* Wheaton, Ill.: Tyndale House Publishers, 1995.
LaHaye, Tim, and Beverly LaHaye. *The Act of Marriage.* Grand Rapids, Mich.: Zondervan, 1998.
Lambert, Frank. *The Founding Fathers and the Place of Religion in America.* Princeton, N.J.: Princeton University Press, 2003.

Lando, Barry M. *Web of Deceit: The History of Western Complicity in Iraq, from Churchill to Kennedy to George W. Bush.* New York: Other Press, 2007.

Ledeen, Michael Arthur. *The First Duce: D'Annunzio at Fiume.* Baltimore, Md.: Johns Hopkins University Press, 1977.

———. *Machiavelli on Modern Leadership: Why Machiavelli's Iron Rules Are as Timely and Important Today as Five Centuries Ago.* New York: Truman Talley Press, 1999.

———. *The War Against the Terror Masters: Why It Happened. Where We Are Now. How We'll Win.* New York: Truman Talley Books, 2003.

Levinson, Jerome I. *Who Makes American Foreign Policy?* Gaithersburg, Md.: Witches Brew, 2004.

Lévy, Bernard-Henri. *American Vertigo: Traveling America in the Footsteps of Toqueville.* New York: Random House, 2006.

Levy, Paul. *The Madness of George W. Bush: A Reflection of Our Collective Psychosis.* Bloomington, Ind.: AuthorHouse, 2006.

Lewis, Bernard. *What Went Wrong?: The Clash Between Islam and Modernity in the Middle East.* New York: Oxford University Press, 2002.

Lief, Michael S., and Mitchell Calwell. *And the Walls Came Tumbling Down: Greatest Closing Arguments Protecting Civil Liberties.* New York: Scribner, 2006.

Lind, Michael. *Made in Texas: George W. Bush and the Southern Takeover of American Politics.* New York: Basic Books, 2003.

Lipartito, Kenneth J., and Joseph A. Pratt. *Baker & Botts in the Development of Modern Houston.* Austin: University of Texas Press, 1991.

Loftus, John, and Mark Aarons. *The Secret War Against the Jews: How Western Espionage Betrayed the Jewish People.* New York: St. Martin's Press, 1994.

MacLean, Nancy K. *Behind the Mask of Chivalry: The Making of the Second Ku Klux Klan.* New York: Oxford University Press, 1995.

Makiya, Kanan. *Republic of Fear: The Politics of Modern Iraq.* Berkeley: University of California Press, 1989.

Malkin, Michelle. *In Defense of Internment: The Case for "Racial Profiling" in World War II and the War on Terror.* Washington, D.C.: Regnery Publishing, 2004.

Mann, James. *Rise of the Vulcans: The History of Bush's War Cabinet.* New York: Viking, 2004.

Mansfield, Stephen. *The Faith of George W. Bush.* New York: Penguin, 2004.

Marcus, Greil. *The Shape of Things to Come: Prophecy and the American Voice.* New York: Farrar, Straus and Giroux, 2006.

Marsden, George M. *Fundamentalism and American Culture: The Shaping of Twentieth-Century Evangelicalism: 1870–1925.* New York: Oxford University Press, 1982.

———. *Understanding Fundamentalism and Evangelicalism.* Grand Rapids, Mich.: Wm. B. Eerdmans Publishing Company, 1991.

Marsden, Peter. *The Taliban: War and Religion in Afghanistan.* New York: Zed Books, 2002.

Martin, William. *With God on Our Side: The Rise of the Religious Right in America.* New York: Broadway, 2005.

Marty, Martin E. *The One and the Many: America's Struggle for the Common Good.* Cambridge: Harvard University Press, 1997.

———. *Pilgrims in Their Own Land: 500 Years of Religion in America.* New York: Penguin, 1984.

Marty, Martin E., and R. Scott Appleby. *The Glory and the Power: The Fundamentalist Challenge to the Modern World.* Boston: Beacon Press, 1992.

Marty, Martin E., and Jonathan Moore. *Politics, Religion, and the Common Good: Advancing a Distinctly American Conversation About Religion's Role in Our Shared Life.* San Francisco: Jossey-Bass Publishers, 2000.

McFarlane, Robert C., with Zofia Smardz. *Special Trust.* New York: Cadell & Davies, 1994.

McKelvey, Tara. *Monstering: Inside America's Policy of Secret Interrogations and Torture in the Terror War.* New York: Carroll and Graf, 2007.

Means, Howard. *Colin Powell: Soldier/Statesman.* New York: Donald I. Fine, 1992.

Milbank, Dana. *Smashmouth: Two Years in the Gutter with Al Gore and George W. Bush.* New York: Basic Books, 2001.
Miller, Perry. *The American Puritans: Their Prose and Poetry.* New York: Anchor Books, 1956.
———. *The New England Mind: The Seventeenth Century.* Cambridge, Mass.: Belknap Press, 1982.
———. *The Puritans: A Sourcebook of Their Writings.* Mineola, N.Y.: Dover Publications, Inc., 2001.
Miller, T. Christian. *Blood Money: Wasted Billions, Lost Lives, and Corporate Greed in Iraq.* New York: Little, Brown, 2006.
Minutaglio, Bill. *First Son: George W. Bush and the Bush Family Dynasty.* New York: Crown, 1999.
Mitchell, Elizabeth. *W: Revenge of the Bush Dynasty.* New York: Hyperion, 2000.
Moens, Alexander. *The Foreign Policy of George W. Bush: Values, Strategy, and Loyalty.* Burlington, Vt.: Ashgate Publishing, 2004.
Moore, James, and Wayne Slater. *The Architect: Karl Rove and the Master Plan for Absolute Power.* New York: Crown, 2006.
Moynihan, Daniel Patrick. *A Dangerous Place.* Boston: Little Brown, 1978.
Nasr, Vali. *The Shia Revival: How Conflicts Within Islam Will Shape the Future.* New York: W. W. Norton, 2007.
Nichols, John. *Dick: The Man Who Is President.* New York: The New Press, 2004.
Niebuhr, Reinhold. *The Irony of American History.* New York: Charles Scribner's Sons, 1952.
———. *Moral Man & Immoral Society: A Study in Ethics and Politics.* New York: Scribner, 2001.
Northcott, Michael. *An Angel Directs the Storm: Apocalyptic Religion and American Empire.* London: I. B. Tauris, 2004.
Norton, Anne. *Leo Strauss and the Politics of American Empire.* New Haven, Conn.: Yale University Press, 2004.
O'Huallachain, D.L., and J. Forrest Sharpe, eds. *Neo-Conned: Just War Principles: A Condemnation of War in Iraq.* Vienna, Va.: Light in the Darkness Publications, 2005.
———. *Neo-Conned Again: Hypocrisy, Lawlessness, and the Rape of Iraq.* Vienna, Va.: Light in the Darkness Publications, 2005.
Olasky, Marvin. *Compassionate Conservatism: What It Is, What It Does, and How It Can Transform America.* New York: Free Press, 2000.
Oren, Michael. *Six Days of War: June 1967 and the Making of the Modern Middle East.* New York: Presidio Press, 2003.
Packer, George. *The Assassin's Gate: America In Iraq.* New York: Farrar, Straus and Giroux, 2006.
Page, Cristina. *How the Pro-Choice Movement Saved America: Freedom, Politics, and the War on Sex.* New York: Basic Books, 2006.
Palast, Greg. *Armed Madhouse: Who's Afraid of Osama Wolf? China Floats, Bush Sinks, the Scheme to Steal '08, No Child's Behind Left, and Other Dispatches from the Front Lines of the Class War.* New York: Penguin, 2006.
Parkhurst, Louis Gifford Jr. *Francis Schaeffer: The Man and his Message.* Carol Stream, Ill.: Tyndale House Publishers, 1985.
Parmet, Herbert S. *George Bush: The Life of a Lone Star Yankee.* New Brunswick, N.J.: Transaction, 2001.
Parry, Robert. *Fooling America: How Washington Insiders Twist the Truth and Manufacture Conventional Wisdom.* New York: William Morrow and Company, 1992.
———. *Secrecy & Privilege: Rise of the Bush Dynasty from Watergate to Iraq.* Arlington, Va.: The Media Consortium, 2004.
Persico, Joseph. *Casey: The Lives and Secrets of William J. Casey—from the OSS to the CIA.* New York: Viking, 1990.
Peterson, Merrill D. *Thomas Jefferson: Writings.* New York: The Library of America, 1984.
Phillips, Kevin. *American Theocracy: The Peril and Politics of Radical Religion, Oil, and Borrowed Money in the 21st Century.* New York: Viking, 2006.

Pitt, William Rivers, with Scott Ritter. *War on Iraq: What Team Bush Doesn't Want You to Know.* New York: Context Books, 2002.
Podhoretz, Norman. *Ex-Friends: Falling Out with Allen Ginsberg, Lionel and Diana Trilling, Lillian Hellman, Hannah Arendt, and Norman Mailer.* San Francisco: Encounter Books, 2000.
——. *The Present Danger: "Do We Have the Will to Reverse the Decline of American Power?"* New York: Simon & Schuster, 1980.
Pollack, Kenneth M. *The Threatening Storm: The Case for Invading Iraq.* New York: Random House, 2002.
Prothro, Stephen. *American Jesus: How the Son of God Became a National Icon.* New York: Farrar, Straus and Giroux, 2004.
Putnam, Robert D. *Bowling Alone: The Collapse and Revival of American Community.* New York: Simon & Schuster, 2001.
Rakove, Jack N. ed. *Interpreting the Constitution: The Debate over Original Intent.* Boston: Northern University Press, 1990.
Rampton, Sheldon, and John Stauber. *Weapons of Mass Deception: The Uses of Propaganda in Bush's War on Iraq.* New York: Tarcher/Penguin, 2003.
Rashid, Ahmed. *Taliban: Militant Islam, Oil, and Fundamentalism in Central Asia.* New Haven: Yale University Press, 2000.
Rich, Frank. *The Greatest Story Ever Sold: The Decline and Fall of Truth.* New York: Penguin, 2006.
Ricks, Thomas E. *Fiasco: The American Military Adventure in Iraq.* New York: Penguin, 2006.
Risen, James. *State of War: The Secret History of the CIA and the Bush Administration.* New York: Free Press, 2006.
Ritter, Scott. *Iraq Confidential: The Untold Story of the Intelligence Conspiracy to Undermine the UN and Overthrow Saddam Hussein.* New York: Nation Books, 2005.
——. *Target Iran: The Truth About the White House's Plans for Regime Change.* New York: Nation Books, 2006.
Robertson, Pat. *The New World Order.* Dallas, Tex.: Word Publishing, 1991.
Robinson, Jeffrey. *Yamani: The Inside Story.* London: Simon & Schuster, 1988.
Ross, Dennis. *The Missing Peace: The Inside Story of the Fight for Middle East Peace.* New York: Farrar, Straus and Giroux, 2005.
Rushdoony, Rousas John. *The Institutes of Biblical Law.* Phillipsburg, N.J.: P&R Publishers, 1978.
Russell, Richard L. *Sharpening Strategic Intelligence: Why the CIA Gets it Wrong and What Needs to Be Done to Get It Right.* New York: Cambridge University Press, 2007.
Ruthven, Malise. *A Fury for God: The Islamist Attack on America.* New York: Granta, 2002.
Ryn, Claes G. *America the Virtuous: The Crisis of Democracy and the Quest for Empire.* New Brunswick, N.J.: Transaction Publishers, 2005.
Sager, Ryan. *The Elephant in the Room: Evangelicals, Libertarians, and the Battle to Control the Republican Party.* Hoboken, N.J.: John Wiley and Sons, 2006.
Scahill, Jeremy. *Blackwater: The Rise of the World's Most Powerful Mercenary Army.* New York: Nation Books, 2007.
Schaeffer, Francis. *A Christian Manifesto.* Westchester, Ill.: Crossway Books, 1981.
——. *How Should We Then Live? The Rise and Decline of Western Thought and Culture.* Wheaton, Ill.: Crossway Books, 2005.
Schaeffer, Francis, and C. Everett Koop. *Whatever Happened to the Human Race?* Wheaton, Ill.: Crossway Books, 1983.
Schwartz, Stephen. *The Two Faces of Islam: The House of Sa'ud from Tradition to Terror.* New York: Doubleday, 2002.
Schweizer, Peter. *Victory: The Reagan Administration's Secret Strategy That Hastened the Collapse of the Soviet Union.* New York: Atlantic Monthly Press, 1994.
Schweizer, Peter, and Rochelle Schweizer. *The Bushes: Portrait of a Dynasty.* New York: Random House, 2004.

Scofield, C. I., ed. *The King James Study Bible Reference Edition.* Uhrichsville, Ohio: Barbour Books, 2000.

Shadid, Anthony. *Night Draws Near: Iraq's People in the Shadow of America's War.* New York: Picador, 2005.

Sharansky, Natan, with Ron Dermer. *The Case for Democracy: The Power of Freedom to Overcome Tyranny and Terror.* New York: Public Affairs, 2004.

Shogan, Robert. *Constant Conflict: Politics, Culture, and the Struggle for America's Future.* Boulder, Colo.: Basic Books, 2004.

Shuck, Glenn. *Marks of the Beast: The Left Behind Novels and the Struggle for Evangelical Identity.* New York: New York University Press, 2004.

Sifry, Micah L., and Christopher Cerf, eds. *The Gulf War Reader.* New York: Times Books, 1991.

Simmons, Matthew R. *Twilight in the Desert: The Coming Saudi Oil Shock and the World Economy.* Hoboken, N.J.: John Wiley and Sons, 2005.

Simon, Merrill. *Jerry Falwell and the Jews.* Middle Village, N.Y.: Jonathan David Publishers, 1984.

Skolfield, Ellis H. *The False Prophet.* Fort Meyers, Fla.: Fish House, 2001.

Slotkin, Richard. *Gunfighter Nation: The Myth of the Frontier in Twentieth-Century America.* New York: HarperPerennial, 1992.

———. *Regeneration Through Violence: The Mythology of the American Frontier: 1600–1860.* Norman: University of Oklahoma Press, 1973.

Slotkin, Richard, and James K. Folsom. *So Dreadful a Judgment: Puritan Responses to King Philip's War, 1676–77.* Middletown, Conn.: Wesleyan University Press, 1979.

Stewart, Rory. *The Prince of the Marshes: And Other Occupational Hazards of a Year in Iraq.* New York: Harcourt, 2006.

Stone, I. F. *Underground to Palestine and Reflections Thirty Years Later.* New York: Pantheon Books, 1978.

Strauss, Leo. *Natural Right and History.* Chicago: University of Chicago Press, 1965.

Suskind, Ron. *The One Percent Doctrine: Deep Inside America's Pursuit of Its Enemies Since 9/11.* New York: Simon & Schuster, 2006.

———. *The Price of Loyalty: George W. Bush, the White House, and the Education of Paul O'Neill.* New York: Pocket Books, 2004.

Sykes, Christopher. *Two Studies in Virtue.* New York: Knopf, 1953.

Teicher, Howard, and Gayle Radley Teicher. *Twin Pillars to Desert Storm: America's Flawed Vision in the Middle East from Nixon to Bush.* New York: William Morrow and Company, 1993.

Tenet, George. *At the Center of the Storm: My Years at the CIA.* New York: HarperCollins, 2007.

Tivnan, Edward. *The Lobby: Jewish Political Power and American Foreign Policy.* New York: Simon & Schuster, 1987.

Toobin, Jeffrey. *Too Close to Call: The Thirty-Six-Day Battle to Decide the 2000 Election.* New York: Random House, 2001.

Tuchman, Barbara W. *Bible and Sword: England and Palestine from the Bronze Age to Balfour.* New York: Ballantine, 1984.

Tuveson, Ernest Lee. *Redeemer Nation: The Idea of America's Millennial Role.* Chicago: University of Chicago Press, 1980.

Unger, Craig. *House of Bush, House of Saud: The Secret Relationship Between the World's Two Most Powerful Dynasties.* New York: Scribner, 2004.

Vidal, Gore. *Gore Vidal: Sexually Speaking: Collected Sex Writings.* San Francisco: Cleis Press, 2001.

Viguerie, Richard A., and David Franke. *America's Right Turn: How Conservatives Used New and Alternative Media to Take Power.* Chicago: Bonus Books, 2004.

Wallach, Janet. *Desert Queen: The Extraordinary Life of Gertrude Bell: Adventurer, Adviser to Kings, Ally of Lawrence of Arabia.* New York: Anchor Books, 1996.

Wallis, Jim. *God's Politics: Why the Right Gets It Wrong and the Left Doesn't Get It.* San Francisco: HarperSanFrancisco, 2005.

Walsh, Lawrence E. *Iran-Contra, the Final Report.* New York: Times Books, 1994.

Weber, Eugen. *Apocalypses: Prophecies, Cults, and Millennial Beliefs Through the Ages.* Cambridge: Harvard University Press, 1999.

Weber, Timothy P. *On the Road to Armageddon: How Evangelicals Became Israel's Best Friend.* Grand Rapids, Mich.: Baker Academic, 2005.

Weinstein, Michael L., and Davin Seay. *With God on Our Side: One Man's War Against an Evangelical Coup in America's Military.* New York: Thomas Dunne Books, 2006.

Weldon, Curt. *Countdown to Terror: The Top-Secret Information That Could Prevent the Next Terrorist Attack on America . . . and How the CIA Has Ignored It.* Washington, D.C.: Regency Publishing, 2005.

Wilentz, Sean. *The Rise of American Democracy: Jefferson to Lincoln.* New York: W. W. Norton and Co., 2005.

Willan, Phillip. *Puppetmasters: The Political Use of Terrorism in Italy.* New York: Author's Choice Press, 2002.

Wilson, Joseph. *The Politics of Truth: Inside the Lies That Led to War and Betrayed My Wife's CIA Identity.* New York: Carroll and Graf, 2004.

Wojcik, Daniel. *The End of the World as We Know It: Faith, Fatalism, and Apocalypse in America.* New York: New York University Press, 1997.

Wolfe, Tom. *The Purple Decades: A Reader.* New York: Farrar, Straus and Giroux, 1982.

Woodward, Bob. *Bush at War.* New York: Simon & Schuster, 2002.

———. *The Commanders.* New York: Simon & Schuster, 1991.

———. *Plan of Attack.* New York: Simon & Schuster, 2004.

———. *State of Denial: Bush at War, Part III.* New York: Simon & Schuster, 2006.

———. *Veil: The Secret Wars of the CIA, 1981–1987.* New York: Simon & Schuster, 1987.

Wright, Jonathan A. *Shapers of the Great Debate on the Freedom of Religion: A Biographical Dictionary.* Westport, Conn.: Greenwood Press, 2005.

Wright, Lawrence. *The Looming Tower: Al-Qaeda and the Road to 9/11.* New York: Knopf, 2006.

Wright, Robin. *Sacred Rage: The Wrath of Militant Islam.* New York: Touchstone, 1985.

Wurmser, David. *Tyranny's Ally: America's Failure to Defeat Saddam Hussein.* La Vergne, Tenn.: AEI Press, 1999.

Yergin, Daniel. *The Prize: The Epic Quest for Oil, Money, and Power.* New York: Simon & Schuster, 1991.

INDEX

Page numbers beginning with 357 refer to endnotes.